# AEROSPACE
# NAVIGATION SYSTEMS

# AEROSPACE NAVIGATION SYSTEMS

*Edited by*

**Alexander V. Nebylov**
**Joseph Watson**

*Library of Congress Cataloging-in-Publication Data*

Names: Nebylov, A. V. (Aleksandr Vladimirovich), editor. | Watson, Joseph, 1931– editor.
Title: Aerospace navigation systems / [edited by] Alexander V. Nebylov, Joseph Watson.
Description: First edition. | Chichester, West Sussex, United Kingdom : John Wiley & Sons Ltd., 2016. |
    Includes bibliographical references and index.
Identifiers: LCCN 2015047272 (print) | LCCN 2015049979 (ebook) | ISBN 9781119163077 (cloth) |
    ISBN 9781119163039 (Adobe PDF) | ISBN 9781119163046 (ePub)
Subjects: LCSH: Navigation (Aeronautics) | Aids to air navigation. | Navigation (Astronautics)
Classification: LCC TL586 .A28 2016 (print) | LCC TL586 (ebook) | DDC 629.135/1–dc23
LC record available at http://lccn.loc.gov/2015047272

A catalogue record for this book is available from the British Library.

Set in 10/12pt Times by SPi Global, Pondicherry, India
Printed and bound in Singapore by Markono Print Media Pte Ltd

1   2016

# Contents

# The Editors

**Alexander V. Nebylov** graduated with honors as an engineer in Missile Guidance from the Leningrad Institute of Aircraft Instrumentation in 1971. He led many R&D projects in aerospace instrumentation, motion control systems, and avionics, and he is a scientific consultant for various Russian design bureaus and research institutes. He was awarded the Candidate of Science degree in 1974 and the Doctor of Science degree in 1985, both in information processing and control systems, from the State Academy of Aerospace Instrumentation. He achieved the academic rank of full professor in the Higher Attestation Commission of the USSR in 1986.

For the past two decades, Dr. Nebylov has been with the State University of Aerospace Instrumentation in St. Petersburg as Professor and Chairman of Aerospace Devices and Measuring Complexes, and Director of the International Institute for Advanced Aerospace Technologies. He is the author of 15 books and numerous scientific papers and has also been a member of the leadership of the IFAC Aerospace Technical Committee since 2002. He has also been a member of the Presidium of the International Academy of Navigation and Motion Control since 1995. In 2006, the title of Honored Scientist of the Russian Federation was bestowed on Dr. Nebylov.

**Joseph Watson** is an electrical engineering graduate of the University of Nottingham, England, and the Massachusetts Institute of Technology, where he held a King George VI Memorial Fellowship. A fellow of the IET, senior member of the IEEE, and a member of Sigma Xi (the Scientific Research Society of America), he has published books and papers in various areas including electronic circuit design, nucleonics, biomedical electronics and gas sensors. He is a former Associate Editor for the *IEEE Sensors Journal*. He is also a qualified pilot.

Dr. Watson has served as Visiting Professor at the University of Calgary, Canada, and the University of California at both Davis and Santa Barbara; was External Examiner to the London University Postgraduate Department of Medical Electronics and Physics at St. Bartholomew's Hospital; and has held various consultancies with firms in the United States, Canada and Japan. Since retirement from Swansea University in the United Kingdom, he has continued as President of the UK-based Gas Analysis and Sensing Group.

# Acknowledgments

The editors wish to thank the distinguished authors of this book, all of whom spent valuable time writing their chapters despite the already huge workloads dictated by their normal activities. The combination of their dedicated efforts has helped create a book that endeavors to cover as much as possible of such a rapidly developing and knowledge-based branch of technology as aerospace navigation systems and many of its current problems. We are extremely grateful to each member of this highly qualified international team of specialists, and hope that readers will also appreciate the sharing of their invaluable in-depth knowledge.

Especial gratitude is due to members of the IFAC Aerospace Technical Committee, particularly its present chairman, Prof. Shinichi Nakasuka, its former chairman Professor Houria Siguerdidjane, and Professor Klaus Schilling who is a member of the Steering Committee along with A.V. Nebylov.

The numerous colleagues of A.V. Nebylov, professors and researchers at the State University of Aerospace Instrumentation in St. Petersburg, also merit profound gratitude. They, along with several generations of students, have provided fruitful dialogues that have helped realize this vision of the problems inherent in the development and perfection of aerospace navigation systems.

We are also indebted to the Wiley editorial and production staff for their valued suggestions and constructive advice during the preparation of the manuscript.

Finally, we would like to thank our respective families, because work on the book has often been performed at the expense of time otherwise available for domestic companionship. Our wives, Elena and Margaret, respectively, have endured these difficulties with considerable forbearance.

Any criticism regarding the contents and material of the book and the quality of its presentation will be accepted with gratitude.

Alexander V. Nebylov
Joseph Watson

# List of Contributors

**Bijay Agarwal**
Sattva E-Tech India Pvt Ltd.
Bangalore, India

**Georgy V. Antsev**
Concern Morinformsystem-Agat JSC
Moscow, Russia and
Radar-mms Research Enterprise JSC
Saint Petersburg, Russia

**Anatoly V. Balov**
Institute of Radionavigation and Time
Saint Petersburg, Russia

**Michael S. Braasch**
School of Electrical Engineering and Computer Science
Avionics Engineering Center
Ohio University
Athens, OH, USA

**Sergey P. Faleev**
State University of Aerospace Instrumentation
Saint Petersburg, Russia

**Walter Geri**
Magneti Marelli S.p.A., Bologna, Italy

**Evgeny A. Konovalov**
Late of the Russian Tsiolkovsky Academy of Cosmonautics
Moscow, Russia

**Alexander V. Nebylov**
State University of Aerospace Instrumentation
Saint Petersburg, Russia

**Ron T. Ogan**
Captain, U.S. Civil Air Patrol and Senior Member, IEEE
USA

**Vladimir Y. Raspopov**
Tula State University
Tula, Russia

**Valentine A. Sarychev**
Radar-mms Research Enterprise JSC
Saint Petersburg, Russia

**Sukrit Sharan**
Sattva E-Tech India Pvt Ltd.
Bangalore, India

**Boris V. Shebshaevich**
Institute of Radionavigation and Time
Saint Petersburg, Russia

**J. Paul Sims**
East Tennessee State University
Johnson City, TN, USA

**Oleg A. Stepanov**
Concern CSRI Elektropribor, JSC
ITMO University
Saint Petersburg, Russia

**Joseph Watson**
Swansea University (retd.)
Swansea, UK

**Matteo Zanzi**
Department of Electrical, Electronic and Information Engineering
"Guglielmo Marconi" (DEI)
University of Bologna, Bologna, Italy

**Sergey P. Zarubin**
Institute of Radionavigation and Time
Saint Petersburg, Russia

# Preface

The generic term "aerospace vehicles" refers to both air and space vehicles, which is logical because they both imply the possibility of three-dimensional controlled motion, high maximum attainable speeds, largely similar methods and parameters of motion, and the need for accurate location measurement. This last aspect was the key factor in determining the title and content of this book.

The main difference between aerospace navigation sensors and systems is actually in the level of complexity. Usually, a system has several sensors and other integral elements (like an Inertial Navigation System or INS), or a set of airborne and ground components (like a radio navigation system); or a set of airborne and space-based elements (like a Satellite Navigation System or SNS), or is constructed using radar or photometry principles (like correlated-extremal systems or map-matching systems). However, the *raison d'être* for navigation sensors and navigation systems is the same—to give valuable and reliable information about changes in the navigation parameters of any aerospace vehicle. In fact, in an integrated navigation complex, all sensors and systems are elements of equal importance, each contributing to the navigational efficiency determined not by the complexity of the design, but by the dynamic properties and spectral characteristics of the measurement error.

It is precisely such properties of navigation systems that make them vastly important to investigate for ensuring high precision in navigation complexes—hence it is this wide-ranging material that is the main subject of the book. When selecting optimum sets of onboard navigational information sources, it is important to know their error properties, their reliabilities, their masses and indicator dimensions, and of course their cost effectiveness. The book also contains material on future prospects for the development of different types of navigation system and opportunities for improving their performances.

Seven different systems are considered, only one of which provides complete autonomous navigation under all conditions, though unfortunately for short periods only. This is the INS, which is based on a set of gyroscopes and accelerometers, various types of which are amongst the most well developed of all devices and are generically termed *inertial sensors*. These provide very wide choices for the INS designer from some extremely accurate versions such as fiber-optic and electrostatic gyros, to some very small and cheap sensors based on MEMS. Principles of INS design, algorithms of INS functioning, and estimations of achievable accuracy are described in Chapter 1, and special attention is paid to the strapdown INS: the most widely used of all aerospace navigation complexes.

Chapter 2 describes SNSs, which in recent years, have become the most common means of determining the positions and velocities of aerospace objects. Global SNS was literally a revolution in navigation and fundamentally changed the available capabilities. It became possible to reduce navigation errors in favorable cases down to several meters, and even (in phase mode) to decimeters and centimeters. However, because of the risk of integrity loss in the satellite measurements and the low interference immunity of satellite navigation, it is not possible to consider the use of SNS as a cardinal approach to satisfying the rapidly increasing requirements for accuracy and reliability in navigation measurements. For personal navigation in large cities, and especially for indoor navigation, the SNS may be complemented by local navigation systems based on electronic maps of Wi-Fi network signal availability. However, for most aerospace vehicles the needed addition to SNS is the use of time-proven classic radio navigation systems.

Chapter 3 describes long-range navigation systems that are extremely reliable but not very accurate in comparison with SNS. Networks for long-range radio navigation cover almost all conceivable aviation routes, making it possible to solve the problem of aircraft, including helicopters, determining their *en route* positions without an SNS receiver, or after losing the working capacity of an SNS.

Chapter 4 is devoted to short-range navigation systems for the very accurate and reliable positioning of aircraft in specific areas with high requirements for precision motion control, usually near airports. Such methods can also be used in aircraft or spacecraft rendezvous applications.

Chapter 5 describes the landing navigation systems that allow aircraft to accurately maintain descending and landing paths under the control of all the necessary movement parameters. Course, glide slope, and marker beacons within the VOR/DME system allow the generation of radio fields for the trajectory control of landing aircraft. The use of these well-established landing systems provides a high level of safety, and they are in the mandatory list of equipment for all higher category aerodromes.

All the radio navigation systems considered in Chapters 2–5 require the deployment of a set of numerous ground radio beacons for creating the artificial radio-fields that permit perfect aircraft navigation. However, in nature there also exist various natural fields exhibiting different physical parameters of the solid underlying surface of the Earth and of other planets, and which can also be used for accurate navigation. The height of the surface relief of the earth under a flying aircraft and its terrain shape are certainly informative parameters for navigation, and it is currently possible to measure such parameters with the help of onboard instruments and to compare their values with pre-prepared maps. There are also some other physical parameters, the actual values of which can be compared with the map values, whence the resulting information will give coordinates for an aircraft's location. These approaches are usually implemented by the principle of correlation-extreme image analysis. Navigation systems based on this principle are described in Chapter 6.

Chapter 7 is devoted to the homing systems that solve problems connected with the docking of two aerospace vehicles where, as for missile guidance, information about the relative position of one vehicle with respect to the other is needed. This information can be obtained on the basis of the principles of active or passive location in different electromagnetic radiation frequency ranges. Guidance systems are rapidly improving performance and becoming "smarter," these trends also being described in Chapter 7.

Chapter 8 describes different approaches to the design of the filtering algorithms used in integrated navigation systems with two or more sensors having different physical properties and principles of operation. The output signals of such sensors invariably need to be subjected to filtering in order to more effectively suppress the measurement error of each sensor. In the case of two different sensors, their outputs are usually passed through a low-pass and high-pass filter, respectively. However, the specific parameters of these filters must be chosen in accordance with the theory of optimal linear filtering; and recently, even nonlinear filtering has been quite frequently used. The synthesis of integrated navigation systems is one of the most popular testing procedures for developing and for checking the reliability of methods for optimal and suboptimal filtering, and it is this that dictated the advisability of devoting a complete chapter to the subject. Hence, the main variants of the filtering problem statement and the algorithms used for their solutions are included in this chapter.

Chapter 9 describes the modern navigational displays that are able to provide effective exchange of information between the crew and the automatic navigation systems of the aircraft. Such displays actually show the result of the entire piloting and navigation system operations. Both hardware and structural means of implementing these displays are shown for a wide class of aircraft including commercial, military, and general aviation categories to illustrate cockpit avionic systems of varying complexities.

Finally, Chapter 10 deals with the navigational requirements of unmanned aerospace vehicles (UAVs)—rapidly becoming generically known as "drones." It is intended to provide a basis for the understanding of new developments of this burgeoning field, which encompasses both civil and military applications.

The systems described throughout the book include those representing the complex and advanced types of technical innovation that made possible the remarkably high levels of development in navigation and motion control systems that occurred near the turn of the century. They are widely used in both civil and military aircraft as well as partially in space technology. In civil aviation, standards for the use of these instruments are determined by the ICAO, and common approaches are used in practically all countries of the world. In military aviation, such complete uniformity does not exist, but many of the design principles used by different developers are similar to each other because of parallel development resulting from the need to find the best technical solutions according to the basic physical principles utilized in the equipment operation. For example, the American GPS, the Russian GLONASS, and the European GALILEO are rather similar in their principles of construction.

Actually, the development of major hardware for aerospace navigation takes place largely in the United States, the European Union, and Russia, and each of these is represented by the present authors. The idea of the book was born during discussions amongst experts at meetings of the IFAC Aerospace Technical Committee, of which A.V. Nebylov is a member. Also, for many years he has been at the State University of Aerospace Instrumentation (SUAI) in St. Petersburg, Russia, which has made it convenient for him to choose several of the authors from the many scientific and manufacturing centers in that city. It was here where the fully automatic and very successful landing systems of the *Buran* (Snowstorm) aerospace plane (Russia's equivalent of America's Space Shuttle) were designed. The Western authors were located by J. Watson, who was also responsible for the

English language formatting of the entire volume. This second program of cooperation between the editors arose naturally from that resulting in a previous and related volume, *Aerospace Sensors* (Momentum Press, 2013).

The primary purpose of the book is to present the fundamentals of design, construction, and application of numerous aerospace navigation and guidance systems to engineers, designers, and researchers in the area of control systems for various aerospace vehicles including aircraft, UAVs, space planes, and missiles. However, it may also be used as a study guide for both undergraduate and graduate students and for postgraduates in aerospace engineering, avionics, aeronautics, astronautics, and various related areas. Finally, the editors hope that it might also be found useful by many other people wishing to satisfy their general interest in modern aerospace technology.

# 1

# Inertial Navigation Systems

Michael S. Braasch

*School of Electrical Engineering and Computer Science, Avionics Engineering Center, Ohio University, Athens, OH, USA*

## 1.1   Introduction

Inertial Navigation Systems (INSs) are modern, technologically sophisticated implementations of the age-old concept of dead reckoning. The basic philosophy is to begin with a knowledge of initial position, keep track of speed and direction, and thus be able to determine position continually as time progresses. As the name implies, the fundamental principle involved is the property of inertia. Specifically, a body at rest tends to stay at rest and a body in motion tends to stay in motion unless acted upon by an external force.

From Newton's second law of motion, the well-known relation can be derived:

$$\vec{F} = m\vec{a} \tag{1.1}$$

where "$\vec{F}$" is a force vector, "$m$" is mass, and "$\vec{a}$" is the acceleration vector. Conceptually, it is then possible to measure force and subsequently determine acceleration. This may then be integrated to determine velocity, which in turn may be integrated to determine position.

Each accelerometer described in chapter 5 of the companion volume, *Aerospace Sensors* (Konovalov, 2013), can determine a measure of linear acceleration in a single dimension, from which it follows that multiple accelerometers are needed to determine motion in the general three-dimensional case. However, in addition, it is necessary to determine the direction in which the accelerometers are pointing, which is not a trivial exercise considering that an aircraft can rotate around three axes. If the accelerometers are hard-mounted to the vehicle (as is typically the case), then theoretically they can be oriented in any direction. Double integration of their outputs is useless if this time-varying orientation is not properly taken into account.

The determination of orientation (also known as attitude determination) is accomplished through processing of data from the gyroscopes described in chapter 6 on *Aerospace Sensors* (Branets *et al.*, 2013). These devices measure either angular rate and/or angular displacement.

*Aerospace Navigation Systems*, First Edition. Edited by Alexander V. Nebylov and Joseph Watson.
© 2016 John Wiley & Sons, Ltd. Published 2016 by John Wiley & Sons, Ltd.

So-called navigation-grade gyros can measure angular rate with accuracies on the order of 0.01°/h. Such sensors are needed to determine attitude to permit velocity and position determination in a given reference frame.

There are two main types of INSs: gimbaled and strapdown (Lawrence, 1998). In a gimbaled system, the accelerometers are mounted on a platform that is rotationally isolated from the vehicle. This platform is maintained in a local-level orientation so that there are two accelerometers in the horizontal plane for providing the data needed to compute horizontal velocity and position. In a strapdown system, the accelerometers are hard-mounted to the vehicle itself. Gimbaled systems are mechanically complex but do not require intensive computational capability. Strapdown systems are mechanically simple but require more intensive computations to deal with the time-varying orientation of the vehicle. Throughout the remainder of this chapter, strapdown operation will be assumed because modern processors are in no way challenged by the computational requirements. Furthermore, the simplified mechanical design (along with modern optical gyros) yields systems with Mean Time Between Failures (MTBFs) measured in tens of thousands of operational hours (versus the hundreds of hours that were typical of gimbaled systems in the late 1970s).

## 1.2   The Accelerometer Sensing Equation

Equation 1.1 stated the well-known relationship between force and acceleration. Solving for acceleration yields the following equation:

$$\frac{\vec{F}}{m} = \vec{a} \tag{1.2}$$

The quantity on the left is referred to as "specific force" and is what an accelerometer inherently measures. However, in the presence of a gravitational field, the specific force measured by an accelerometer is not equal to the classic Newtonian acceleration. As an example, the pendulous accelerometers described by Konovalov in chapter 5 of *Aerospace Sensors* have pendulums that displace under acceleration because of inertia. However, even if a sensor is stationary, the pendulum will still displace due to the effects of gravity. Specifically, the pendulum will displace downward just as it would have if the device were actually accelerating upward. Accelerometers thus measure the combination of Newtonian acceleration and the reaction force to gravity:

$$\vec{f} = \vec{a} - \vec{g} \tag{1.3}$$

where $\vec{f}$ is the measured specific force vector, $\vec{a}$ is the Newtonian acceleration vector, and $\vec{g}$ is the earth's gravity vector. This is the so-called accelerometer sensing equation. The desired Newtonian acceleration vector is thus given by the following equation:

$$\vec{a} = \vec{f} + \vec{g} \tag{1.4}$$

Although the equation is quite simple, it reveals a fundamental concept in inertial navigation. In order to determine acceleration from specific force, the inertial system needs to determine the value of the gravity vector at its current position because the value of this gravity vector varies significantly with position and especially with altitude. A variety of gravity models have been developed over the past 60 years, and the ones typically utilized in inertial navigation have accuracies on the order of 5–10 μg (i.e., five to ten one-millionths the average magnitude of gravity at the earth's surface) (Hsu, 1998).

## 1.3   Reference Frames

A variety of coordinate (or reference) frames are utilized in INSs (Britting, 1971). Accelerometers and gyros form their measurements with respect to inertial space (i.e., the inertial frame), but they output their data with respect to frames that are fixed to the vehicle. Furthermore, velocity is typically expressed in east/north/up components (i.e., a local-level reference frame), and position is frequently expressed in terms of latitude and longitude (i.e., a reference frame fixed to the Earth). It is therefore important to understand these various reference frames and then to understand how to take quantities expressed in one frame and convert them to another.

### *1.3.1   True Inertial Frame*

The true inertial frame is given by any reference frame that does not accelerate with respect to the stars in the galaxy in which Earth resides—the Milky Way. Classical physics has established these "fixed stars" as an acceptable inertial reference frame (i.e., a frame in which Newton's laws are valid).

### *1.3.2   Earth-Centered Inertial Frame or i-Frame*

The Earth-Centered Inertial Frame (ECI), or i-frame, is a frame that is centered in the Earth but does not rotate with it—its orientation with respect to the stars remains fixed. This frame is not truly inertial since the Earth revolves around the Sun, and hence the i-frame undergoes nonzero acceleration with respect to the stars. However, for navigation tasks on or near the surface of the Earth, this acceleration can be considered negligible.

### *1.3.3   Earth-Centered Earth-Fixed Frame or e-Frame*

The Earth-Centered Earth-Fixed (ECEF) frame, or e-frame, is also a frame that is centered in the Earth, but is also fixed to it. Hence, it rotates along with the Earth. Two of the axes are in the equatorial plane, and the third axis is aligned with the nominal pole of rotation of the Earth. Although the choice for the equatorial plane axes is arbitrary, one of the most common specifies an $x$-axis that intersects the prime meridian (through Greenwich, UK).

### *1.3.4   Navigation Frame*

The navigation frame, or n-frame, is centered in the vehicle, but it does not rotate with the vehicle—it remains at a local level. The choice of the horizontal plane axes is arbitrary, but one of the most common specifies an $x$-axis that points north, a $y$-axis that points east, and a $z$-axis that points "down" (i.e., approximately along the gravity vector). This particular

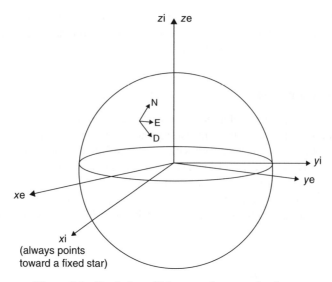

**Figure 1.1**   Depiction of i-frame, e-frame, and n-frame

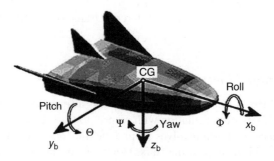

**Figure 1.2**   Depiction of b-frame (reprinted, with permission from Depiction of b-frame reprinted by permission of Nebylov, A. V., chapter 1, p. 7, "Introduction," *Aerospace Sensors*, Momentum Press, LLC, New York, 2013)

convention is sometimes referred to as the "north–east–down" or "NED" frame. However, in some cases it is desirable not to force the local-level frame to have an axis pointing north. When the local-level frame is permitted to rotate (around the vertical) away from north, the resulting frame is sometimes referred to as a "wander frame." The angle of rotation around the vertical is known as the "wander angle" and is typically denoted by $\alpha$. The i-, e-, and n-frames are depicted in Figure 1.1.

## 1.3.5   Body Frame

The body frame, or b-frame, is centered in the vehicle and is fixed to it (Figure 1.2) (Nebylov, 2013). It is typically specified to coincide with the principal axes of the vehicle. For an aircraft, the convention is chosen such that the $x$-axis aligns with the longitudinal axis of the vehicle (positive out of the nose), the $y$-axis points out of the right wing, and the $z$-axis points out of the bottom of the vehicle (i.e., "down" if the vehicle is level). This convention is therefore such that positive rotations about these axes correspond to positive Euler angles (roll, pitch, and yaw).

### 1.3.6 Sensor Frames (a-Frame, g-Frame)

The sensor frames are specified by orthogonal sensor triads. An orthogonal set of three accelerometers forms the a-frame, and an orthogonal set of gyroscopes forms the g-frame. The a- and g-frames are not coincident since it is physically impossible to mount both the accelerometers and the gyros in exactly the same space. Furthermore, the sensor triads are typically not mounted in perfect alignment with the b-frame. In fact, for installation convenience, it is not uncommon to have large angular offsets between the sensor frames and the b-frame. For the remainder of this chapter, however, it will be assumed that the offsets have been determined during installation and that the sensor outputs have been converted to their b-frame equivalents.

## 1.4 Direction Cosine Matrices and Quaternions

A variety of coordinate frame conversions are utilized in inertial navigation. For example, accelerometer outputs expressed in the b-frame need to be converted to the n-frame before velocity and position can be computed. It is assumed the reader is already familiar with the concept of rotation matrices, so only the key relationships for inertial navigation will be summarized here. For a more detailed discussion of the fundamental concepts, the reader is directed to Kuipers (2002).

A vector expressed in the b-frame may be converted to the n-frame as follows:

$$\vec{f}^n = C_b^n \vec{f}^b \tag{1.5}$$

where the so-called body-to-navigation direction cosine matrix (DCM) is given by the following equation:

$$C_b^n = \begin{bmatrix} \cos\theta\cos\psi & -\cos\phi\sin\psi + \sin\phi\sin\theta\cos\psi & \sin\phi\sin\psi + \cos\phi\sin\theta\cos\psi \\ \cos\theta\sin\psi & \cos\phi\cos\psi + \sin\phi\sin\theta\sin\psi & -\sin\phi\cos\psi + \cos\phi\sin\theta\sin\psi \\ -\sin\theta & \sin\phi\cos\theta & \cos\phi\cos\theta \end{bmatrix} \tag{1.6}$$

where:

$\phi$ = roll angle
$\theta$ = pitch angle
$\psi$ = yaw angle

This DCM defines the orientation (attitude) of the vehicle by specifying the rotational difference between the local-level n-frame and the fixed-to-the-vehicle b-frame; and the initialization and updating of this matrix will be discussed later in this chapter. Although there are nine elements in the matrix, only six of them need to be updated (the other three can be derived from the orthonormality property of the matrix). As discussed in Kuipers (2002), an equivalent attitude representation is given by the quaternion. Quaternions have only four elements and thus are more efficient than DCMs that need at least six elements to be specified uniquely. Again, both representations are equivalent and the remainder of this chapter will consider only DCMs for attitude representation.

The rotational differences between the Earth frame and the n-frame are given by the latitude, longitude, and (optionally) the wander angle and are expressed in the Earth-to-nav DCM:

$$C_e^n = \begin{bmatrix} -\cos\alpha\sin\text{Lat}\cos\text{Lon} - \sin\alpha\sin\text{Lon} & -\cos\alpha\sin\text{Lat}\sin\text{Lon} + \sin\alpha\cos\text{Lon} & \cos\alpha\cos\text{Lat} \\ \sin\alpha\sin\text{Lat}\cos\text{Lon} - \cos\alpha\sin\text{Lon} & \sin\alpha\sin\text{Lat}\sin\text{Lon} + \cos\alpha\cos\text{Lon} & -\sin\alpha\cos\text{Lat} \\ -\cos\text{Lat}\cos\text{Lon} & -\cos\text{Lat}\sin\text{Lon} & -\sin\text{Lat} \end{bmatrix}$$

(1.7)

where:

$\alpha$ = wander angle

Lat = latitude

Lon = longitude

Since DCMs are orthonormal matrices, their inverse is equal to their transpose:

$$C_n^b = \left(C_b^n\right)^{-1} = \left(C_b^n\right)^T$$
$$C_n^e = \left(C_e^n\right)^{-1} = \left(C_e^n\right)^T$$

## 1.5  Attitude Update

The accelerometer outputs expressed in the b-frame need to be resolved into a frame of interest for subsequent updating of the vehicle velocity and position states. For aircraft applications, the frame of interest is usually the n-frame. As described in the previous section, the body-to-nav DCM can be used to perform this conversion. However, this DCM needs to be updated continuously to account for rotations of both the b-frame and the n-frame.

The two rotations are driven by two corresponding angular rates:

$\vec{\omega}_{ib}^b$ = angular rate of the b-frame, relative to the i-frame, expressed in b-coordinates

$\vec{\omega}_{in}^n$ = angular rate of the n-frame, relative to the i-frame, expressed in n-coordinates

The differential equation governing the rate of change of the body-to-nav DCM is given by the following equation (Titterton and Weston, 2005):

$$\dot{C}_b^n = C_b^n\left[\vec{\omega}_{ib}^b \times\right] - \left[\vec{\omega}_{in}^n \times\right]C_b^n$$

(1.8)

where:

$$\left[\vec{\omega}_{ib}^{b} \times\right] = \text{skew symmetric matrix form of the body-rate vector } \vec{\omega}_{ib}^{b}$$

$$= \begin{bmatrix} 0 & -\omega_{ib}^{b}(z) & \omega_{ib}^{b}(y) \\ \omega_{ib}^{b}(z) & 0 & -\omega_{ib}^{b}(x) \\ -\omega_{ib}^{b}(y) & \omega_{ib}^{b}(x) & 0 \end{bmatrix}$$

$$\left[\vec{\omega}_{in}^{n} \times\right] = \text{skew symmetric matrix form of the n-frame-rate vector } \vec{\omega}_{in}^{n}$$

$$= \begin{bmatrix} 0 & -\omega_{in}^{n}(z) & \omega_{in}^{n}(y) \\ \omega_{in}^{n}(z) & 0 & -\omega_{in}^{n}(x) \\ -\omega_{in}^{n}(y) & \omega_{in}^{n}(x) & 0 \end{bmatrix}$$

Since the angular rate of the body can be very high (e.g., hundreds of degrees per second for aerobatic aircraft) and the n-frame rate is very low (as will be discussed later), the DCM update is performed in two parts: one for the b-frame update and the other for the n-frame update.

## 1.5.1 Body Frame Update

The rotation of the b-frame relative to the inertial frame is measured by the gyros. The discrete-time update of the body-to-nav DCM accounting for body rotation is given by the following equation (Titterton and Weston, 2005):

$$C_{b[k+1]}^{n} = C_{b[k]}^{n} e^{[\vec{\sigma} \times]} \tag{1.9}$$

where:

$k = $ time index

$\left[\vec{\sigma} \times\right] = $ skew symmetric matrix form of the rotation vector $\vec{\sigma}$

$$= \begin{bmatrix} 0 & -\sigma_z & \sigma_y \\ \sigma_z & 0 & -\sigma_x \\ -\sigma_y & \sigma_x & 0 \end{bmatrix}$$

The components of the rotation vector are computed from the gyro outputs. If the body angular-rate vector had a fixed orientation, the rotation vector could be computed very simply as follows:

$$\vec{\sigma} \approx \int_{t_k}^{t_{k+1}} \vec{\omega}_{ib}^{b} dt \tag{1.10}$$

However, in practice the body angular-rate vector changes orientation, at least in part, because of vibration. On a larger scale, this change can also occur when an aircraft is performing so-called S-turns (Kayton and Fried, 1997). The change of angular rate vector orientation is referred to as "coning." The term refers to the geometric figure of a cone being swept out by the axis of rotation. When this occurs during the interval in which the angular rate is being integrated, an erroneous angular displacement will be computed. As a result, a coning compensation algorithm is applied in the computation of the rotation vector (Ignagni, 1996).

Before continuing, it should be noted that the matrix exponential in Equation 1.9 is typically approximated using the first few terms of a Taylor series expansion.

## 1.5.2   Navigation Frame Update

The rotation of the n-frame, relative to the inertial frame, is a function of two rotations: Earth rate and transport rate:

$$\vec{\omega}_{in}^n = \vec{\omega}_{ie}^n + \vec{\omega}_{en}^n \tag{1.11}$$

This rotation rate of the n-frame is also known as "spatial rate." The two components will now be described.

### 1.5.2.1   Earth Rate

Even if the vehicle is stationary relative to the Earth, the n-frame must rotate relative to the i-frame, at the earth's rotation rate, in order for it to stay locally level.

$$\vec{\omega}_{ie}^n = C_e^n \vec{\omega}_{ie}^e$$
$$\vec{\omega}_{ie}^e = \begin{bmatrix} 0 & 0 & \omega_{ie} \end{bmatrix}^T$$
$$\omega_{ie} \approx 7.292115e-5 \text{ rad/s}$$

### 1.5.2.2   Transport Rate

Furthermore, if the vehicle is moving relative to the Earth, then to stay at a local level the n-frame must also rotate to account for the motion of the vehicle over the surface of the curved Earth. This is obviously a function of the horizontal velocity components. For north-pointing mechanizations, the transport-rate vector is given by (Kayton and Fried, 1997; Titterton and Weston, 2005):

$$\vec{\omega}_{en}^n = \begin{bmatrix} \dfrac{V_E}{R_P + h} & -\dfrac{V_N}{R_M + h} & -\dfrac{V_E \tan \text{Lat}}{R_P + h} \end{bmatrix}^T \tag{1.12}$$

where:
  $V_E$ = east component of velocity
  $V_N$ = north component of velocity
  $h$ = altitudue
  $R_M$ = radii of curvature of the earth in the north-south direction (also known as the meridian radius of curvature)
$$= \dfrac{a\left(1-e^2\right)}{\left(1-e^2 \sin^2 \text{Lat}\right)^{3/2}}$$

$R_p$ = radii of curvature of the earth in the east-west direction (also known as the prime radius of curvature)

$$= \frac{a}{\left(1 - e^2 \sin^2 \text{Lat}\right)^{1/2}}$$

a = semi-major axis of the earth ellipsoid (i.e., Equatorial radius)
e = eccentricity of the earth ellipsoid

It should be noted that the vertical component of transport rate has a singularity at the earth's poles. Specifically, the term "tan(Lat)" approaches infinity as latitude approaches ±90°. This is not an issue for vehicles that stay in mid-latitude or equatorial regions. For polar or near-polar crossings, however, north-pointing mechanizations must be replaced with a wander-azimuth mechanization that does not force one of the n-frame axes to point in the north direction (Jekeli, 2000).

### 1.5.2.3   Navigation-Frame Update Algorithm

The magnitude of the spatial rate vector is very small. Even for commercial jets traveling at 500 knots, it can be shown that the magnitude of transport-rate is approximately half of Earth rate. As a result, a simple trapezoidal approximation of the discrete-time update is generally acceptable:

$$C_b^{n[k+1]} = \left\{ I - \frac{1}{2} \left( \left[ \vec{\omega}_{in}^n (t_k) \times \right] + \left[ \vec{\omega}_{in}^n (t_{k+1}) \times \right] \right) \Delta t \right\} C_b^{n[k]} \tag{1.13}$$

where:

$\left[ \vec{\omega}_{in}^n (t_k) \times \right]$ = skew symmetric matrix form of $\vec{\omega}_{in}^n$ at time $t_k$

$\Delta t = t_{k+1} - t_k$

## 1.5.3   *Euler Angle Extraction*

Once the body-to-nav DCM has been updated for both b-frame and n-frame rotations, it may be used to convert accelerometer outputs as shown in Equation 1.5. It should also be noted that the updated DCM inherently contains the updated representation of the attitude of the vehicle or inertial system. From Equation 1.6, it follows that the Euler angles can be extracted from the updated DCM as follows:

$$\phi = \arctan2\left(C_b^n (3,2), C_b^n (3,3)\right)$$
$$\theta = \arcsin\left(-C_b^n (3,1)\right) \tag{1.14}$$
$$\psi = \arctan2\left(C_b^n (2,1), C_b^n (1,1)\right)$$

Note: the four-quadrant arc-tangent functions are needed to preserve the full range of roll and yaw angles.

## 1.6   Navigation Mechanization

The differential equation governing the rate of change of the velocity vector is given by

$$\dot{v}_e^n = C_b^n \vec{f}^b - \left[ 2\vec{\omega}_{ie}^n + \vec{\omega}_{en}^n \right] \times \vec{v}_e^n + \vec{g}_{eff}^n \qquad (1.15)$$

where the cross-product term accounts for the effects of Coriolis on the rotating Earth and n-frames and where

$$\vec{g}_{eff}^n = \vec{g}^n - \vec{\omega}_{ie}^n \times \vec{\omega}_{ie}^n \times \vec{R}$$

is known as "effective gravity," "apparent gravity," "local gravity," or "plumb-bob gravity." It is the combination of the earth's mass attraction (first term) and centripetal acceleration due to Earth rotation (vector triple-product term $\vec{R}$ is the position vector with origin at the center of the Earth). To the first order, effective gravity is aligned with the local vertical (note that since the Earth is ellipsoidal, extension of the local vertical down to the equatorial plane does not intersect the mass center of the Earth, unless one is located at the poles or on the equator).

Conceptually, the Coriolis term may be understood as follows: Consider a vehicle that starts at the North Pole and travels due south along a meridian (path of constant longitude). Relative to the Earth, the horizontal path is "straight" (i.e., no east–west motion). However, relative to the i-frame, the path is clearly curved due to the rotation of the Earth. The inertial sensors detect the actual curved path in space, but the Coriolis term allows the overall equation to compute a velocity relative to the Earth, not relative to the inertial frame.

The discrete-time update associated with 1.14 is very straightforward:

$$v_e^n[k+1] = v_e^n[k] + C_b^n \overline{\Delta V}_k^b + \left\{ -\left[ 2\vec{\omega}_{ie}^n + \vec{\omega}_{en}^n \right] \times \vec{v}_e^n[k] + \vec{g}_{eff}^n[k] \right\} \Delta t \qquad (1.16)$$

where:

$$\overline{\Delta V}_k^b \approx \int_{t_k}^{t_{k+1}} \vec{f}^b \, dt$$

The integration of the specific force vector can be performed numerically if necessary, but some accelerometers perform this integration as part of their normal operation. For high-accuracy applications, two corrections need to be included in this specific force integration: sculling and size effect. Sculling error arises when an accelerometer is swinging back and forth at the end of a pendulum. This combination of rotation and acceleration results in a nonzero net sensed acceleration even though the average displacement of the accelerometer is zero. In practice, this motion occurs due to vibration. Sculling error can be corrected with a compensation algorithm that utilizes both the accelerometers' outputs and the coning-compensated gyro outputs (Mark and Tazartes, 1996; Savage, 1998). Alternatively, sculling error may be avoided by transforming the accelerometer outputs to the n-frame at a rate significantly higher than the highest anticipated vibration frequencies (Kayton and Fried, 1997).

Size-effect errors result from the nonzero lever arms between the three orthogonal accelerometers. Since the three sensors are not physically located in exactly the same point, they will sense different accelerations when the vehicle is rotating as well as accelerating. For high accuracy applications, a size-effect correction must be applied to the accelerometer outputs (Savage, 2009).

## 1.7   Position Update

The nav-to-Earth DCM is the transpose (also inverse) of the Earth-to-nav DCM. The differential equation governing the rate of change of the nav-to-Earth DCM is given by the following equation:

$$\dot{C}_n^e = C_n^e \left[ \vec{\omega}_{en}^n \times \right] \tag{1.17}$$

The discrete-time update is then given by the following equation:

$$C_n^e[k+1] = C_n^e[k] e^{\left[ \vec{\zeta} \times \right]} \tag{1.18}$$

$$\left[ \vec{\zeta} \times \right] = \text{skew} - \text{symmetric matrix form of } \vec{\zeta}$$

$$= \begin{bmatrix} 0 & -\zeta_z & \zeta_y \\ \zeta_z & 0 & -\zeta_x \\ -\zeta_y & \zeta_x & 0 \end{bmatrix}$$

$$\vec{\zeta} = \int_{t_k}^{t_{k+1}} \vec{\omega}_{en}^n dt$$

Again, the matrix exponential is typically approximated using the first few terms of the Taylor series expansion. The integration of the transport rate vector is typically performed either with rectangular or trapezoidal integration.

Once the DCM has been updated, the transpose is taken:

$$C_e^n = \left( C_n^e \right)^T$$

Finally, based on Equation 1.7, the position angles may be extracted as follows:

$$\text{Lon} = \arctan2\left(-C_e^n(3,2), -C_e^n(3,1)\right)$$
$$\text{Lat} = \arcsin\left(-C_e^n(3,3)\right) \tag{1.19}$$
$$\alpha = \arctan2\left(-C_e^n(2,3), C_e^n(1,3)\right)$$

Over short intervals of time, altitude may be determined by the following equation:

$$h_{k+1} = h_k + \int_{t_k}^{t_{k+1}} -v_e^n(z)\,dt \tag{1.20}$$

As will be discussed later, however, the vertical channel of an INS is inherently unstable and hence is unusable over long periods of time.

## 1.8   INS Initialization

The discrete-time updates for attitude, velocity, and position were given in Equations 1.9, 1.13, 1.16, 1.17, and 1.19. Each equation assumes that the quantity to be updated was known at the previous time step. Hence, these equations do not explain how attitude, velocity, and position are initialized.

Initial position is generally determined in one of two ways. For stationary vehicles, initial position can be determined by parking the vehicle at a known, surveyed location. At international airports, for example, travelers may see signs (visible to pilots from the cockpit) specifying the latitude, longitude, and altitude of a given gate. For vehicles that are in motion, initial position typically is provided through the use of a radio navigation aid such as GPS or GLONASS. For stationary vehicles, initial velocity may be specified as zero and, again, for vehicles in motion, external radio navigation aids may be utilized.

The most challenging aspect of initialization lies in the determination of the body-to-nav DCM. This amounts to the determination of the local level plane (from which roll and pitch are determined) and the determination of the direction of north (from which yaw or heading is determined).

Conceptually, the process of attitude determination is most easily understood in the context of a gimbaled INS. For a stationary vehicle, the only force sensed by the accelerometers is the force of gravity and, to first order, the gravity vector lies along the local vertical. The leveling process in a gimbaled system involves rotating the platform until the two accelerometers with sensitive axes in the platform plane are outputting zero (and thus are orthogonal to the gravity vector). In practice, all accelerometers output noise regardless of the value of the true sensed specific force, which is therefore averaged over a finite interval of time. For strapdown inertial systems, leveling is performed in two parts: coarse leveling and fine leveling.

Coarse leveling amounts to the estimation of pitch and roll by processing accelerometer outputs that have been averaged over a brief interval of time (i.e., a few seconds) (Kayton and Fried, 1997):

$$\phi \approx a\tan\left(\frac{A_y^b}{A_z^b}\right)$$

$$\theta \approx a\tan\left(\frac{A_x^b}{\sqrt{\left(A_y^b\right)^2 + \left(A_z^b\right)^2}}\right) \tag{1.21}$$

where "A" is the averaged accelerometer output.

Given an initial estimate of level, the direction of north (or equivalently, platform heading or yaw angle) can be achieved by exploiting the fact that a stationary vehicle only experiences rotation due to the rotation of the Earth, that is, the gyros will only sense Earth rate. This is significant because Earth rate only has north and vertical components (i.e., there is no east component of Earth rate). The procedure works as follows. With the coarse estimation of roll and pitch, a coarse body-to-local-level DCM can be formed:

$$C_b^{LL}\left(\theta,\phi\right)\Big|_{\psi=0} = \begin{bmatrix} \cos\theta & \sin\phi\sin\theta & \cos\phi\sin\theta \\ 0 & \cos\phi & -\sin\phi \\ -\sin\theta & \sin\phi\cos\theta & \cos\phi\cos\theta \end{bmatrix}$$

With this coarse body-to-local-level DCM, the gyro outputs (which are only sensing Earth rate) can be transformed from the b-frame to the local-level frame:

$$\vec{\omega}_{ie}^{LL} = C_b^{LL}\,\vec{\omega}_{ie}^b$$

The level components of the sensed Earth rate are actually components of the north component of Earth rate (there being no east component of Earth rate). Thus, heading or yaw can be determined by (Kayton and Fried, 1997):

$$\psi \approx \arctan2\left(\frac{-\omega_{ie}^{LL}(y)}{\omega_{ie}^{LL}(x)}\right) \tag{1.22}$$

Again, the gyro outputs are very noisy, so averaging is necessary before Equation 1.24 is computed.

The aforementioned procedure provides coarse initialization of roll, pitch, and yaw (or equivalently, the body-to-nav DCM). Fine leveling and yaw/azimuth determination typically involve the use of a Kalman filter and the exploitation of the knowledge that the velocity (of a stationary vehicle) is zero. The fine initialization Kalman filter is generally of the same type as that used to integrate external radio navigation aid information while the vehicle is in motion.

## 1.9   INS Error Characterization

INSs are subject to a wide variety of errors. These include mounting errors, initialization errors, sensor errors, gravity model errors, and computational errors, all of which are briefly described in the following text.

### 1.9.1   Mounting Errors

These include nonorthogonality of the sensor triads, unknown angular offsets between the sensor triads and reference points on the system case (that houses the sensors), and unknown angular offsets between the system case and the vehicle in which it is installed. The bulk of the nonorthogonality of the sensor triads and the angular offsets between the triads and the system case are determined during factory calibration, and the raw sensor outputs are compensated accordingly. Determination of the angular offsets between the inertial system case and the vehicle (i.e., the b-frame) is referred to as "boresighting" and is performed when the inertial system is initially mounted in the vehicle. Boresighting errors affect the accuracy of the Euler angles (roll, pitch, and yaw).

### 1.9.2   Initialization Errors

Initialization errors are the errors in the determination of initial position, velocity, and attitude. In the absence of external aiding, navigation-grade inertial systems (i.e., inertial systems capable of stand-alone positioning with drift rates on the order of 1 nautical mile per hour of operation) require accuracy of initial pitch and roll to be better than 0.1 mrad, initial azimuth better than 1 mrad, and initial velocity better than 0.5 m/s.

### 1.9.3   Sensor Errors

A variety of errors affect the performances of accelerometers and gyros and are discussed more fully in Konovalov and Branets (chapters 5 and 6 of *Aerospace Sensors* 2013). The primary sensor errors affecting INS performance are residual biases. These are the bias errors that remain after calibration/compensation and are different each time a given sensor is powered on. A secondary effect is scale factor error. The inherent noise in the sensors primarily drives the amount of filtering/averaging required during initialization.

### 1.9.4   Gravity Model Errors

A variety of closed-form gravity models exist (Hsu, 1998), which provide accuracies on the order of 5–10 µg. However, none is able to characterize the so-called deflection of the vertical because the actual direction of the gravity vector at the surface of the Earth is not perfectly orthogonal to the surface of the reference ellipsoid (i.e., the imaginary surface that defines zero altitude). The true gravity vector has nonzero horizontal components with magnitudes on the order of a few micro-gs. These values are position-dependent and vary greatly, for example, in mountainous regions. Though precise characterization of the gravity field can be accomplished using extensive look-up tables, for the vast majority of inertial navigation applications the closed-form models are sufficiently accurate.

## 1.9.5   Computational Errors

As discussed earlier in this chapter, inertial navigation algorithms require a variety of numerical methods including transcendental function approximation and numerical integration. When strapdown systems were being developed in the 1970s, computational power was limited, and the approximations needed to achieve real-time operation had a nontrivial impact on system performance. With the capabilities of modern processors, however, computational errors can be made negligible.

## 1.9.6   Simulation Examples

A detailed analysis of error sources may be found in Titterton and Weston (2005). Simulation results will be presented in this section to illustrate the impact of key error sources. The INS is simulated in a vehicle located at 45° North latitude, is level, and is pointed northward.

### 1.9.6.1   Accelerometer Bias Simulation

In this example, a 100 μg bias is simulated on the body $x$-axis accelerometer. Since the vehicle is level and is pointed northward, this accelerometer is sensitive in the north direction (Figures 1.3 and 1.4).

As would be expected with a bias on an accelerometer that is pointed north, the dominant errors are north position error and north velocity error along with pitch (since the vehicle is level and is also pointed north). The periodicity of the errors is known as the "Schuler period" and is approximately 84.4 min long. It is named after Maximilian Schuler (Pitman, 1962). In 1923, Schuler published a paper regarding the properties of a mechanical system (the specific example was a gyrocompass) tuned to maintain a local-level/locally vertical frame of reference regardless of external applied forces. Schuler's work was later shown to apply to INSs since they also implement a local-level/locally vertical frame of reference (e.g., the n-frame).

As the simulation goes on, errors start to build up in the east–west direction. This cross-coupling results from incorrect application of Earth rate (which results from an erroneous computation of latitude) in the n-frame update. If the simulation is extended, the magnitudes (i.e., envelope) of the east–west and north–south errors will slowly oscillate with a period that is inversely proportional to the sine of the latitude (Titterton and Weston, 2005). This is known as the Foucault oscillation.

### 1.9.6.2   Horizontal Gyroscope Bias Simulation

In this example a 0.01°/h bias is simulated on the body $y$-axis gyroscope. Since the vehicle is level and pointed north, the gyro is sensitive in the east direction (Figures 1.5 and 1.6).

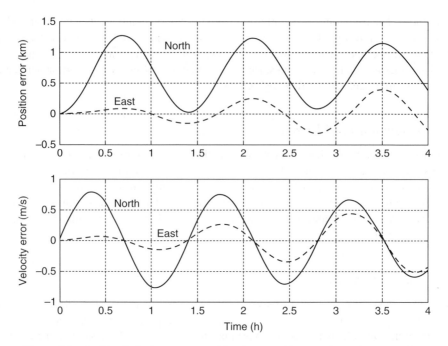

**Figure 1.3**   Position/velocity errors resulting from a 100 µg north accelerometer bias

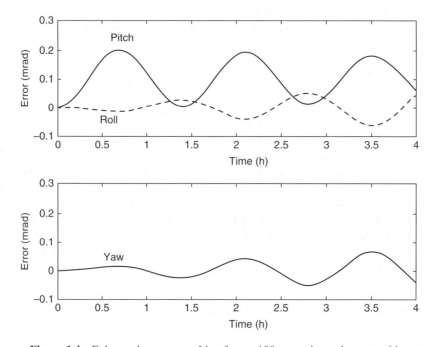

**Figure 1.4**   Euler angle errors resulting from a 100 µg north accelerometer bias

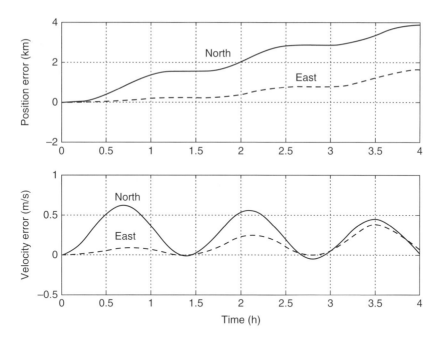

**Figure 1.5**   Position/velocity errors resulting from a 0.01°/h east gyroscope bias

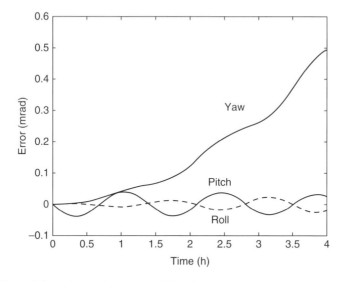

**Figure 1.6**   Euler angle errors resulting from a 0.01°/h east gyroscope bias

In this case, the Schuler periodicity is still present in the position error, but the dominant effect is a longer-term growth trend that is approximately linear over the first few hours. Although somewhat counterintuitive, the horizontal gyro bias induces a longer-term error growth trend in the yaw angle, whereas the bounded Schuler oscillations are present in the roll and pitch angles.

### 1.9.6.3 Vertical Gyroscope Bias Simulation

In this example, a 0.01°/h bias is simulated on the body $z$-axis gyroscope. Since the vehicle is level, the gyro is sensitive to yaw motion in the vertical direction (Figures 1.7 and 1.8).

In this case, the linear error growth trend is observed on the velocity components, and the position error growth is then quadratic. Nevertheless, the position error growth is still slower for the vertical gyro bias than was the case for the horizontal gyro bias.

### 1.9.6.4 Azimuth Initialization Error Simulation

In this example, a 1 mrad error in the initial azimuth (yaw) determination is simulated (Figures 1.9 and 1.10).

It is noted that the position, velocity, and attitude error characteristics for an initial azimuth error are very similar to those for a horizontal gyro bias.

### 1.9.6.5 Velocity Initialization Error Simulation

In this example, a 0.1 m/s error in the initial north velocity determination is simulated (Figures 1.11 and 1.12).

Virtually pure Schuler oscillations are present, and it is noted that all the errors in this case are zero mean over the long term. The magnitudes of the errors scale with the magnitude of the initial velocity error. An initial velocity error of 1 m/s would therefore result in a peak position error of approximately 0.8 km.

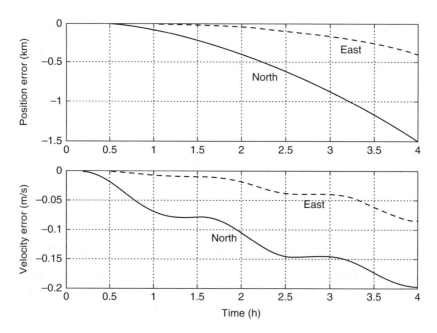

**Figure 1.7**   Position/velocity errors resulting from a 0.01°/h vertical gyroscope bias

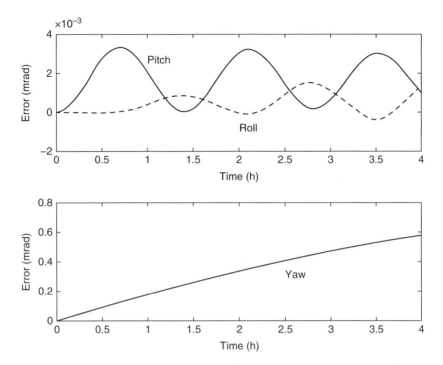

**Figure 1.8**  Euler angle errors resulting from a 0.01°/h vertical gyroscope bias

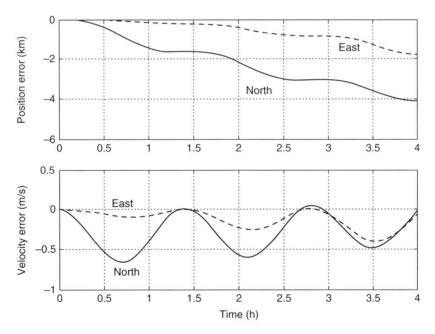

**Figure 1.9**  Position/velocity errors resulting from a 1 mrad initial azimuth error

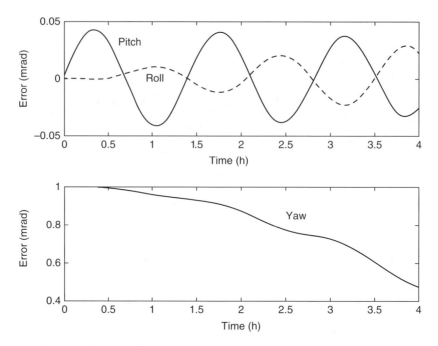

**Figure 1.10**    Euler angle errors resulting from a 1 mrad initial azimuth error

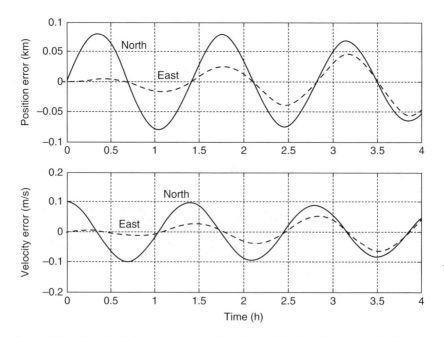

**Figure 1.11**    Position/velocity errors resulting from a 0.1 m/s initial north velocity error

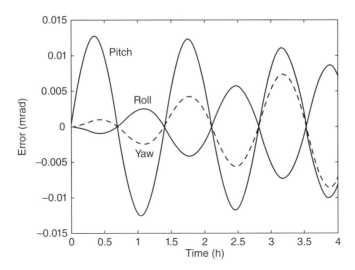

**Figure 1.12**   Euler angle errors resulting from a 0.1 m/s initial north velocity error

### 1.9.6.6   Combination Simulation of All Error Sources

In this example, all the previously simulated single error sources are simulated simultaneously: 100 µg north accelerometer bias, 0.01°/h east gyro bias, 0.01°/h vertical gyro bias, 1 mrad initial azimuth error, and 0.1 m/s initial north velocity error (Figures 1.13 and 1.14).

The error characteristics for the combination appear to be similar to those for the horizontal gyro bias and initial azimuth error cases. To show that this is not a coincidence, the next plot depicts the north component of position error for each error source along with the combination (Figure 1.15).

Although the long-term error is clearly dominated by the horizontal gyro bias and initial azimuth error, it would appear that the accelerometer bias plays a more significant role in the first few minutes after initialization. Figure 1.16 is a zoomed view of Figure 1.15 to show that this is indeed the case.

The figure clearly illustrates that initial short-term error growth is dominated by the horizontal accelerometer bias along with initial velocity error.

### 1.9.6.7   Vertical Instability

At this point, the reader may have noted that all of the simulation results presented in this chapter depict only horizontal position and velocity errors. This was not an accident. It can be shown that the so-called vertical channel of an INS has an inherent instability. That is, any perturbation in the vertical position and/or velocity determination will result in a positive feedback and thus an unbounded error growth. Thus, altitude determination cannot be performed by a stand-alone INS over long periods of time (i.e., more than a couple of minutes). In practice, the vertical channel of an inertial system is always integrated with an external source of altitude information. Historically, this external source has been a barometric altimeter (Ausman, 1991) although modern satellite-based navigation systems are now also used (Groves, 2008; Grewal *et al.*, 2013).

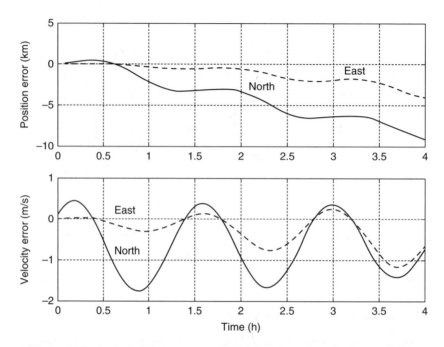

**Figure 1.13**    Position/velocity errors resulting from combination of error sources

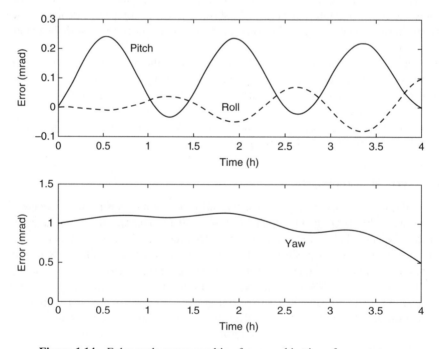

**Figure 1.14**    Euler angle errors resulting from combination of error sources

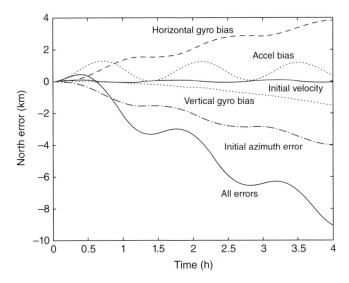

**Figure 1.15**   North position error resulting from various error sources

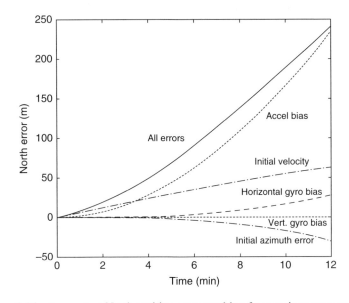

**Figure 1.16**   *Zoom view*: North position error resulting from various error sources

## 1.10   Calibration and Compensation

Although the details of sensor calibration and compensation are beyond the scope of this chapter, it must be noted that such procedures and algorithms are essential to provide the most accurate system performance (Rogers, 2007). A top quality gyroscope with a specified bias error of less than 0.01°/h will only achieve this specification after the repeatable component of the bias error has been calibrated and after the sensor output has been compensated for temperature-dependent variation.

## 1.11   Production Example

Navigation-grade INSs are produced by several companies around the world. Examples include Thales and Sagem in France; iMAR in Germany; and Honeywell, Kearfott, L-3, and Northrop-Grumman in the United States. Unit prices are generally negotiated with individual customers and are a function of the volume of the production. At the time of writing, unit prices were on the order of US$50 000–US$100 000.

**Figure 1.17**   Northrop Grumman LN-251 INS/GPS (courtesy of Northrop Grumman)

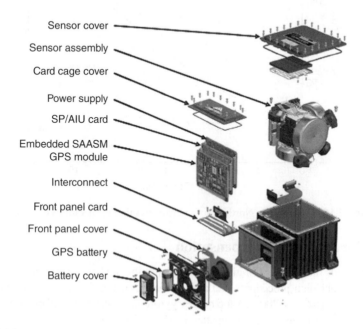

Sensor cover
Sensor assembly
Card cage cover
Power supply
SP/AIU card
Embedded SAASM
GPS module
Interconnect
Front panel card
Front panel cover
GPS battery
Battery cover

**Figure 1.18**   Exploded view of Northrop Grumman LN-251 (courtesy of Northrop Grumman)

An example of navigation-grade product offered by Northrop Grumman is the LN-251, which is a fiber-optic-gyro-based INS with an integrated GPS (Volk, Lincoln, and Tazartes, 2006). This unit is shown in Figure 1.17, and its dimensions are approximately $26 \times 19 \times 14$ cm. The weight is approximately $5.8$ kg, and it has an MTBF that exceeds that exceeds 20,000 h. Figure 1.18 shows an exploded view of the unit.

# References

J. S. Ausman, "A Kalman Filter Mechanization for the Baro-Inertial Vertical Channel," Proceedings of the 47th Annual Meeting of the Institute of Navigation, Williamsburg, VA, USA, June 1991.

V. N. Branets, B. E. Landau, U.N. Korishko, D. Lynch, and V. Y. Raspopov, "Gyro Devices and Sensors," in *Aerospace Sensors*, edited by A. Nebylov, Momentum Press, LLC, New York, 2013.

K. Britting, *Inertial Navigation System Analysis*, Wiley-Interscience, New York, 1971.

M. Grewal, A. Andrews, and C. Bartone, *Global Navigation Satellite Systems, Inertial Navigation, and Integration*, 3rd edition, Wiley-Interscience, New York, 2013.

P. Groves, *Principles of GNSS, Inertial, and Multi-Sensor Integrated Navigation Systems*, Artech House, Boston, 2008.

D. Hsu, "An Accurate and Efficient Approximation to the Normal Gravity," 1998 IEEE Position, Location and Navigation Symposium (PLANS '98), Palm Springs, CA, USA, April 1998. IEEE, Piscataway, NJ, pp. 38–44.

M. Ignagni, "Efficient Class of Optimized Coning Compensation Algorithms," *AIAA Journal of Guidance, Control and Dynamics*, 1996, 19(2) 424–429.

C. Jekeli, *Inertial Navigation Systems with Geodetic Applications*, Walter De Gruyter Inc., Berlin, 2000.

M. Kayton and W. Fried, *Avionics Navigation Systems*, Wiley-Interscience, New York, 1997.

S. F. Konovalov, "Devices and Sensors for Linear Acceleration Measurement," in *Aerospace Sensors*, edited by A. Nebylov, Momentum Press, LLC, New York, 2013.

J. B. Kuipers, *Quaternions and Rotation Sequences: A Primer with Applications to Orbits, Aerospace and Virtual Reality*, Princeton University Press, Princeton, NJ, 2002.

A. Lawrence, *Modern Inertial Navigation Technology – Navigation, Guidance and Control*, 2nd edition, Springer-Verlag, New York, 1998.

J. Mark and D. Tazartes, "On Sculling Algorithms," Third St. Petersburg International Conference on Integrated Navigation Systems, St. Petersburg, Russia, May 1996.

A. Nebylov, "Introduction," in *Aerospace Sensors*, edited by A. Nebylov, Momentum Press, LLC, New York, 2013.

G. R. Pitman, Jr., (1962). *Inertial Guidance*, John Wiley & Sons, Inc., New York, 1962.

R. Rogers, *Applied Mathematics in Integrated Navigation Systems*, 3rd edition, American Institute of Aeronautics and Astronautics, Reston, 2007.

P. Savage, "Strapdown Inertial Navigation Integration Algorithm Design Part 2: Velocity and Position Algorithms," *AIAA Journal of Guidance, Control and Dynamics*, 1998, 21(2), 208–221.

P. Savage, "Computational Elements for Strapdown Systems," NATO Lecture Series RTO-EN-SET-116-2009, Low-Cost Navigation Sensors and Integration Technology, 2009.

D. Titterton and J. Weston, *Strapdown Inertial Navigation Technology*, 2nd edition, American Institute of Aeronautics and Astronautics, Reston, VA, 2005.

C. Volk, J. Lincoln and D. Tazartes, "Northrop Grumman's Family of Fiber-Optic Based Inertial Navigation Systems," IEEE/ION Position, Location and Navigation Symposium (PLANS 2006), San Diego, CA, April 2006.

# 2

# Satellite Navigation Systems*

Walter Geri[1], Boris V. Shebshaevich[2] and Matteo Zanzi[3]

[1] *Magneti Marelli S.p.A., Bologna, Italy*
[2] *Institute of Radionavigation and Time, Saint Petersburg, Russia*
[3] *Department of Electrical, Electronic and Information Engineering "Guglielmo Marconi" (DEI), University of Bologna, Bologna, Italy*

## 2.1 Introduction

This chapter describes the Global Navigation Satellite Systems (GNSS) used for positioning and navigation applications. Although a satellite navigation system is very complex, it is common to speak of a "satellite navigation system sensor" when referring to the tools that an end user needs in order to exploit the functionalities of a satellite navigation system, and this chapter will give a general survey of this topic. Firstly, the positioning problem within the framework of a positioning satellite system will be described. Then, general GNSS descriptions and prospects will be presented, outlining the main features of the American GPS–GNSS "gold standard," the Russian GLONASS (revived and modernized a few years ago), and the European Galileo and the Chinese BeiDou (formerly called Compass), both successfully developing, along with the Japanese QZSS and the Indian Regional Navigation Satellite System (IRNSS). Following these will be sections that include sources of errors and receiver structures common to all GNSS. Modernization and augmentation (GBAS and SBAS) of satellite systems are treated in a specific paragraph. Finally, some brief outlines of the aerospace applications of satellite systems will conclude the chapter.

Major attention has been paid to the principles of satellite positioning, descriptions of the measurements that can be obtained from satellite signals, the way they are processed, the effects that affect the measurements, and how they propagate. These choices were motivated by the fact that their principles are common to all the GNSS and can provide the reader with the tools to approach the following subject matter. Many references are also

---

*The authors are very grateful to Prof. Gianni Bertoni from the University of Bologna and Prof. Valery Ipatov from the Russian Institute of Radionavigation and Time (Saint Petersburg) for their valuable advice.

---

*Aerospace Navigation Systems*, First Edition. Edited by Alexander V. Nebylov and Joseph Watson.

given in the bibliography for the study of particular topics in depth, but the most recent novelties are quickly outlined because the relevant research is still ongoing and new applications are emerging.

## 2.2   Preliminary Considerations

Steady research in the field of radio navigation by means of Earth-orbiting artificial satellites commenced in the early 1950s, several years before the first satellite was launched. The possibility of exploiting satellite signals for positioning was initially suggested by the observed Doppler effect on the ground-satellite data link signal that allowed for trajectory tracking of satellites (Kotelnikov *et al.*, 1958; Shebshaevich, 1958, 1971; Parkinson and Spilker, 1996; Kayton and Fried, 1997). The use of satellites as powerful and global radio-navigation beacons for terrestrial navigation was a subject of experimentation in the 1960s, and later, marine positioning applications were considered, closely followed by aircraft navigation in the early 1970s. More recently, the miniaturization of electronic components has led to the wide availability of compact and reduced cost receivers and, at present, even most new general aviation aircraft feature onboard satellite navigation receivers that allow for the determination of three-dimensional (3D) position and accurate time. The International Civil Aviation Organization (ICAO) has defined a general system that includes each existing and future satellite system (GPS, GLONASS, Galileo, etc.) together with geostationary overlay satellites and has named it the Global Navigation Satellite System (GNSS). A GNSS receiver is a device for positioning and navigating based on the integration of many satellite-based positioning services. It is essentially a passive sensor that only receives and does not transmit and is based on the measurement of range and change of range between several satellites and the aircraft.

Major advantages of GNSS are that they provide all-weather worldwide navigation capability. However, a disadvantage is that the satellite-receiver link is long distance and hence the power of the received signal is low, which results in high vulnerability to interference.

## 2.3   Navigation Problems Using Satellite Systems

A GNSS configuration consists of the following three segments:

1. A satellite constellation that transmits radio signals that represent both range (distance between a satellite and a user) and range-rate information sources. This may be called the *Space Segment*.
2. The entire onboard equipment consisting of the receiver and the processing devices that compute the navigational solutions. This is the *User Segment*.
3. A set of ground-based stations that monitor the transmitted satellite signals. Each of these consists of a master control station for performing computations on the monitored data and an upload station where the results obtained are uploaded to the satellites via ground antennas. This is the *Control Segment*.

It should be noted that in addition to performing computations on satellite navigation data, another primary task of the master control station is to recognize necessary satellite orbit corrections and clock adjustments.

Satellite navigation systems are based on the concept of direct ranging that involves the direct propagation of radio signals from transmitter to receiver (like terrestrial radio-navigation systems such as Loran-C, Decca, and Omega).

The ranging capability is related to the propagation time, since the speed of radio signals is assumed constant and equal to the speed of light in a vacuum. Using this principle, the receiver measurements are based on the time of arrival of the satellite radio signals at the flying user vehicle, for which reason it is necessary to install sufficiently stable onboard clocks at relatively low cost.

The main advantage of satellite navigation systems compared with ground-based radio-navigation systems is that they provide a worldwide navigation solution with better accuracy and less dependency on atmospheric conditions. On the other hand, the main disadvantages are:

- A strong dependence of the overall performance on signal-in-space interference.
- The possibility of the navigation solution being unavailable owing to the lack of a sufficient number of tracked satellite signals. This problem can arise either because of signal masking or because not enough satellites are in view.
- The necessity for external augmentation equipment in order to fulfill the accuracy, integrity, and availability requirements of particular applications.

### 2.3.1 The Geometrical Problem

In order to adopt the time of arrival of all the received signals as range information, a common reference time system encompassing the entire satellite constellation is required. This is ensured by the Control Segment of the satellite system.

As a first attempt, assume that the receiver clock is synchronized to the common reference time system and that the emission times of all the received satellite signals are known. The time of arrival of each signal can then be easily translated into the value of the range between the corresponding satellite and the receiver antenna. In the two-dimensional case (i.e., when the description of the user position on a plane surface is sufficient), each computed range defines a circumference, centered in the corresponding satellite, as a set of candidate user locations. Figure 2.1 represents the particular configuration in which two ranges, $R1$ and $R2$, corresponding, respectively, to satellites Sat1 and Sat2, are available. In this case the user position is represented by points A and B, both of which satisfy the range constraints. In order to solve this ambiguity there are two possibilities:

1. Exploiting the range information from at least one other satellite
2. Using further geometric constraints to discard one of the two candidate solutions

In the 3D case each received range reduces the receiver candidate positions to the points on a spherical surface centered at the corresponding satellite. The set of user locations compatible with two satellite ranges can be obtained by means of the intersection between the two spherical surfaces. The result is the circumference shown in Figure 2.2.

It is worth noting that the circumference of the candidate user positions shown in Figure 2.2 is placed on a plane perpendicular to a line connecting Sat1 and Sat2 and, moreover, that the

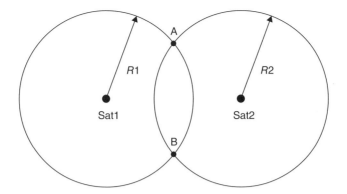

**Figure 2.1**    Two-dimensional positioning on the basis of two satellite ranges

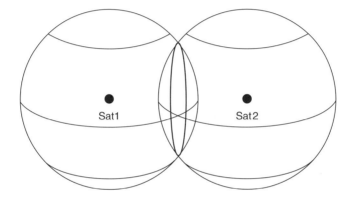

**Figure 2.2**    Three-dimensional positioning on the basis of two satellite ranges

intersection between the plane and the line determines the circumference center. By introducing range information from a third satellite Sat3, the candidate user positions reduce to two points A and B as shown in Figure 2.3, where Π is the plane defined by the positions of Sat1, Sat2, and Sat3, while Γ is the circumference defined by the intersection between the spherical surfaces of Sat1 and Sat2.

It should be noted that the resulting candidate user positions are specular points of one another with respect to the plane Π of the satellites. Regarding typical applications, there is no ambiguity since users are always under the plane Π, that is, between the satellites and the Earth's surface. On the other hand, to avoid ambiguity in space navigation applications, there is need of further geometric constraints or at least another satellite range.

## 2.3.2   Reference Coordinate Systems

To perform user position computation by means of satellite ranging, as described earlier, it is necessary to adopt a common reference coordinate system to represent both user and satellite positions. (Note that similar considerations can be stated for the user velocity.)

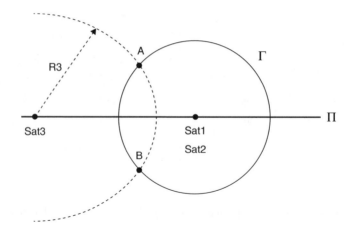

**Figure 2.3**   Three-dimensional positioning on the basis of three satellite ranges

### 2.3.2.1   The ECI Coordinate System

*Earth-Centered Inertial* (ECI) systems are right-handed orthogonal coordinate systems (with $x$, $y$, $z$ axes) that do not rotate with respect to the fixed stars and have their origin fixed at the Earth's center of mass. Although ECI coordinate systems can be defined in several ways (Kovalevsky *et al.*, 1989), usually the $x$, $y$ axes are taken on the Earth's equatorial plane.

ECI coordinate systems are typically used as inertial reference frames to express the dynamics of satellites: in fact, in this case the Earth's rotational effects are neglected during the formulation of the equations of motion. Since the velocity vector of the Earth's center of mass and its spin axis is not constant, the ECI systems are not truly inertial. In order to overcome this problem, the orientation of the ECI frame axes can be defined with reference to the Earth's equatorial plane at a particular instant of time.

### 2.3.2.2   The ECEF Coordinate System

Air navigation applications require the obtaining of user position and velocity with respect to an Earth-referenced frame. Earth-Centered Earth-Fixed (ECEF) frames are right-handed orthogonal coordinate systems ($x$, $y$, $z$ axes) with the origin placed on the Earth's center of mass. The usual definition of the ECEF coordinate system assumes that both the $x$- and $y$-axes are on the Earth's equatorial plane (as for the ECI) but the $x$-axis crosses the Greenwich Meridian, which rotates with the Earth. It is worth observing that computations of the user position and velocity using the ECEF system require that all the available quantities should be expressed in the same reference frame, including the satellite orbits. In order to change the coordinate system from ECI to ECEF, a proper time-variant elementary rotation around the $z$-axis has to be applied to the satellite position and velocity vectors (as described in Long *et al.* (1989)).

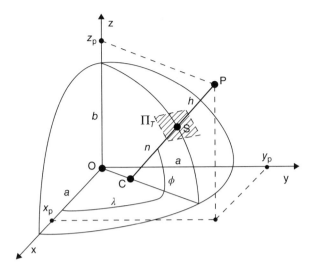

**Figure 2.4**  Earth model as ellipsoidal shape

### 2.3.2.3  The World Geodetic System (WGS84)

Typical navigation applications refer to user position in terms of latitude, longitude, and altitude with respect to a standard physical model of the Earth. The most widely used model is that defined in NIMA (2000) and indicated as the World Geodetic System 1984 (WGS84). The Earth is modeled as a particular ellipsoidal shape in WGS84, where cross sections parallel to the equatorial plane (ECEF $x$, $y$ plane) are circular, while cross sections containing the ECEF $z$-axis are elliptical.

In Figure 2.4, let $a$ and $b$ be the major semiaxis and the minor semiaxis of the elliptical shape, respectively. Then, its eccentricity is defined as

$$e = \sqrt{1 - \frac{b^2}{a^2}}$$

It can be noted that for each point P in the tridimensional space, both a set of $(x_p, y_p, z_p)$ ECEF coordinates and a set of $\phi$, $\lambda$, and $h$ WGS84 geodetic coordinates can be equivalently assumed. In particular $\phi$, $\lambda$, and $h$ represent the latitude, longitude, and altitude, respectively. The relations among the $\phi$, $\lambda$, and $h$ parameters and the Cartesian coordinates $(x_p, y_p, z_p)$ are represented in Figure 2.4, where $\Pi_T$ is the local tangent plane to the WGS84 reference ellipsoid at the point S, which represents the orthogonal projection of P to the ellipsoidal surface, while $n$ is the geodetic normal to the local tangent plane $\Pi_T$.

In more detail, the following expressions provide the Cartesian coordinates corresponding to a known set of geodetic ones:

$$\begin{cases} x_p = \left( \left\| \overrightarrow{OC} \right\| + \left\| \overrightarrow{CS} \right\| \cos\phi \right) \cos\lambda + h \cos\phi \cos\lambda \\ y_p = \left( \left\| \overrightarrow{OC} \right\| + \left\| \overrightarrow{CS} \right\| \cos\phi \right) \sin\lambda + h \cos\phi \sin\lambda \\ z_p = \left\| \overrightarrow{CS} \right\| \sin\phi + h \sin\phi \end{cases} \tag{2.1}$$

with the following positions that are well known from elementary geometry:

$$\begin{cases} \left\| \overrightarrow{OC} \right\| = d\, e^2 \cos\phi \\ \left\| \overrightarrow{CS} \right\| = \left(1 - e^2\right) \\ d = \dfrac{a}{\sqrt{1 - e^2 \sin^2 \phi}} \end{cases} \tag{2.2}$$

Here, $d$ represents the length of the segment joining the point S and the $z$-axis along the direction of the geodetic normal $n$, in accordance with Figure 2.4.

Conversely, the transformation from Cartesian coordinates to geodetic ones is more involved but is nevertheless very important in navigation based on satellite systems. The proposed approach is based on iterative computations leading to a fast convergence. The solution algorithm is described by the following steps:

1. If $\left(x_p^2 + y_p^2\right) = 0$, then the longitude $\lambda$ cannot be defined. Moreover,

$$\begin{cases} \phi = \pi/2, & \text{if } z_p > 0 \\ \phi = -\pi/2, & \text{if } z_p < 0 \\ h = z_p - b, & \text{if } z_p > 0 \\ h = -z_p - b, & \text{if } z_p < 0 \end{cases}$$

and the algorithm stops here.

2. Directly compute the longitude $\lambda$ as follows (assuming $x_p^2 + y_p^2 > 0$ in this case):

$$\begin{cases} \lambda = \arctan\left(y_p/x_p\right), & \text{if } x_p \geq 0 \\ \lambda = \pi + \arctan\left(y_p/x_p\right), & \text{if } x_p < 0 \text{ and } y_p \geq 0 \\ \lambda = -\pi + \arctan\left(y_p/x_p\right), & \text{if } x_p < 0 \text{ and } y_p < 0 \end{cases} \tag{2.3}$$

3. Setting up at will the sensitivity levels $\varepsilon_\phi$ and $\varepsilon_h$ in order to define the convergence condition of the iterative process and initializing the algorithm with the following starting values:

$$\phi_s = \frac{\pi}{2}, \quad d_s = \frac{a}{\sqrt{1-e^2}}, \quad h_s = 0.$$

4. Computing the new values related to the latitude

$$\phi_n = \arctan\left( \frac{z_p}{\sqrt{x_p^2 + y_p^2}} \frac{d_s + h_s}{d_s\left(1-e^2\right) + h_s} \right), \quad d_n = \frac{a}{\sqrt{1 - e^2 \sin^2 \phi_n}}. \tag{2.4}$$

5. Computing the new altitude value

$$
\begin{cases}
h_{n} = \dfrac{\sqrt{x_{p}^{2} + y_{p}^{2}}}{\cos\phi_{n}} - \delta_{n}, & \text{if } -\dfrac{\pi}{4} \leq \phi_{n} \leq \dfrac{\pi}{4} \\[4mm]
h_{n} = \dfrac{z_{p}}{\sin\phi_{n}} - \delta_{n}\left(1 - e^{2}\right), & \text{if } \phi_{n} > \dfrac{\pi}{4} \text{ or } \phi_{n} < -\dfrac{\pi}{4}
\end{cases}
\tag{2.5}
$$

6. Evaluating the convergence of the obtained results

$$\text{if } |\phi_{n} - \phi_{s}| < \varepsilon_{\phi} \text{ and } |h_{n} - h_{s}| < \varepsilon_{h}$$
then
$$\phi = \phi_{n}, \quad h = h_{n}, \text{ and the algorithm stops here.}$$
else
$$\phi_{s} = \phi_{n}, \quad \delta_{s} = \delta_{n}, \quad h_{s} = h_{n}, \text{ and go back to the step 4.}$$

As a remark, the quick convergence of the algorithm depends on the choice of the particular latitude update expression $\phi_{n}$. In fact, the sensitivity of $\phi_{n}$ to $d_{s}$ and $h_{s}$ errors is very low, as can be easily verified from Equation 2.4.

As described in Zhu (1994), in addition to the previously described algorithm (the greater the required accuracy, the higher the number of iterations), there are several closed-form solutions for the geodetic latitude $\phi$ and altitude $h$ with respect to the Cartesian coordinates.

### 2.3.3 The Classical Mathematical Model

As previously stated, in order to determine the user position, the availability of both a minimum set of satellite ranges and the related satellite positions is necessary.

Regarding the satellite orbits, the classical mathematical model is based on the following equation:

$$
\frac{d^{2}\mathbf{r}_{s}}{dt^{2}} = \left(\nabla V\right)^{\mathrm{T}} + \mathbf{a}_{d}
\tag{2.6}
$$

Referring to an ECI reference frame, $\mathbf{r}_{s}$ and $\mathbf{a}_{d}$ are the vectors representing, respectively, the satellite position and the perturbing acceleration, while $V$ is the Earth's gravitational potential. The term $\mathbf{a}_{d}$ models additional accelerations due to the gravity of the moon and sun, the solar radiation pressure, and satellite maneuvers. This *gravitational potential V* must be modeled as a spherical harmonic series in order to take into account the ellipsoidal shape and nonhomogeneity of the mass distribution of the Earth (NIMA, 2000).

Now, a set of $n$ satellite range measurements $r_{i}$ (with $i = 1,\ldots,n$) is assumed to be available at the user receiver without any errors. Furthermore the positions of the $n$ satellites are assumed to be known by means of the received ephemeris, and the vector $\mathbf{s}_{i} = \left(X_{i}\ Y_{i}\ Z_{i}\right)^{\mathrm{T}}$ (expressed in the ECEF coordinate system) represents the $i$th satellite position to which $r_{i}$

refers. The relationship between the range measurements along with the user and satellite positions is then as follows:

$$r_i = \|\mathbf{s}_i - \mathbf{u}\| = \sqrt{(X_i - x_u)^2 + (Y_i - y_u)^2 + (Z_i - z_u)^2}, \quad i = 1,\ldots,n \tag{2.7}$$

where $\mathbf{u} = (x_u \ y_u \ z_u)^T$ is the unknown user position whose components represent the ECEF Cartesian coordinates. In this framework, if the user receiver has at least $n = 3$ range measurements, it is possible to compute $x_u$, $y_u$, $z_u$ by means of an iterative procedure based on the linearization of Equation 2.7, as widely treated in Hofmann-Wellenhof et al. (1997).

Furthermore, a set of $n$ satellite range-rate measurements $\dot{r}_i$ (with $i = 1,\ldots,n$) is also assumed to be available at the user receiver without any errors. The relationships between range-rate measurements and user and satellite velocities can be obtained by differentiating (2.7) as follows:

$$\dot{r}_i = \mathbf{l}_i^T (\mathbf{v}_i - \mathbf{v}_u) = \left(l_{xi} \ l_{yi} \ l_{zi}\right) \begin{pmatrix} \dot{X}_i - \dot{x}_u \\ \dot{Y}_i - \dot{y}_u \\ \dot{Z}_i - \dot{z}_u \end{pmatrix}, \quad i = 1,\ldots,n \tag{2.8}$$

where

$$\mathbf{l}_i^T = \left(l_{xi} \ l_{yi} \ l_{zi}\right) = \left( \frac{X_i - x_u}{r_i} \quad \frac{Y_i - y_u}{r_i} \quad \frac{Z_i - z_u}{r_i} \right), \quad i = 1,\ldots,n \tag{2.9}$$

represents the line of sight (unitary vector) between the user and the $i$th satellite. The directional cosines $l_{xi}, l_{yi}, l_{zi}$ are known if the user position has been computed as previously shown. Moreover the $i$th satellite velocity components $\dot{X}_i, \dot{Y}_i, \dot{Z}_i$ are known by means of the received ephemeris. In this framework, $n = 3$ range-rate measurements are sufficient to determine the $\dot{x}_u, \dot{y}_u, \dot{z}_u$ ECEF Cartesian components of the user velocity vector $\mathbf{v}_u$, and in this case an iterative procedure to solve the problem is not necessary. In fact, from Equation 2.8 with $n = 3$, and provided the satellites have different lines of sight from one another, a nonsingular square system of linear equations is defined.

Clock onboard satellites are very accurate and keep themselves constantly synchronized so that the *common satellite time system* can be taken as the *reference time system*. However, the clock in the receiver equipment on board the user vehicle cannot be considered to be aligned with that reference time system. Consequently, the error in the navigation solution would generally be too high with respect to the required accuracy. In fact, typical trade-offs between cost and performance lead to receiver clocks that currently employ crystal oscillators, so that only good short-term stability can be achieved. For this reason, the time offset $\Delta t_u$ of the user clock and the frequency offset of the user oscillator (leading to a drift $\dot{t}_u$ of the user time with respect to the reference time) have to be included in the equations of range (2.7) and range rate (2.8) when receiver measurements (pseudorange $\rho_i$ and pseudorange-rate $\dot{\rho}_i$) are involved. Moreover, it is worth observing that the available measurements are affected by several other error sources, such as atmospheric effects, multipath received signal reception, and noise introduced by the receiver. Therefore, the classical models of both pseudorange and pseudorange-rate measurements point to the need for an additive term representing the measurement error.

In this case $\Delta t_u$, defined as the advance of the receiver clock with respect to the reference time system, is one more unknown, whence instead of Equation 2.7, a new set of $n$ equations provides the pseudorange model, as follows:

$$\rho_i = \sqrt{(X_i - x_u)^2 + (Y_i - y_u)^2 + (Z_i - z_u)^2} + c\Delta t_u + \xi_i, \quad i = 1,\ldots,n \qquad (2.10)$$

where $c$ is the speed of light in a vacuum and $\xi_i$ represents the pseudorange measurement error. The classical model of $\xi_i$ is very simple and is given by a stationary random variable with zero mean value and variance $\sigma_{\xi_i}^2$. This makes possible the adoption of the least squares (LS) method in order to obtain the best estimation of the user position coordinates $x_u$, $y_u$, $z_u$ and of the user clock bias $\Delta t_u$ (Farrell and Barth, 1999) by means of an iterative algorithm as follows.

This algorithm is based on the model of Equation 2.10 linearized around $\mathbf{P}_0 = (x_{u0}\ y_{u0}\ z_{u0}\ \Delta t_{u0})$ as follows:

$$\tilde{\rho}_i = \begin{pmatrix} l_{xi} & l_{yi} & l_{zi} & 1 \end{pmatrix} \begin{pmatrix} \tilde{x}_u \\ \tilde{y}_u \\ \tilde{z}_u \\ -c\Delta \tilde{t}_u \end{pmatrix} + \xi_i', \quad i = 1,\ldots,n \qquad (2.11)$$

where

$$l_{xi} = \frac{\partial r_i}{\partial x_u}\bigg|_{\mathbf{P}_0} = \frac{X_i - x_u}{r_i}\bigg|_{\mathbf{P}_0}, \quad l_{yi} = \frac{\partial r_i}{\partial y_u}\bigg|_{\mathbf{P}_0} = \frac{Y_i - y_u}{r_i}\bigg|_{\mathbf{P}_0}, \quad l_{zi} = \frac{\partial r_i}{\partial z_u}\bigg|_{\mathbf{P}_0} = \frac{Z_i - z_u}{r_i}\bigg|_{\mathbf{P}_0},$$

$$\tilde{\rho}_i = \rho_i - \rho_{i0}, \quad \tilde{x}_u = x_u - x_{u0}, \quad \tilde{y}_u = y_u - y_{u0}, \quad \tilde{z}_u = z_u - z_{u0}, \quad \Delta \tilde{t}_u = \Delta t_u - \Delta t_{u0}$$

in which $\rho_{i0}$ is computed from the chosen linearization point $\mathbf{P}_0$ and $\xi_i'$ embeds the measurement and linearization errors.

Instead of Equation 2.8, a new set of $n$ equations represents the pseudorange-rate model where, in addition to $\dot{x}_u$, $\dot{y}_u$, $\dot{z}_u$, the drift of the user time with respect to the reference time $\dot{t}_u$ is one more unknown, as follows:

$$\dot{\rho}_i = \begin{pmatrix} l_{xi} & l_{yi} & l_{zi} \end{pmatrix} \begin{pmatrix} \dot{X}_i - \dot{x}_u \\ \dot{Y}_i - \dot{y}_u \\ \dot{Z}_i - \dot{z}_u \end{pmatrix} + c\dot{t}_u + v_i, \quad i = 1,\ldots,n \qquad (2.12)$$

Here, $v_i$ is the pseudorange-rate measurement error, and its classical model in this case is given by a stationary random variable with zero mean value and variance $\sigma_{v_i}^2$. It should be noted that instead of the pseudorange rate, the measurement of the Doppler shift on the carrier of the transmitted satellite signal is usually adopted (Kayton and Fried, 1997). Doppler shift $\Delta f_i$ and pseudorange-rate $\dot{\rho}_i$ are related as follows:

$$c\frac{\Delta f_i}{f_0} = -\dot{\rho}_i \qquad (2.13)$$

where $f_0$ is the nominal value of the transmitted satellite carrier frequency.

### 2.3.3.1 Computation of the User Position and Velocity

The classical algorithm used to compute the user position, under the hypothesis that all the error components $\xi_i'$ have the same variance and are mutually uncorrelated, is described by the following steps:

7. Setting up, at will, the sensitivity level $\varepsilon$ in order to define the convergence condition of the iterative process and also setting up the following starting values:

$$x_u = 0, \quad y_u = 0, \quad z_u = 0, \quad \Delta t_u = 0$$

8. Letting $\hat{x}_u = x_u$, $\hat{y}_u = y_u$, $\hat{z}_u = z_u$, $\Delta\hat{t}_u = \Delta t_u$, and computing the following terms for all the $n$ tracked satellites:

$$\hat{r}_i = \sqrt{(X_i - \hat{x}_u)^2 + (Y_i - \hat{y}_u)^2 + (Z_i - \hat{z}_u)^2}, \quad \hat{\rho}_i = \hat{r}_i + c\Delta\hat{t}_u, \quad i = 1,\ldots,n$$

$$\hat{l}_{xi} = \frac{X_i - \hat{x}_u}{\hat{r}_i}, \hat{l}_{yi} = \frac{Y_i - \hat{y}_u}{\hat{r}_i}, \qquad \hat{l}_{zi} = \frac{Z_i - \hat{z}_u}{\hat{r}_i}, \quad i = 1,\ldots,n$$

9. From Equation 2.2, the displacement of the solution according to the LS estimator is given by

$$\begin{pmatrix} \tilde{x}_u \\ \tilde{y}_u \\ \tilde{z}_u \\ -c\Delta\tilde{t}_u \end{pmatrix} = \hat{\mathbf{H}}^{\dagger}\tilde{\rho} = \begin{pmatrix} \hat{l}_{x1} & \hat{l}_{y1} & \hat{l}_{z1} & 1 \\ \hat{l}_{x2} & \hat{l}_{y2} & \hat{l}_{z2} & 1 \\ \vdots & \vdots & \vdots & \vdots \\ \hat{l}_{xn} & \hat{l}_{yn} & \hat{l}_{zn} & 1 \end{pmatrix}^{\dagger} \begin{pmatrix} \hat{\rho}_1 - \rho_1 \\ \hat{\rho}_2 - \rho_2 \\ \vdots \\ \hat{\rho}_n - \rho_n \end{pmatrix} \qquad (2.14)$$

where $\hat{\mathbf{H}}^{\dagger}$ is the Moore–Penrose pseudoinverse of $\hat{\mathbf{H}}$.

10. Computing the new solution

$$x_u = \hat{x}_u + \tilde{x}_u, \quad y_u = \hat{y}_u + \tilde{y}_u, \quad z_u = \hat{z}_u + \tilde{z}_u, \quad \Delta t_u = \Delta\hat{t}_u + \Delta\tilde{t}_u$$

11. Evaluating the convergence of the obtained solution:

$$\text{if } \left\| \tilde{x}_u \ \tilde{y}_u \ \tilde{z}_u \ c\Delta\tilde{t}_u \right\| < \varepsilon$$

then the algorithm stops here.
else go back to the step 2.

As a remark, the system of nonlinear equations (Eq. 2.10) can be solved in several other ways, for example, by adopting closed-form solutions as described in Chaffee and Abel (1994).

Regarding the user velocity computation, from Equations (2.12), (2.13), and a known user position, the classical LS estimation (under the hypothesis that all the error components $\nu_i$ have the same variance and are mutually uncorrelated) is as follows:

$$\begin{pmatrix} \dot{x}_u \\ \dot{y}_u \\ \dot{z}_u \\ -c\dot{t}_u \end{pmatrix} = \mathbf{H}^\dagger \mathbf{f_d} = \begin{pmatrix} l_{x1} & l_{y1} & l_{z1} & 1 \\ l_{x2} & l_{y2} & l_{z2} & 1 \\ \vdots & \vdots & \vdots & \vdots \\ l_{xn} & l_{yn} & l_{zn} & 1 \end{pmatrix}^\dagger \begin{pmatrix} l_{x1}\dot{X}_1 + l_{y1}\dot{Y}_1 + l_{z1}\dot{Z}_1 + \dfrac{c}{f_0}\Delta f_1 \\ l_{x2}\dot{X}_2 + l_{y2}\dot{Y}_2 + l_{z2}\dot{Z}_2 + \dfrac{c}{f_0}\Delta f_2 \\ \vdots \\ l_{xn}\dot{X}_n + l_{yn}\dot{Y}_n + l_{zn}\dot{Z}_n + \dfrac{c}{f_0}\Delta f_n \end{pmatrix}. \tag{2.15}$$

A deeper analysis of the solution accuracy (limited to the position in this framework) is generally carried out by means of a factor separation as follows (Kaplan and Hegarty, 2006):

$$\varepsilon_{sol} = \text{UERE} \cdot \text{DOP} \tag{2.16}$$

where $\varepsilon_{sol}$ is the resulting error on the position solution, user equivalent range error (UERE) represents the overall effect of the error sources on each pseudorange measurement, and the dilution of precision (DOP) represents the sensitivity of the solution to the measurement errors. (Further details on UERE will be provided in the following sections.)

Furthermore, the DOP depends on the relative geometry between the user and the satellite constellation. Moreover, as shown in Kaplan and Hegarty (2006), it is possible to define several DOP parameters that can be directly related to the $\mathbf{H}$ matrix properties, where $\mathbf{H}$ has been presented in the linearized equation (Eq. 2.14) by means of $\hat{\mathbf{H}}$ as a specific case. In particular, the commonly used geometric dilution of precision (GDOP) is defined as follows:

$$\text{GDOP} = \sqrt{\frac{1}{\lambda_{s1}^2} + \frac{1}{\lambda_{s2}^2} + \frac{1}{\lambda_{s3}^2} + \frac{1}{\lambda_{s4}^2}} = \sqrt{\text{tr}\left\{\left(\mathbf{H}^\mathrm{T}\mathbf{H}\right)^{-1}\right\}} \tag{2.17}$$

where $\lambda_{si}$ (with $i = 1,\ldots,4$) represent the nonnull (positive) singular values of the matrix $\mathbf{H}$.

## 2.4   Satellite Navigation Systems (GNSS)

The *Global Positioning System* (GPS) and *Global Navigation Satellite System* (GLONASS) were designed as dual-purpose systems, and free access to their open signals made them available to users worldwide.

More than 30 years of successful GPS and GLONASS general usage, along with the strategic importance of global navigation satellite systems, have convinced the European Union, China, Japan, and India to design their own global or regional systems. The designers state rather promising characteristics of these future systems, so the United States and Russia—providers of the GPS and GLONASS, respectively—are upgrading their systems to keep their leading positions in the industry. New technologies are being applied to greatly improve the accuracy, availability, and interference immunity of their systems and to enhance the quality of navigational services, so minimizing the vulnerability aspects of global satellite navigation.

The appearance of new participants in the global navigation sphere enforces all GNSS providers to coordinate their activities on a worldwide basis. On the one hand, this coordination ensures interoperability between all GNSS, which results in tangible benefits for users who combine several systems. On the other hand, it is also necessary to provide compatibility between the systems so that different GNSS do not interfere with each other, which is important for the navigational independence of owner countries.

### 2.4.1   The Global Positioning System

The Navstar GPS is a satellite positioning and navigation system conceived as a US Department of Defense (DoD) multiservice program in 1973 (Parkinson and Spilker, 1996; Kayton and Fried, 1997). The US Air Force manages the overall program at the GPS Joint Program Office (JPO), which is located at the Space Division Headquarters in Los Angeles, California. The result of this program is an all-weather radio-navigation system that allows any user equipped with a GPS receiver to determine its 3D position, velocity, and time (PVT) information anywhere on the surface of the Earth and in the air.

GPS is a ranging system, which means that it provides user-to-satellite range measurements together with satellite position-related data, to be processed by the user equipment to compute four-dimensional navigation solutions (user position and time) in a dynamic environment. It also gives precise Doppler range rate-of-change information. It is a passive means of navigation: a GPS user receiver does not transmit anything, but only receives signals from GPS satellites. All the GPS signals share the same carriers, and the signal separation is based on code division multiple access (CDMA). This results in reduced complexity in the radio receiving equipment.

The main advantage of GPS is that it provides an accurate all-weather worldwide navigation capability that remains constant with time and is therefore not affected by any kind of drift. Its main disadvantage is that satellite signals are vulnerable to intentional and unintentional interference and temporary unavailability due to signal masking or lack of visibility coverage: these effects can cause a temporary loss of navigation solutions that can be unacceptable for some applications. Augmentations and/or data fusion with other sensors is therefore necessary in order to overcome these problems.

### 2.4.1.1   GPS Architecture

GPS was designed to provide two positioning services: the Precise Positioning Service (PPS) and the Standard Positioning Service (SPS). Available free of charge to any user worldwide, SPS was initially based on the tracking of coarse/acquisition (C/A) code modulated on an L1 frequency carrier in the GPS signal. The typical accuracy of the SPS was several metres. Nevertheless, degradation of the SPS in order to deny accuracy to an unfriendly force (and, unfortunately, the civil community) was imposed by the introduction of Selective Availability (SA), an intentional error that affected the satellite clock phase and ephemeris data, and could be added at any time by the DoD, so decreasing the SPS accuracy to about 100 m. However, the United States switched off SA in May 2000.

Denied to unauthorized users, PPS was initially based on the acquisition and tracking of Code P, modulated over both L1 and L2 carriers. PPS is intended to be a highly accurate and antispoofing (A–S) service. Consequently, P-code was always filtered by an encryption process (modulo-2 addition of W code) and transformed to Y-code, the actual code used in the PPS. Y-code cannot be spoofed nor used by unauthorized users that are not crypto-capable. Only encryption-keyed GPS receivers, typically NATO and military users of US-friendly countries, can benefit from the PPS high-performance service, which has a positioning accuracy of a few decimetres. Moreover, differently from SPS users, SA can be removed by PPS users. The current GPS SPS and PPS are based on traditional and modern signals providing new capabilities and quality.

The GPS consists of three segments: the Space Segment (SS), the Operational Control Segment (OCS), and the User Equipment Segment. The standard SS consists of a constellation of 24 operative satellites plus three spares. The 24 satellites are positioned in six Earth-centered orbital planes with four satellites in each plane. The orbits are nearly circular, with an inclination of 55° with respect to the equatorial plane and equally spaced from one another at 60° around the Earth's rotational axis. The orbits are designated by the first six letters of the alphabet, A, B, C, D, E, and F. The altitude of each satellite is approximately 20 200 km from the Earth's surface. The nominal orbital period is 11 h and 58 min, that is, one-half of a sidereal day (23 h, 56 min). Hence, a GPS satellite makes two revolutions around the Earth in a sidereal day, but in terms of a solar day of 24 h, each satellite is in the same position in the sky about 4 min earlier each day.

Several notations are used to refer to GPS satellites. A letter followed by a number indicates the orbital plane and which satellite is in that orbit (e.g., B3 stands for satellite #3 on orbit B). The second notation is that of the US Air Force: the Satellite Vehicle Number (SVN) and Navstar satellite number, for example, SVN 2 refers to satellite 2. The third notation indicates a satellite by the unique pseudorandom (PRN) code generator on board a satellite, for example, PRN 5.

The navigation payload of each satellite consists of a transmitting antenna with a field-of-view coverage beam of 28.6° along with RF and digital processing equipment and redundant atomic clocks. The latter are cesium and rubidium clocks characterized by a stability of about 1–2 parts in $10^{14}$. A good frequency standard is at the heart of the GPS navigation concept; hence redundancy in the generation of frequency as obtained by the combination of more than two atomic clocks is essential in order to assure functional reliability and accuracy.

Seven generations of satellites—Block I, IA, II, IIA, IIR, IIR-M, and IIF—have been developed since the GPS project began. Currently (March 2016), GPS SS includes 32 satellites (see Figure 2.1), including Block II-R—12, Block IIR-M—8, and Block II-F—12 satellites.

Development of advanced generation GPS-IIR-M may be considered as the beginning of active modernization of the system. Significant improvements have been implemented to block IIR-M satellites: a second civil signal L2C providing higher positioning reliability and accuracy, two new military signals L1M and L2M providing improved antispoofing protection and antijamming immunity, and an upgraded antenna panel to increase the transmitted signal power. The Block IIR-M satellite commissioning increased the GPS accuracy from 1.6 m in 2001 up to 0.9 m in 2009 (SIS URE). The third civil signal of the L5 band and a reprogrammable onboard processor increased the active life cycle to 12 years and will be implemented for the next GPS-IIF generation.

Currently, the design of the GPS-III generation satellites is ongoing (Revnivykh, 2012a), and this step in incremental complication will provide:

- Independent operation, launch windows not being required—satellites can be launched all year-round into any orbit plane.
- High-speed and high-precision satellite cross-links (Ka-band 22.5–23.55 GHz; V-band 59.3–64 GHz, laser communication).
- High-power transmission antennas operating in narrow beam mode (power increase up to 27 dB) to implement the Navwar function, this being the American DoD initiative to provide navigation services to custom users in specified military operation areas:
  - A search and rescue (SAR) function
  - An additional L4 band frequency (1379.91 MHz) to compensate for ionosphere error
  - An L1C signal with BOC (1.1) modulation compatible and complementary to the L1 Galileo signal
  - Onboard integrity monitoring capabilities

The OCS has responsibilities for maintaining the satellites and ensuring their proper functioning. This includes keeping the satellites in their proper orbital positions and in their correct attitudes, monitoring the health and status subsystems, and maintaining the proper functioning of the navigation payload. In particular, the OCS has to update each satellite clock in order to exactly model its present drift and to synchronize it to the GPS time. Moreover, the OCS updates the ephemeris and almanac in order to store on the satellite its exact position. These data represent the most important information that every satellite continuously broadcasts to the users via the navigation messages.

To accomplish the aforementioned functions, the OCS consists of four subsystems: the *Master Control Station* (MCS), the *Backup Master Control Station* (BMCS), the network of *Ground Antennas* (GA), and the network of globally distributed *Monitor Stations* (MSs) (GPS Standard Positioning Service (SPS) Performance Standard, 4th edition, September 2008, GPS Modernization Fact Sheet, http://www.gps.gov/technical/ps/2008-SPS-performance-standard.pdf).

The MSs are georeferenced reception stations spread around the Earth in order to maximize the visibility of the GPS satellites (see Figure 2.5). Initially, this network comprised five sites located in Hawaii (United States), Colorado Springs (United States), Ascension Island (South Atlantic), Diego Garcia (Indian Ocean), and Kwajalein (North Pacific). The number of stations in the network was increased to 6 in 2001 by the addition of Cape Canaveral (United States) and up to 17 in 2005–2006 by the inclusion of Adelaide (Australia), Buenos Aires (Argentina), Hermitage (United Kingdom), Manama (Bahrain), Quito (Ecuador), Washington, DC (United States), Fairbanks (Alaska, United States), Osan (Republic of Korea), Papeete (Tahiti),

**Figure 2.5**   The GPS orbit constellation

Pretoria (South Africa), and Wellington (New Zealand). According to station configuration, each satellite is monitored continuously by not less than three MSs.

The main activity of these stations consists of satellite signal code and carrier-phase tracking as the satellites pass overhead and the computation of ranges and delta ranges. These data, together with the navigation message and the local weather data, are transmitted to the MCS at Falcon Air Force Base, Colorado, where they are processed and used to form satellite clock corrections and ephemeris and almanac data for each satellite. The MCS then sends its output to the GAs that are colocated at four MSs—Ascension Island, Cape Canaveral, Diego Garcia, and Kwajalein. From each GA, the data within the navigation messages, and also command telemetry from the MCS, are uploaded to the satellites via an S-band data communication uplink. The OCS is capable of uploading the data to each satellite every 8 h, but it is usually done just once each day.

The new-generation ground OCS is based on new antijamming technologies, forecasting algorithms, and more frequent orbit and time data updates, to support gradual capacity building for the control of GPS Block III satellites having extra functions. For example, Navwar will replace the existing OCS.

The User Equipment Segment consists of all the users worldwide who are equipped with a GPS receiver. This receiving equipment processes the L-band signals broadcast by the satellites to determine the user navigation solution, that is, its PVT. There are many applications using GPS-computed PVT—for example, aerial navigation applications need different performances for each phase of flight. The advent of GPS has led to a definition for the Required Navigation Performance (RNP) concept, which is a set of criteria by which the aerospace user segment can assess all the performances available by satellite navigation and, more generally, by whichever navigation system is utilized.

Since 1987, the GPS has used the World Geodetic System WGS84, developed by the US DoD, which is a unified terrestrial reference system for position and vector referencing. The GPS ephemeris is linked to the position of the satellite antenna phase center in the WGS84 reference frame, and the user receiver coordinates will be expressed in the same ECEF frame. The more recently refined realizations WGS84(G730) and WGS84(G873), using more accurate coordinates for the monitor stations, approximate to some International Terrestrial Reference System (ITRS) realizations and correspond to ITRF92 and ITRF94, respectively. The refined frame WGS84(G250) was introduced in 2002, which agrees with ITRF2000 at the centimetre level.

### 2.4.1.2 GPS Signals

As already pointed out, the GPS SS has evolved through seven phases of enhancement with the gradual appearance of new ranging signals. Actually, to keep existing navigators operable, the original ranging signals are not simply being replaced by new ones, but are to exist in parallel with these over a rather long transitional period. To distinguish between the original and the new signals in GPS literature, the former are usually referred to as *legacy signals*. It is therefore appropriate to begin the description of GPS signals with the legacy format.

### *Modulation*

GPS satellites broadcast two L-band signals: $s_{L1}(t)$ and $s_{L1}(t)$. The first is obtained by a modulation of an L1 carrier centered at a frequency of 1575.42 MHz, while the second utilizes an L2 carrier frequency of 1227.6 MHz. A detailed description of GPS signal structure can be found in Kayton and Fried (1997), Kaplan and Hegarty (2006), and Parkinson and Spilker (1996) and, of course, in the GPS Interface Control Document (ICD) (IS-GPS-200F, 202). Both these frequencies are multiples of the fundamental clock frequency $f_0 = 10.23$ MHz: $f_1 = 154 f_0$ for L1 and $f_2 = 120 f_0$ for L2.

The availability of two carriers at different frequencies facilitates ionosphere correction, and each is modulated by two types of digital (a sequence of 1's and 0's) modulation signal, the PRN codes (PRN P-code and PRN C/A code), and also a data signal (D) called the *navigation message*. The modulation is a binary phase shift keying (BPSK) that changes the phase of the carrier by 180° for each PRN or data signal change. Since a 180° phase shift simply changes the sign of the carrier, the BPSK is equivalent to a polarity modulation. The resulting mathematical expression for the signals is as follows:

$$s_{L1}(t) = AP(t)D(t)\cos(2\pi f_1 t + \varphi_{01}) + \sqrt{2}AC(t)D(t)\sin(2\pi f_1 t + \varphi_{01})$$
$$s_{L2}(t) = \frac{A}{\sqrt{2}}P(t)D(t)\cos(2\pi f_2 t + \varphi_{02})$$

$$(2.18)$$

where $A$ is the L1 carrier amplitude, $P(t) = \pm 1$ and $C(t) = \pm 1$ are the P and C/A PRN sequences, $D(t) = \pm 1$ is the data bit sequence, and $\varphi_{01}$ and $\varphi_{02}$ are the L1 and L2 carrier initial phases. Note that in the initial GPS segment version, the L2 carrier was modulated only by P-code and the data bits, while the in-phase L1 component was modulated by P-code along with data, and the quadrature L1 component was modulated by C/A code, also along with data.

Relativistic effects influence the nominal frequency $f_0$. The velocity of a satellite causes the onboard time to slow down, while the difference between the gravitational field strength on the Earth's surface and at the satellite orbit causes the onboard time to speed up. The net relativistic effect requires that the output of each SV frequency clock be tuned to $f_0 = 10.22999999543$ MHz in order to appear to a user on the ground as $f_0 = 10.23$ MHz.

### Legacy PRN Codes

The PRN codes are fixed length digital sequences of 1's and 0's that repeat periodically. The relevant property of these sequences is that they are near-orthogonal, meaning that the correlation between different sequences under arbitrary mutual time shift is almost zero. Also their autocorrelations are impulse-like, that is, nearing zero at any nonzero time shift. These features of PRN codes give the PRN modulated signals two fundamental functionalities:

1. Their near-orthogonality makes CDMA possible. With this technique, signals transmitted from many satellites do not interfere with each other even if they share the same carrier frequency and are time synchronous. The signal transmitted from each satellite can be distinguished among the others because a specific PRN sequence is assigned to each satellite.
2. Despite signal time continuity, the pulse-like autocorrelation of PRN code allows the ranging process as if the signal itself were a pulse. In fact, by means of correlating a PRN sequence with its time-shifted local replica, it is possible to determine the phase difference between the two sequences and then the mutual time delay and the spatial distance. This process is discussed in more detail in the section on receivers.

There are two types of ranging signals and, correspondingly, PRN codes: the C/A and the precise (P) types. The former is used for supporting the SPS, while the latter, in the encrypted version (Y), is adopted for the PPS. Each of the ranging signals is composed of repeated elementary symbols called *chips*, which are BPSK-modulated by PRN codes. Chip rates of the C/A and P signals are 1.023 and 10.23 MHz, respectively. Both ranging signals are periodic: the C/A has a real-time period of 1 ms, while the period of the P signal is 1 week.

C/A PRN codes belong to a quite popular family of binary Gold sequences existing for any length $N = 2^n - 1$ with natural $n$. Every Gold sequence is produced by element-wise modulo-2 summation of two basic maximal-length (or *m*-) sequences of the above length. The first one may be arbitrary, the second being chosen to be a result of *d*-decimation (picking every *d*th symbol) of the first one, where $d = 2^s + 1$ and $s$ is either relatively prime to $n$ (odd $n$) or has most common divisor 2 with $n$ ($n$ even) (Ipatov, 2005). Technically, each of two *m*-sequences is generated by an *n*-stage linear feedback shift register (LFSR), the initial state of the second specifying the Gold sequence obtained. Since there exist $N = 2^n - 1$ initial LFSR states (all-zero state excluded), the total of $N$ different Gold sequences can be formed in this way. With two original *m*-sequences also included, there are $N + 2 = 2^n + 1$ Gold sequences altogether.

The reason for the ubiquitous appearance of Gold sequence sets in CDMA applications is their remarkable correlation properties. The correlation peak $\rho_{max}$ of a Gold set, that is, the greatest among all undesired correlations (autocorrelation sidelobes and cross-correlation levels), is defined as

$$\rho_{max} = \begin{cases} \dfrac{\sqrt{2(N+1)}+1}{N}, n = 1 \bmod 2, \\ \dfrac{2\sqrt{N+1}+1}{N}, n = 2 \bmod 4. \end{cases} \tag{2.19}$$

It is well known (Ipatov, 2005) that odd-$n$ Gold sets are optimal in the sense that no set of binary sequences of the same length and volume can have a smaller value of $\rho_{max}$. As is seen from Equation (1.19), with a large length $N \gg 1$, the correlation peak of the Gold set becomes close to either $\sqrt{2/N}$ for odd $n$ or $2/\sqrt{N}$ for even $n$.

For arranging an SPS Gold set of length, $N = 1023$ was chosen. In this case $n = 10$ so that two 10-stage LFSR were needed to create basic $m$-sequences. It is common to describe a feedback circuit of LFSR by polynomial notation, a nonzero coefficient of the term $t^x$ pointing out that the $l$th LFSR stage left-to-right output is connected to the modulo-2 adder. Specifically, GPS C/A code feedback polynomials of two basic LFSR sequences are

$$g_1(x) = x^{10} + x^3 + 1, \qquad g_2(x) = x^{10} + x^9 + x^8 + x^6 + x^3 + x^2 + 1.$$

Both LFSR are initialized by all-ones states. In order to imitate different initial states of the second LFSR (necessary to form different Gold sequences), the shift-and-add property of the $m$-sequence can be used: the element-wise modulo-2 sum of an $m$-sequence and its time-shifted replica produces another time-shifted replica of the same $m$-sequence (Ipatov, 2005). Thus, modulo-2 combined outputs of two stages of the second LFSR make it possible to obtain 45 time-shifted replicas of the second sequence as though 45 different initial states were used. The patterns of such circuitry are tabulated in interface documentations (ICD-GPS-200F, 202). Among 45 Gold sequences generated accordingly, 37 are chosen for SV, the rest being reserved for testing and similar purposes.

It follows from the previous equations that when $n = 10$, the correlation peak of the Gold set is about −23.9 dB. However, allowing for mutual Doppler shift, this value can increase up to near −21 dB. This means that, processing the signal of the selected SV, the user's terminal can experience interference from some side SV (multiple access interference (MAI)) whose power is only −21 dB lower than that of the useful signal. Though correlation spikes at the maximal level are rather infrequent, and the average one-signal C/A MAI power is about −30 dB, the length of the new SPS codes is designed in order to gradually replace current C/A code with future GPS codes 10 times larger, resulting in remarkable MAI reduction.

As has already been pointed out, the chip rate of P-code is 10 times that of C/A at 10.23 MHz. The P-code itself is obtained as a chip-wise modulo-2 sum of two sequences X1 and X2, the first, in its turn, being a chip-wise modulo-2 sum of sequences X1A and X1B, while the second being produced by modulo-2 summation of sequences X2A and X2B. The X1A, X1B, X2A, and X2B are truncated $m$-sequences generated by four 12-stage LFSR, whose feedback polynomials are

$$g_{X1A} = x^{12} + x^{11} + x^8 + x^6 + 1,$$
$$g_{X1B} = x^{12} + x^{11} + x^{10} + x^9 + x^8 + x^5 + x^2 + x + 1,$$
$$g_{X2A} = x^{12} + x^{11} + x^{10} + x^9 + x^8 + x^7 + x^5 + x^4 + x^3 + x + 1,$$
$$g_{X2B} = x^{12} + x^9 + x^8 + x^4 + x^3 + x^2 + 1.$$

When the GPS week begins, all LFSR are set to their initial states. The natural cycles of X1A and X2A are shortened by omitting the last three chips so that the sequences produced have a repetition period of 4092 chips instead of 4095. In a similar manner, the periods of the X1B and X2B are truncated to 4093 chips instead of 4095. As will be seen, the cycle lengths 4092 and 4093 are relatively prime, so with no more manipulation done, X1 would have a period of 16 748 556 chips. In fact, this period is truncated in the following way: 3750 X1A cycles, that is, $3750 \times 4092 = 15\,345\,000$ chips, make up an epoch (period) of X1 spanning a time interval of 1.5 s. After 3749 X1B cycles or $3749 \times 4093 = 15\,344\,657$ chips, X1B LFSR is halted for 343 chips to equate its epoch to that of X1A.

Similarly controlled LFSR X2A and X2B produce the X2 sequence with a period of 15\,345\,037 chips, which is 37 chips longer than that of X1. Since epoch lengths of X1 and X2 are relatively prime, the modulo-2 sum of these sequences would have an extremely long period of $15\,345\,037 \times 15\,345\,037 \approx 2.3547 \times 10^{14}$ chips or more than 38 weeks. However, as indicated earlier, at the beginning of every week, all registers are reset to their initial states, so that the actual P-code length is 7 days, corresponding to around $6.1871 \times 10^{12}$ chips. A unique P-code for each SV is assigned by a proper time shifting of X2 relative to X1 preceding modulo-2 summation. As in the case of C/A code, there are 37 available P-codes altogether, 32 being allocated to operational SV and the rest reserved for some other purposes.

The reasons behind using such enormous P-code lengths are to attain antispoofing ability in addition to preventing unauthorized access to the PPS resource. If the P-code repeated every split second or several seconds, a potential eavesdropper could intercept and store a one-period segment of it. Then a replica so obtained could serve either as a correlator reference to implement PPS or as an instrument for arranging spoofing. Actually however, a huge length of P-code cannot in itself secure high antispoofing ability because the code structure is presented in the public domain. Due to this, an overlay encryption of the P-code by way of its modulo-2 summation with a secret Y-code is provided, resulting in the code denoted P(Y). The Y-code structure is described only in the strictly classified documents and keys necessary for its generation and is available only to a very limited number of users and with the permission of the body running the system. It is known, nevertheless, that the chip rate of the Y-code is 20 times lower than that of the P-code, meaning that one Y-chip spans 20 P-chips.

### Navigation Message

The navigation message is a digital data stream at a rate of 50 bits per second (bps) that is transmitted by each satellite. The bits in the stream are synchronized and added modulo-2 with the PRN codes and then modulated onto the L1 and L2 carriers. The results of the modulation process are briefly exemplified in Figure 2.6.

The navigation message contains all the information necessary to allow the user to compute the position coordinates of the satellite and to correct the offset, drift, and drift rate of the satellite atomic clock with respect to the GPS time. The GPS time starts at midnight between Saturday and Sunday, and the largest unit of time is 1 week, defined as 604\,800 s. GPS time is maintained by the OCS within a maximum difference (modulo-1 second) of 1 μs from the Universal Time Coordinated (UTC). The navigation message also contains other information about the health status of the satellite, the week number, and the ionosphere delay model parameters. Moreover, it provides the almanacs of the other satellites, that is, the parameters the user needs in order to compute a coarse position estimation of those other satellites. In particular, the navigation message consists of a 30 s frame subdivided into five 300-bit subframes of 6 s each. Each subframe starts with two 30-bit words, the Telemetry Word (TLM Word) and the Handover Word (HOW). The TLM Word is used in order to synchronize the C/A code with the subframes and hence to establish the time of reception: in fact C/A code duration is 1 ms, while the transit time from satellite to user is about 70 ms, so the received C/A code epoch is 1 out of 70. TLM allows for resolving this ambiguity. The HOW contains the Z-count, a number providing the user time information required to hand over to P-code.

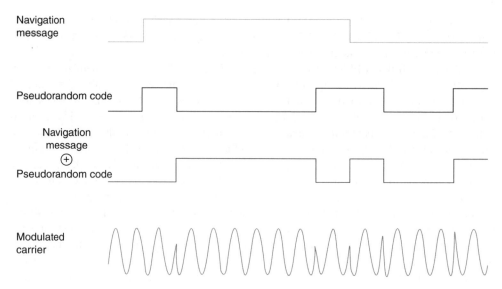

Navigation message

Pseudorandom code

Navigation message
$\oplus$
Pseudorandom code

Modulated carrier

**Figure 2.6**   The carrier modulation process

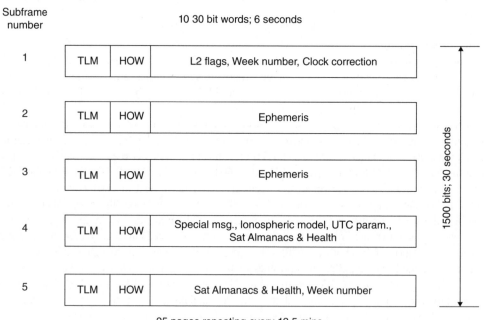

Subframe number

10 30 bit words; 6 seconds

| 1 | TLM | HOW | L2 flags, Week number, Clock correction |
| 2 | TLM | HOW | Ephemeris |
| 3 | TLM | HOW | Ephemeris |
| 4 | TLM | HOW | Special msg., Ionospheric model, UTC param., Sat Almanacs & Health |
| 5 | TLM | HOW | Sat Almanacs & Health, Week number |

1500 bits; 30 seconds

25 pages repeating every 12.5 mins

**Figure 2.7**   The GPS navigation message frame

**Table 2.1**   Orbital parameters

| Parameter | Description |
|---|---|
| $M_0$ | Mean anomaly |
| $\Delta n$ | Mean motion difference |
| $e$ | Eccentricity |
| $\sqrt{A}$ | Semimajor axis square root |
| $\Omega_0$ | Longitude of ascending node |
| $i_0$ | Inclination angle |
| $\omega$ | Argument of perigee |
| $\dot{\Omega}$ | Rate of right ascension |
| $di/dt$ | Rate of inclination angle |
| $C_{uc}$ | Amplitude of cos harmonic correction term to the argument of latitude |
| $C_{us}$ | Amplitude of sin harmonic correction term to the argument of latitude |
| $C_{rc}$ | Amplitude of cos harmonic correction term to the orbit radius |
| $C_{rs}$ | Amplitude of sin harmonic correction term to the orbit radius |
| $C_{ic}$ | Amplitude of cos harmonic correction term to the angle of inclination |
| $C_{is}$ | Amplitude of sin harmonic correction term to the angle of inclination |
| $t_{oe}$ | Reference time for ephemeris |
| IODE | Issue of data ephemeris |

The structure of the navigation message frame and its five subframes is shown in Figure 2.7.

Subframes 1, 2, and 3 repeat every 30 s for about 1 h, while subframes 4 and 5 change 25 times each, so that a complete data message requires the transmission of 25 full 1500-bit frames, or 12.5 min.

Subframes 2 and 3 contain the ephemeris parameters. These are the Kepler orbital parameters that describe a curve fitting the trajectory of a satellite during a certain time interval of 4 h or more. These parameters are summarized in Table 2.1.

In Table 2.1, in addition to the classical Kepler parameters, some others are indicated and used in satellite position determination to take into account also the orbit perturbations due to the sun and moon gravitational forces and the solar radiation pressure. Table 2.2 shows the algorithm that is widely used to compute the satellite position from the orbital parameters.

### Modernized GPS Air Interface Signals

The legacy GPS signals were designed more than four decades ago, so that their philosophy reflected both user demands and the technology advancements anticipated for the last quarter of twentieth century. After the millennium, it became evident that novel and radical improvements could be introduced into the GPS air interface architecture to bring it into accordance with new performance standards and technological possibilities. Among the requirements dictated by dramatically extended GNSS applications, there are higher positioning accuracies for nonmilitary navigators, the possibility of user operation under poor signal conditions, better spoofing immunity, and antijam resistance, to name but a few.

To attain such targets, several initiatives proved to be fruitful:

- The introduction of a special ranging signal called a *pilot signal*, which being free of data modulation, provides better phase lock tracking robustness, so enabling a user to operate in weak signal scenarios.

**Table 2.2**  Satellite position computation

| Equation | Description |
| --- | --- |
| $t_k = t - t_{0e}$ | Time from reference $t_{0e}$ |
| $n = n_0 + \Delta n$ | Corrected mean motion |
| $M_k = M_0 + n t_k$ | Mean anomaly at time $t_k$ |
| $E_k = M_k + e \sin(E_k)$ | Eccentric anomaly at time $t_k$ |
| $v_k = \operatorname{atan}\left(\dfrac{\sqrt{1-e^2}\,\sin(E_k)}{\cos(E_k)-e}\right)$ | True anomaly at time $t_k$ |
| $\Phi_k = v_k + \omega$ | Argument of latitude at time $t_k$ |
| $\delta u_k = C_{us}\sin(2\Phi_k) + C_{uc}\cos(2\Phi_k)$ | Second harmonic perturbation to argument of latitude at time $t_k$ |
| $\delta r_k = C_{rs}\sin(2\Phi_k) + C_{rc}\cos(2\Phi_k)$ | Second harmonic perturbation to orbit radius at time $t_k$ |
| $\delta i_k = C_{is}\sin(2\Phi_k) + C_{ic}\cos(2\Phi_k)$ | Second harmonic perturbation to inclination angle at time $t_k$ |
| $u_k = \Phi_k + \delta u_k$ | Corrected argument of latitude at time $t_k$ |
| $r_k = A\left(1 - e\cos(E_k)\right) + \delta r_k$ | Corrected orbit radius at time $t_k$ |
| $i_k = i_0 + \delta i_k + di/dt \cdot t_k$ | Corrected inclination angle at time $t_k$ |
| $x_{1k} = r_k \cos(u_k)$ | Position in orbital plane at time $t_k$ |
| $y_{1k} = r_k \sin(u_k)$ | Position in orbital plane at time $t_k$ |
| $\Omega_k = \Omega_0 + (\dot{\Omega} - \dot{\Omega}_e)t_k - \dot{\Omega}_e t_{0e}$ | Corrected longitude of ascending node at time $t_k$ accounting for Earth's rotation rate $\dot{\Omega}_e$ |
| $x_k = x_{1k}\cos(\Omega_k) - y_{1k}\sin(\Omega_k)\cos(i_k)$ | ECEF coordinates at time $t_k$ |
| $y_k = x_{1k}\sin(\Omega_k) + y_{1k}\cos(\Omega_k)\cos(i_k)$ | ECEF coordinates at time $t_k$ |
| $z_k = y_{1k}\sin(i_k)$ | ECEF coordinates at time $t_k$ |

- The transmission of ranging signals on a newly introduced carrier at 1176.45 MHz (L5 band) in order to allow for the demands of civil aero navigation (and rejecting L2 band as being overloaded).
- An increase in the PRN code length, which provides better interference suppression.
- The design of a new military signal separated in the frequency domain from the civil versions, so allowing a noticeable rise in power of the military signal with no deterioration in any civil terminal performance.

Some specific information now follows:

**The M-code**: Block IIR-M satellites, the first of which was launched in September 2005 along with legacy signals, now transmit a new military signal called *M-code*, and future generation SVs will certainly transmit this signal, too. Unfortunately, little can be said about its fine structure, which is described only in highly classified sources and encrypted in a sophisticated manner. However, what is in the public domain (Kaplan and Hegarty, 2006; Groves, 2008) reveals that this signal is designed to be autonomous, that is, searching of it is direct, without any preliminary acquisition of C/A or other open signal. Also known is that modulation of the M-code is of the *binary offset carrier* (BOC) type. This modulation mode, long established as the Manchester code in GPS documents, is denoted by BOC(*m,n*), which means that a plain BPSK signal with chip rate $nf_c$ is multiplied with a meander wave of frequency $mf_c$, where $m \geq n$ and $f = 1.023$ MHz is a chip frequency of C/A code. Specifically, M-code is BOC(10,5) modulated so that its spectrum main lobes are separated by a 20 MHz space within which the spectra of the rest of the signals are located. Thus, the M-code spectrum is isolated from the rest of the spectra, so making it possible to introduce (with launching Block III satellites) a new GPS operation mode termed *spot beam* with negligible damage to the SPS. In such a regime, a high-gain directional antenna provides, within a limited area (hundreds of kilometres), M-code power 20 dB higher than that delivered by a standard (all-Earth) antenna. While a standard M-code power is anticipated to be $-158$ dB W, the spot beam mode power should be about $-138$ dB W.

**L2C**: In the original GPS architecture the L2 band was not intended for exploitation by nonmilitary users, so that only P signal was broadcast by the space segment. However, a real GPS role in civil applications has grown drastically since the beginning of the twenty-first century, making obvious the necessity of access by a nonmilitary terminal to the accuracy resource warranted by two-frequency operation. After some testing it was decided that a better option is to refrain from doubling C/A code in the L2 band in favor of a new signal L2C (C meaning "civil"). The L2C signal came on air with the launch of the first GPS Block IIR-M satellite in September 2005.

An initial novelty in L2C compared with C/A is that the former contains two components: CM (M for "moderate") and CL (L for "long"). CM code has a length of 10.230 chips—10 times longer than C/A code—which potentially improves the suppression of interference (MAI, intersystem, and narrowband) by about 10 dB. The length of CL code may even be 75 times greater, equaling 767.250 chips, giving a further gain in interference immunity of over 18 dB.

Navigation data at a net rate of 25 bit/s before modulating ranging code are encoded by a rate 1/2 and constraint length 7 convolutional FEC code. Two times slower data rates, along with quite efficient FEC, make an L2C data channel much more reliable than a C/A channel. Then, the data stream of rate 50 bit/s modulates only CM code, leaving CL data-free, thereby serving as a pilot signal. Both these codes have chip rates of 511.5 kHz and are time-multiplexed: after CM chip, CL chip follows, then again CM chip, etc. Thus, the total chip rate of L2C remains the same as that of the C/A code: 1.023 MHz. Thus, a real-time period of CM component is 20 ms, while for CL code it is 1.5 s.

Structurally, both CM and CL codes are simply truncated $m$-sequences produced by 27-stage LFSR as described by the polynomial

$$g(x) = x^{27} + x^{24} + x^{21} + x^{19} + x^{16} + x^{13} + x^{11} + x^9 + x^6 + x^5 + x^4 + x^3 + 1$$

Specific to an individual SV is just an initial LFSR state to which the CM register is reset every 20 ms and the CL register every 1.5 s. The predicted power level of this signal near the Earth's surface is $-160$ dB W.

More detailed information on L2C signal can be found in Interface Specification (IS-GPS-200F, 2Q2).

**L5**: This signal is designed to meet rigorous requirements for safety-of-life (SOL) transportation and other applications of a similar kind. It is broadcast by satellites of generation Block II-F, the first of which was launched in May 2010. The chip rate of the L5 signal is the same as that for P-code, that is, 10.23 MHz. It thereby allows a nonmilitary user to position with a tenth of the noise error compared with the legacy civil signals. There are two L5 components, I5 and Q5, modulating a carrier of 1176.45 MHz in quadrature. Each of sequences I5 and Q5 is formed by modulo-2 summation of two subsequences XA and XB, being in turn generated by a properly reset 13-stage shift register. The feedback polynomials of these LFSR are

$$g_{XA}(x) = x^{13} + x^{12} + x^{10} + x^9 + 1, \quad g_{XB} = x^{13} + x^{12} + x^8 + x^7 + x^6 + x^4 + x^3 + x + 1.$$

The period of I5 and Q5 is 10 230 chips, that is, 1 ms. Since 13-stage LFSR would naturally output $m$-sequence of length 8191, some resets are used to come to the above period. Specifically, the natural cycle (8191) of XA is reduced to 8190. In other words, one chip before LFSR XA comes to its initial state (all ones) is set to this state forcibly, the XB register continuing its normal cycle. After that, the sequence XA advances XB by one chip until 1 ms point arrives when both LFSRs are returned to their initial states. Individual I5 and Q5 codes for every SV are arranged by an appropriate choice of initial state of LFSR XB.

As usual, a data stream is modulo-2 added to the ranging code. The net data rate for L5 signal is 50 bit/s, but after convolutional FEC (the same as for L2C signal), the encoded stream rate grows two times up to 100 bit/s. Data bits manipulate only I5 code, leaving Q5 data-free to serve as a pilot signal.

There is one more new feature in L5 that is absent in C/A code. To make a transition from a 1 ms code period to data symbol boundaries easier, the ranging codes I5 and Q5 are overlaid by modulo-2 added codes with a chip duration 1 ms. For example, in I5 branch is the Neuman–Hoffman code of length 10 : 0000110101, so that its real-time period is exactly the data symbol duration of 10 ms. An overlay code of Q5 is the Neuman–Hoffman code of length 20 : 00000100110101001110.

As estimated, an Earth surface power of L5 signal should be slightly above −155 dB W. For further details, see Interface Specification (Navstar, 2011).

**L1C**: Deploying SVs of the Block III series is due to start in 1014 and will bring into existence an enhanced civil signal of L1 band (1575.42 MHz) called L1C, which is designed to replace C/A code in the long term. Similarly to L2C, it will have two components, one of them, L1C$_p$, being a data-free pilot signal and the other, L1C$_D$, being data-modulated in the usual form. To better separate modernized and legacy L1 signals (to be used in parallel for a decade or so) for the L1C BOC(1,1), a modulation format has been chosen. A ranging code of L1C$_D$ having length 10 230 chips is obtained by extending a Weil sequence of length 10 223 with a common 7-chip inset whose position is dependent on the SV number. In its turn, the Weil sequence is a result of modulo-2 summation of the Legendre sequence (Ipatov, 2005) with its time-delayed replica. Varying delay, specific Weil sequences with predictable correlation features are obtained for assignment to different SVs. A navigation message of net rate 50 bit/s is FEC-encoded by a half-rate, length 1200, LDPC code to produce data a stream of rate 100 bit/s modulo-2 added to L1C$_D$.

The pilot L1C$_p$ component has a layered (cascaded) structure. The primary code which is again an extended Weil sequence of length 10 230 is overlaid by an SV-specific external (overlay) code of length 1800 and 10 ms chip duration modulo-2 added to the primary one. This results in L1C$_p$ sequence of a total period 18 s. Overlay codes themselves are just truncated $m$-sequences or Gold sequences of natural length 2047.

Earth surface received powers of L1C components are expected to be −158.25 dB W for a pilot code L1C$_p$ and −163 dB W for a data-modulated part L1C$_D$. (See Interface Specification IS-GPS.)

## 2.4.2  GLONASS

The GLONASS is a satellite navigation system developed by the former USSR (now the Russian Federation). The first GLONASS satellite was launched on October 12, 1982, and by 1993 the system consisted of 12 satellites and was formally declared operational for limited usage. By December 1995 the full constellation of 24 satellites was fully operational. As a result, the system performance reached the level of the American GPS that had achieved full operational status a year earlier. The economic problems of 1990s and the short 3-year active life cycles of the first-generation GLONASS satellites led to an orbit constellation degradation down to six satellites in orbit in 2001. Due to the Governmental Program adopted in 2001, the system was completely restored and deeply modernized by 2012. There are many similarities with GPS but also important implementation differences (Kayton and Fried, 1997).

### 2.4.2.1  GLONASS Architecture

GLONASS was designed to provide two positioning services: the *Standard Accuracy* (SA) service and the *High Accuracy* (HA) service, which in Russian documents are abbreviated to ST and VT, respectively (GLONASS ICD, 2008). SA is available free of charge to any user worldwide and is based on the acquisition and tracking of SA code modulated over L1-band and L2-band carriers in FDMA GLONASS air interface. Degradation of the SA service by SA introduction is not used.

The HA service, which is more accurate and better protected against nonintentional and intentional interference, is based on the acquisition and tracking of HA code modulated over L1-band and L2-band carriers in FDMA GLONASS air interface. The continuous HA service is intended only for authorized users, which is why the encryption procedures are employed.

In the modernized GLONASS service system, the *open* (O) access service and the *sanctioned* (S) access service will be based on FDMA(F) and CDMA(C) L1-band, L2-band, and L3(5)-band signals including L1OF, L2OF, L1SF, L2SF (equal to L1SA, L2SA, L1HA, L2HA) and L1OC, L2OC, L1SC, L2SC, L3OC. The L1OCM and L5OC, compatible with L1C and L5 GPS, may complete this set of signals to reinforce GNSS interoperability (Revnivykh, 2012b).

The GLONASS comprises three components: an orbiting constellation of satellites (the *Space Segment*), ground-based measurement and control facilities (*Ground Control Segment*) to maintain open access and sanctioned access basic navigation services, and user equipment (the *User Segment*).

Actually, Space and Ground-Based Augmentation Segments are becoming integral parts of the GLONASS. These provide GLONASS with differential accuracy and integrity monitoring for local and regional services, precise orbit and clock determination systems for global-level services differential accuracy in postprocessing or real time, and a fundamental segment to maintain the geodetic reference frame, universal time, and an Earth rotation service.

The GLONASS orbit constellation is composed of 24 satellites placed in groups of eight on three orbital planes. These three planes are separated by 120° and have inclination angles of 64.8° with respect to the equatorial plane. The revolution period of the satellites is 675.73 min (for GPS it is 717.94 min) with an orbital semimajor axis a little smaller than GPS (25 510 km). Due to these implementation differences, in some respects the GLONASS orbital constellation suites are more suitable than the GPS suites: firstly, for positioning in high northern or southern latitudes and secondly because the satellite orbital positions are

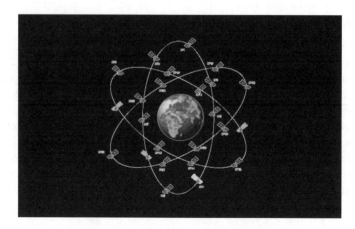

**Figure 2.8**   GLONASS orbit constellation

more stable and so do not need any orbital corrections, the satellite rotations over the Earth and the Earth's rotation being asynchronous (Figure 2.8).

Historically, three satellite generations GLONASS, GLONASS-M, and GLONASS-K were developed. Currently (March 2016), the GLONASS orbital constellation includes 28 satellites: 26 GLONASS-M and 21 GLONASS-K. Typical satellite active life cycle estimations are about 3 years for GLONASS, 7 years for GLONASS-M, and 10 years for GLONASS-K generations. GLONASS navigation accuracy is based on precise onboard clock synchronization and time-keeping. The clocks include sets of three cesium frequency standards with instability $5 \times 10^{-13}$ for GLONASS and $1 \times 10^{-13}$ for GLONASS-M and two cesium plus two rubidium standards, both with instability $5 \times 10^{-14}$. A reserved timekeeping system is also present.

Future constellation modernization involves new open and sanctioned access CDMA signal implementation in three GLONASS frequency bands. Also, systems for SAR and for cross-links in radio and optical bands for data transmission and intersatellite measurements and for improving ephemerides and clock data will be implemented.

The GLONASS Ground Control Segment is responsible for the proper operation of the GLONASS. It consists of a System Control Center (SSC); a network of five Telemetry, Tracking, and Command (TT&C) centers; the Central Clock (CC); three Upload Stations (UL); two Satellite Laser Ranging Stations (SLR); and a network of four Monitoring and Measuring Stations (MS), all distributed over the territory of the Russian Federation. Six additional monitoring and measurement stations are to begin operation in the near future, and this network is complemented by additional measurement stations, also on the same territory, and equipped with laser ranging and other monitoring facilities. The construction of further MS outside Russian territory is ongoing. Synchronization of all the processes in the GLONASS is very important for its proper operability, and the Central Synchronizer is a highly precise hydrogen atomic clock that forms the entire GLONASS time scale.

The main differences from GPS concern the time and the reference coordinate systems, so the GLONASS Ground Control Segment provides corrections such that GLONASS time is related to UTC within 1 µs. Moreover, differently from GPS time, it follows the leap second corrections that UTC occasionally makes.

The GLONASS ephemerides are given in the Parametry Zemli 1990 (Parameters of the Earth, 1990) (PZ-90) reference frame. As for the WGS84, this is an ECEF frame with a set of associated fundamental parameters (Table 2, GLONASS ICD, 2008). The PZ-90.02 reference system was refined and updated on all operational GLONASS satellites on September 20, 2007, and now uses the ECEF reference frame, which is an updated version of PZ-90 closest to the ITRF2000. The transformation from PZ-90.02 to ITRF2000 contains only an origin shift vector, but no rotations nor scale factor.

### 2.4.2.2  GLONASS Legacy Signals and Navigation Message

GLONASS satellites broadcast two L-band spread-spectrum signals, L1 and L2. The main difference compared with GPS is that in the original version of GLONASS, air interface satellites used FDMA instead of CDMA, with each satellite transmitting on a channel with its own frequency. However, like GPS, GLONASS uses two ranging codes that are transmitted in quadrature.

Each satellite was originally assigned a unique L1 frequency according to the following equation (GLONASS ICD, 2008):

$$f_{1i} = 1602 + 0.5625i \text{ MHz} \tag{2.20}$$

where $i$ is an integer ranging from −7 to 6. Antipodal space vehicles never observable simultaneously from anywhere on the Earth's surface are assigned the same $i$.

The original (block GLONASS) satellites transmitted only HA code in the L2 band, while the latter (GLONASS-M) sputniks broadcast both SA and HA codes. For the L2 signal, the frequencies were assigned to each satellite according to

$$f_{2i} = 1246 + 0.4375i \text{ MHz} \tag{2.21}$$

L1 and L2 signals are both modulated with the PRN codes and the navigation data. The PRN code characteristics are as follows:

- The SA code is a 511-bit length $m$-sequence clocked at a rate of 511 kHz. Hence it has a period of 1 ms. Its generator is a standard 9-stage LFSR described by a polynomial (GLONASS ICD, 2008).
- $g(x) = x^9 + x^5 + 1$.
- The HA code is a sequence short-cycled to 5 110 000 bits and clocked at a rate of 5.11 MHz, so that its period is 1 s, and is synchronized with the 1 ms sequence in order to ease the handover from one code to the other.

The navigation data is a 50-bit per second sequence modulo-2 added to both the codes. It is return-to-zero encoded and the modulation is actually 100 symbols per second. It differs from the GPS navigation message, being made up of lines, frames, and superframes. Streams of mixed data with SA and HA BPSK codes modulate, respectively, in-phase and quadrature components of both the L1 and L2 carriers. The navigation message provides users with the necessary information for positioning, that is, the GLONASS satellite coordinates, their clock offsets, and various other system parameters.

| Frame number | String number | 2 s | | | |
| | | 1.7 s | | | 0.3 s |
| --- | --- | --- | --- | --- | --- |
| I | 1 | 0 | Immediate data | Kx | TM |
| | 2 | 0 | for | Kx | TM |
| | 3 | 0 | transmitting | Kx | TM |
| | 4 | 0 | satellite | Kx | TM |
| | 5 | 0 | Non immediate data | 0 | TM |
| | : | : | (almanac) for | : | TM |
| | 15 | 0 | five satellites | Kx | TM |
| II | 1 | 0 | Immediate data | Kx | TM |
| | 2 | 0 | for | Kx | TM |
| | 3 | 0 | transmitting | Kx | TM |
| | 4 | 0 | satellite | Kx | TM |
| | 5 | 0 | Non immediate data | 0 | TM |
| | : | : | (almanac) for | : | TM |
| | 15 | 0 | five satellites | Kx | TM |
| III | 1 | 0 | Immediate data | Kx | TM |
| | 2 | 0 | for | Kx | TM |
| | 3 | 0 | transmitting | Kx | TM |
| | 4 | 0 | satellite | Kx | TM |
| | 5 | 0 | Non Immediate data | 0 | TM |
| | : | : | (almanac) for | : | TM |
| | 15 | 0 | five satellites | Kx | TM |
| IV | 1 | 0 | Immediate data | Kx | TM |
| | 2 | 0 | for | Kx | TM |
| | 3 | 0 | transmitting | Kx | TM |
| | 4 | 0 | satellite | Kx | TM |
| | 5 | 0 | Non immediate data | 0 | TM |
| | : | : | (almanac) for | : | TM |
| | 15 | 0 | five satellites | Kx | TM |
| V | 1 | 0 | Immediate data | Kx | TM |
| | 2 | 0 | for | Kx | TM |
| | 3 | 0 | transmitting | Kx | TM |
| | 4 | 0 | satellite | Kx | TM |
| | 5 | 0 | Non immediate data | 0 | TM |
| | : | | (almanac) for four sat | : | |
| | 14 | 0 | Reserved data | Kx | TM |
| | 15 | 0 | Reserved data | Kx | TM |

(Bracket annotations: Frame I spans "30 s"; the five frames span "30 s × 5 = 2.5 min")

**Figure 2.9**   GLONASS navigation message structure (Source: GLONASS-ICD)

The navigation message of the SA signal is broadcast as continuously repeating superframes of 2.5 min duration. Each superframe consists of five frames of 30 s, and each frame consists of 15 strings of 2 s duration (of length 100 bits). Each 2 s string includes the 0.3 s time mark (TM) (Figure 2.9).

The message data is divided into immediate data for the transmitting satellite and nonimmediate data for the other satellites. The immediate data is repeated in the first four strings of every frame. It comprises the ephemerides parameters, the satellite clock offsets, the satellite healthy flag, and the relative difference between the carrier frequency of the satellite and its nominal value. The nonimmediate data is broadcast in strings 5–15 of each frame (almanac for 24 satellites). The frames I–IV contain almanacs for 20 satellites (five per frame), and the V frame almanac for 4 satellites. The last two strings of frame V are reserved bits (the almanac of each satellite uses two strings).

The ephemerides values are predicted from the Ground Control Center (GCC) for a 12 or 24 h period, and the satellite transmits a new set of ephemerides every 30 min. These data differ from GPS data: instead of Keplerian orbital elements, they are provided as ECEF Cartesian coordinates for position and velocity, along with lunar–solar acceleration perturbation parameters. The GLONASS-ICD (2008) provides integration equations based on the fourth-order Runge–Kutta method, which includes the second zonal geopotential harmonic coefficient. The almanac is quite similar to the GPS one, given as modified Keplerian parameters, and it is updated approximately once per day.

The navigation message of the HA signal structure is not officially published.

### 2.4.2.3  Future GLONASS Air Interface Modernization

Despite years of successful exploitation of the GLONASS, from the point of view of further interoperability and complementarity with GPS (as well as Galileo), it is recognized that it would be desirable to introduce the CDMA segment of the air interface functionality in parallel with the legacy FDMA, but gradually replace the latter over time. As a preliminary phase in the GLONASS CDMA era, the first GLONASS-K satellite was orbited in November 2011 and transmitted a test signal on the L3 band (1202.25 MHz) (Urlichich *et al.*, 2012a). This signal is considered to be a prototype of the whole CDMA signal family.

Like the new GPS signals, the complete L3 GLONASS signal contains two openly accessible ranging codes. The first, carrying data stream, is broadcast on an in-phase carrier, while the second, serving as a pilot signal, is transmitted on the quadrature carrier. Both ranging codes have a chipping rate of 10.23 MHz and are Kasami sequences (Ipatov, 2005) of initial length $2^{14}-1$ short-cycled to a length of 10 230 that corresponds to a real-time period of 1 ms. Like Gold codes, every Kasami sequence is generated by modulo-2 summation of two *m*-sequences, but this time of different length. While the (long) first has a period of $N=2^n-1$ chips with $n$ even, the (short) second has a period $N_1=2^{n/2}-1$ and is obtained by $d$-decimation of the long one, where $d=2^{n/2}+1$. Specifically, for the L3 ranging code, long and short sequences are generated by 14-stage and 7-stage shift registers respectively, whose feedback loops are given by polynomials

$$g_1(x)=x^{14}+x^{10}+x^6+x+1,\ g_2(x)=x^7+x+1.$$

Thus formed primary ranging codes are also modulo-2 added by overlay codes of chipping rate 1 kHz. The data-carrying component is of length-5 Barker sequence 00010, and the pilot is of length-10 Neuman–Hoffman sequence 0000110101. Finally the FEC-encoded navigation message bits of rate 100 bits/s are modulo-2 added to the Kasami–Barker ranging sequence to modulate an in-phase carrier.

The navigation message is packed into five string frames, each 3 s duration string starting with a time marker giving a number of the string within a day of the GLONASS time scale. It is proposed that, starting with the launch of the first block GLONASS-K2 satellite, CDMA signals will be added to the legacy FDMA signals in the subbands L1 and L2 also. The proposed L1/L2 CDMA signal structure (Urlichich *et al.*, 2012b) will contain quadrature multiplexed ranging codes with open and authorized access. In turn, an open access (in-phase) component will be a time-multiplexed combination of a data carrying BPSK(1) sequence (whose chip rate equals that of the C/A GPS code, i.e., 1.023 MHz) and a BOC(1,1) data-free pilot code, while an authorized access quadrature component will be of the BOC(5,2.5) type.

It should be stressed that the L1 GLONASS signal design had been strongly dependent on a very tough limitation on the power flux penetrating into the neighboring radio astronomy band (1610.6–1613.8 MHz). In particular, BOC-modulation parameters were adjusted to secure as low as possible a level of GLONASS emission within the radio astronomy window. Still, to hold this level below the required threshold of $-194 \, \text{dB W} / \text{m}^2$, additional 15–18 dB band rejection is necessary. This is why the alternative L1 signal structures have also been discussed based on a spectral-compact (continuous phase) modulation (Ipatov and Shebshaevich, 2010, 2012).

### 2.4.3   Galileo

At the beginning of this century, the European Union decided to initiate a new project for a civilian satellite-based navigation system capable of satisfying defined performances in terms of security, integrity, continuity, and availability. This was to be called Galileo.

In May 2003, the European Commission and European Space Agency (ESA) officially approved the project, agreeing that this system must be open, global, and fully compatible with GPS but independent of it. The system is called "open" because it is available to international participation and "global" because it has worldwide coverage. Further, it is thought to be equipped with a constellation whose distribution will guarantee coverage up to very high latitudes.

Regarding compatibility, the European Union and the United States have subscribed to an agreement by which the Galileo system and the new GPS version will be compatible regarding radio frequencies, that is, each system will be designed so as not to produce interference that might downgrade, in an unacceptable way, the service supplied by the other system; more precisely, the two systems will share the L1 and L5 transmission bands and will be interoperable for nonmilitary applications. Interoperability at user level means that a receiver making use of navigation signals supplied by both systems reaches positioning and timing performances equal or better than a receiver that uses a single system. All this is made possible by an appropriate definition of the structure of the signals in shared bands.

The main characteristics of Galileo are provided both by increasing the accuracy reached in the positioning information and by providing integrity information, which is extremely important for critical applications such as those related to aerospace.

### 2.4.3.1 Galileo Architecture and Signals

Given long-standing world experience of the GPS and GLONASS penetration into various areas of social life, Galileo designers had to solve a complicated task: to offer users clear advantages for their system as compared to other systems. The deployment of additional satellites, innovations in signal structure and its processing methods to attain better design accuracy, and an expanded range of services (e.g., a service for consumers demanding a high safety level SOL and a public regulated service (PRS) providing integrity information, along with a commercial service (CS) providing an encrypted antijamming signal with higher data transmission rate) are among the presented advantages.

In order to meet the specific demands of the different communities of users, it is intended that Galileo will supply the following services:

- Open service (OS): A completely free-of-charge service for positioning, navigation, and time reference. OS data, which are transmitted on the E5a, E5b, and E2-L1-E1 carrier frequencies, are accessible to all users and mainly include navigation and SAR data.
- CS: A service for higher performances than those supplied by OS. This is a charged service and access is controlled by the receiver using access keys. It is supported by two different signals containing data referred to:
  ○ Satellite integrity conditions
  ○ Precise timing
  ○ Parameters for the determination of a more sophisticated model of the ionosphere
  ○ Local differential corrections

The use of two signals instead of one permits a higher data rate and therefore higher service accuracy. CS data are transmitted on the E5b, E6, and E2-L1-E1 carriers:

- SOL: This service has the same accuracy as OS, but it guarantees (with worldwide cover) the reception of messages with the system integrity included in the OS navigation message. This service, access to which is by means of certified receivers operating on two frequencies, is particularly important in cases where system failure has consequences for human life. SOL data are transmitted on the same carriers as for OS data (E5a, E5b, and E2-L1-E1).
- PRS: This service is referred to public utility applications (police, fire brigade, first aid) and supplies data for navigation and timing by utilizing restricted-access signals in addition to signals independent of those supplied by other services. Moreover, this service must be always available and certainly during crisis situations when other kinds of services might have problems; for this reason PRS is characterized by a highly robust signal with respect to jamming and spoofing. PRS data are transmitted on E6 and L1 carrier frequencies.
- At present an in-orbit validation (IOV) stage is being implemented. Demonstration testing of four IOV satellite flight units (two SV launched in October 2011and two more in October 2012), upgraded after launch and operation of the first two experimental satellites GIOVE-A and GIOVE-B, has commenced along with ground infrastructure functional testing (satellite control and mission control segments).
- The Galileo system is designed to be compatible with the GPS but completely independent of it. In fact, the satellite constellation will be completely new, as will the network of ground control stations.

- The Galileo infrastructure (Subirana *et al.*, 2013) will be composed of:
  - A constellation of 30 satellites in medium Earth orbit (MEO) each containing a navigation payload and a SAR transponder
  - A global network of Galileo Sensor Stations (GSS) providing coverage for clock synchronization and orbit measurements
  - Two control centers and two Launch and Early Operations Phase (LEOP) centers
  - A network of Mission Uplink Stations (ULS)
  - Several TT&C stations

This infrastructure is organized in two segments, the SS and the Ground Segment, to be complemented by the user receivers, which comprise the User Segment.

The standard Galileo system constellation includes 30 satellites on circular orbits at an altitude of 23 222 km in three planes inclined at an angle of 56°. One satellite in each orbital plane is a spare (Galileo ICD, 2010). Highly accurate atomic clocks are installed on these satellites, and each will have two types on board, a rubidium and a hydrogen maser clock. The frequencies are about 6 GHz for the rubidium clock and about 1.4 GHz for the hydrogen clock. The system uses the clock frequency as a very stable reference by which other units can generate the accurate signals that the Galileo satellites will broadcast. The broadcast signals will also provide a reference by which the less stable user-receiver clocks can continuously reset their time (Figure 2.10).

The complete constellation will be deployed in two stages:

1. In 2014–2015 an initial operation capability constellation of 18 SV will be deployed (4 IOV SV flight units and 14 standard Galileo SV).
2. In 2016–2020 the final operation capability constellation consisting of 30 SV will be deployed (27 SV will be used as intended and 3 SV will be spares).

**Figure 2.10**   Galileo orbit constellation

The Galileo Ground Segment will include GCC: two main centers (in Fuchino, Italy, and Oberpfaffenhofen, Germany), one hot standby center (in Spain) to support all system functions, five TT&C stations, the global network of 30–40 GSS, and nine Mission ULS. The Galileo Ground Segment is decomposed into the Galileo Control System (GCS) and the Galileo Mission System (GMS). The GCS, responsible for satellite constellation control and management of the satellites, provides the telemetry, telecommand, and control functions for the whole constellation. The GMS is responsible for the determination and uplink of the navigation and integrity data messages needed to provide the navigation and UTC time transfer service.

In the final configuration Galileo will provide 10 navigation signals in right-handed circular polarization (RHCP) over the frequency ranges 264–1215 MHz (E5a and E5b), 1260–1300 MHz (E6), and 1559–1591 MHz (E2-L1-E1), which are part of the Radio Navigation Satellite Service (RNSS) allocation.

Table 2.3 presents an overview of these signals, indicating the nominal carrier frequency, type of modulation, chip rate, ranging code length, and the data rate for each signal. All the Galileo satellites will share the same nominal frequency, making use of CDMA techniques compatible with the GPS approach. A specific feature of the Galileo air interface is that in the E5 band it uses a novel modulation technique called AltBOC. This mode allows the broadcasting of four independent signals (two data carrying and two pilots) on two carriers via a single transmitter–antenna trunk with a constant instantaneous power, that is, highest energy efficiency. Also in the E1 range the aggregated modulation CBOC(1,1,6,1) is implemented, combining additively signals of BOC (1,1) and BOC(6,1) types with allocation of the 10/11 part of the total power to the first.

As to ranging codes, in the public domain (ICD, 2010) only the E5 and E1 ones have been specified so far. Those described have a layered structure with primary component lengths given in Table 2.3. The primary codes can be generated either as modulo-2 sums of truncated $m$-sequences or by reading from the memory ("memory codes"), where they are stored accordingly to tables from ICD.

In order to guarantee interoperability with the GPS and promote their combined use, the agreement foresees that Galileo and GPS realize their own geodetic reference frames as near as possible to the International Navigation System, so reducing as much as possible errors due to differences in the coordinates of the sensor stations, which affect pseudorange errors. Transmission of the offset between the time scales of the two systems in both the navigation messages has been agreed.

Finally, interoperability is also foreseen at the level of the SAR service, both with regard to the definition of the kind of message transmitted by beacons and satellites and for the coordination of the rescue activity put into operation as a response to alarms.

When the Galileo constellation is fully operational, by utilizing the signals coming from its satellites and from the GPS constellation, the availability of the service will be greatly improved. In fact, since at least 5 satellites in the GPS constellation alone are always visible, and in the Galileo constellation 7 are visible, by joint use of the two the visibility of at least 12 satellites will be guaranteed.

The Galileo satellites will broadcast five types of data in four navigation messages: the Freely Accessible Navigation Message (F/NAV) and Integrity Navigation Message (I/NAV), plus the Commercial Navigation Message (C/NAV) and the Governmental Navigation Message (G/NAV). Table 2.4 summarizes the content of the Galileo messages, with an indication of the associated channels and services (table inspired by Hofmann-Wellenhof et al., 2008).

**Table 2.3** Galileo signal parameters

| Frequency bands | E5a | | E5b | | E6 | | | E2-L1-E1 | | |
|---|---|---|---|---|---|---|---|---|---|---|
| Carrier frequency, MHz | 276.450 | | 1207.140 | | 1278.750 | | | 1575.420 | | |
| Channel | I | Q | I | Q | A | B | C | A | B | C |
| Modulation | AltBOC(15,10) | | | | BOC(10,5) | BPSK(5) | BPSK(5) | Flexible | CBOC(1,1,6,1) | CBOC(1,1,6,1) |
| Chip rates (Mcps) | 10.23 | 10.23 | 10.23 | 10.23 | 5.25 | 5.25 | 5.25 | TBD | 2.046 + 12.276 | 2.046 + 12.276 |
| Primary ranging code length | 10230 | 10230 | 10230 | 10230 | — | — | — | TBD | 4092 | 4092 |
| Symbol rates (sps) | 50 | No data (pilot) | 250 | No data (pilot) | TBD | 1000 | No data (pilot) | TBD | 250 | No data (pilot) |
| User minimum received power at 10° elevation (dBW) | −158 | −158 | −158 | −158 | −155 | −158 | −158 | −155 | −160 | −160 |

**Table 2.4** Content of the Galileo message types

| Message-type Galileo services | F/NAV OS | I/NAV Os/CS/SoL | | C/NAV CS | G/NAV PRS | |
|---|---|---|---|---|---|---|
| Channels | E5a-1 | E1B | E5b-1 | E6B | E1A | E6A |
| Data rate (bps) | 25 | 125 | | 500 | 50 | |
| Navigation/positioning | X | X | X | | X | |
| Integrity | | X | X | | X | |
| Supplementary | | | | X | | |
| Public regulated | | | | | X | |
| Search and rescue | | X | | | | |

As represented in the previous table, Galileo will offer several types of service with different purposes to suit the needs of different types of users:

- The *integrity* data will provide time alarms and parameters for computing the integrity risk in supporting SOL applications.
- The *supplementary* data is expected to provide information for supporting different envisaged commercial services as differential corrections for a high-precision positioning service and different kinds of information data such as weather alerts and traffic information. The data is encrypted by the service providers in order to limit access to authorized users.
- The *public regulated* data is under governmental control, and it is devoted to the PRS. The system will guarantee a high continuity of service with controlled access via data encryption.
- The *SAR* data will provide the capability to send acknowledgment SAR messages to a beacon equipped with a suitable Galileo receiver.

Like GPS, Galileo will establish a dedicated Galileo Terrestrial Reference Frame (GTRF) that will be an independent realization of the ITRS. According to Galileo requirements, the 3D differences in position compared to the most recent ITRS should not exceed 3 cm (2-sigma). The realization of the GTRF is the task of the Galileo Geodetic Reference Service Provider (GRSP). Operational GTRF will include all GSS and selected stations of International GNSS Service (IGS). These IGS stations are used for alignment to ITRF (because GSS are not a part of it) and for densification of the network to improve the accuracy of the results. GTRF computation includes two parts: free network adjustment including all stations (GSS and selected IGS) and network alignment to ITRS using the IGS stations. GRSP prototype has been developed by a consortium of leading European geodetic institutions. The network included 131 IGS stations and 13 Galileo Experimental Sensor Stations (GESS) for the GIOVE mission. The initial realization of the GTRF (called GTRF07v01) was already in agreement with ITRF05 up to 0.9, 0.9, and 2.7 mm for north, east, and up, respectively.

## 2.4.4 BeiDou (Compass)

In the year 2000 China began the deployment of its own independent space-based navigation system. Initially it was conceived as a regional system covering the Chinese and neighboring territories and, unlike GPS and GLONASS, using space vehicles on geostationary

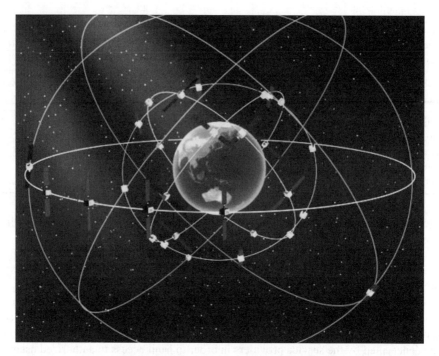

**Figure 2.11**   Compass orbit constellation

and geosynchronous orbits. This phase of the BeiDou is now referred to as BeiDou-1. In 2006 after successful tests of BeiDou-1, it was announced that the future system BeiDou-2, also called Compass, would by 2020 be on a par with already existing GNSS—that is, it would provide continuous all-weather positioning and timing service over the entire world. Development of the system is progressing rather quickly and efficiently: by the beginning of 2013 there had been 16 successful satellite launches. After its full deployment the system is expected to guarantee OS positioning accuracy within 10 m and timing with errors no greater than 20 ns.

The system deployment involved two stages (Chen, 2012). During the first stage (by the end of 2012) a group of 14 SV consisted of:

- Five satellites on geostationary orbit (GEO)
- Five satellites on inclined geosynchronous orbit IGSO (two SV are reserved on the orbit)
- Four satellites on MEO

This group of satellites is capable of providing services to users in the Asia-Pacific region (55° north latitude –55° south latitude and 84° east longitude –160° east longitude).

The second stage is planned to be deployed by 2020. When fully deployed, the BeiDou space segment will contain 35 satellites (see Figure 2.11), including five geostationary, three geosynchronous (orbiting at 55° inclination to the equatorial plane), and 27 MEO satellites. The first two types will be at an altitude of 35 786 km, while the last ones will be at 21 528 km, again with inclinations to the equatorial plane of 55° (BeiDou ICD, 2012).

The positioning accuracy (95% probability) for the Compass system will be equal to:

- Regional system: 20 m minimum
- Global system: 5 m in the horizontal plane and 5 m in the vertical plane on People's Republic of China territory and 10 m in any point on the Earth's surface

Due to the absence of ground control stations outside the People's Republic of China, it is problematical to implement control and data transmission. The control arc is small, so the accuracy of SV orbit determination decreases and consequently positioning accuracy reduces concomitantly.

Compass system options may consist of a short messages service, cosmic particle recorders and laser reflectors installed on some satellites, and a cross-link (possibly in the X band).

Along with GPS, Galileo, and modernized GLONASS, the BeiDou is CDMA based. Like all its cousins BeiDou will provide two kinds of service: open and authorized. The respective ranging codes are, as usual, quadrature multiplexed. Details are published only on signal B1I of the OS (BeiDou ICD, 2012).

The B1I ranging code is transmitted on the in-phase component of a 1561.098 MHz carrier frequency. For every space vehicle it is a Gold sequence of initial length 2047 short-cycled by one chip to come to a final length of 2046. Each Gold sequence is generated as a symbol-wise modulo-2 sum of two $m$-sequences formed by 2-stage shift registers with linear feedback given, respectively, by polynomials

$$g_1(x) = x^{11} + x^{10} + x^9 + x^8 + x^7 + x + 1,$$

$$g_2(x) = x^{11} + x^9 + x^8 + x^5 + x^4 + x^3 + x^2 + x + 1.$$

The chip rate of B1I code is 2.046 MHz so that real-time period of this signal is 1 ms. As far as can be seen from the ICD, no pilot component is incorporated into an OS signal.

The navigation message D1 broadcast by MEO satellites has a rate 50 bit/s and contains all the fundamental data: vehicle state, almanac, time offset against other systems, etc. The geostationary and geosynchronous satellites broadcast a navigation message D2 at a rate of 500 bit/s, including augmentation service information such as system integrity, differential and ionosphere correction grid, etc. The simplest (15,11) Hamming FEC code is used as a means of both D1 and D2 data protection. Naturally, data streams are modulo-2 summed with ranging codes before BPSK modulation of the carrier. The expected signal power of the B1I signal at the Earth's surface is about −163 dB W.

## 2.4.5   State and Development of the Japanese QZSS

By 2020 the Japanese government plans to deploy a four-satellite constellation on quasi-zenith orbits. One of SV—Michibiki—was launched in September 2010 and is used for demonstration testing. The QZSS is designed to solve the following three tasks:

1. Augmentation of the GPS on Japanese territory and adjoining areas of the Asia-Pacific region (due to the increase in GPS availability and full compatibility with existing and advanced GPS signals)

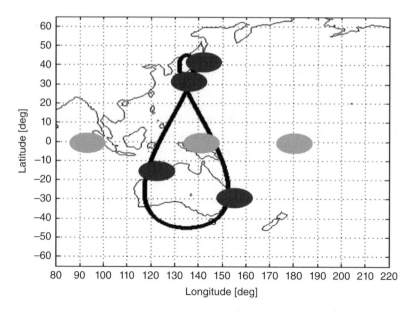

**Figure 2.12**   QZSS SV tracks on the Earth's surface

2. An improvement in navigation service quality for the combined operation of QZSS and GPS (due to a wide range of corrective data including differential corrections, data on integrity, ionosphere corrections, data on state the of SV included in other GNSS at full interoperability with L1C/A GPS signals, and compatibility with SBAS signals)
3. Implementation of the short message service (Figure 2.12)

Experimental synchronization system is one of the QZSS options. During Michibiki SV flight tests, it is planned to study the possibility of timekeeping with no atomic clock on board—a synchronization system integrated with a simple clock onboard will be used instead. This clock operates as a transceiver broadcasting exact time data transmitted remotely via a time synchronization network on the Earth's surface (Fujiwara, 2011).

### 2.4.6   State and Development of the IRNSS

The IRNSS coverage area will include the whole Indian continent and part and the territory about 1500 km outside, including a major part of Indian Ocean (Ganeshan, 2011).

The orbit constellation will consist of seven satellites:

- Satellites on GEO.
- Satellites on geosynchronous orbit with the following parameters: apogee of 24 000 km, perigee of 250 km, and inclination of 29°. Two satellites will cross the equatorial plane at point 55° east longitude and two SV at point 21.5° east longitude (Figure 2.13).

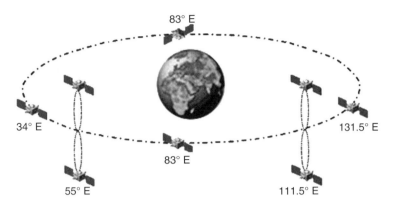

**Figure 2.13**   The IRNSS orbit constellation

The IRNSS will transmit CDMA navigation signals in L5 bands at a center frequency of 1176.45 MHz and in the S band at a center frequency of 2492.028 MHz.

## 2.5   GNSS Observables

The main measurement that is output by a GNSS receiver is the *pseudorange*. This provides the so-called code measurement that is directly related to the PRN code. Pseudoranges have been described in Section 2.3.3 (see Eq. 2.10).

In the present sections the other most important satellite measurements and measurement-related quantities (simply called observables) are briefly presented: the carrier-phase and Doppler frequency observables and single, double, and triple differences. The last three observables are obtained by carrier-phase manipulation, and then greater consideration is given to the carrier-phase section. See also Hofmann-Wellenhof *et al.* (1997).

### 2.5.1   Carrier-Phase Observables

The numerically controlled oscillator (NCO), which controls the carrier frequency/phase tracking loops, provides an indication of the observed frequency shift of the received signal together with its phase. The wavelength of the L1 carrier is about 19 cm, so that very precise measurements can be made from knowledge of the carrier phase. The amount of noise depends largely on the parameters of the carrier tracking loops in the receiver. Usually it ranges from 1% in a stationary receiver to 2% in a navigation-type receiver, where the high dynamics of the vehicle needs a tracking loop with a wider bandwidth in order to maintain proper tracking of the carrier. However, a wider bandwidth contains a higher amount of noise. Nevertheless, the level of noise in carrier measurements is much smaller than in pseudorange measurements; hence carrier measurements are much better than pseudorange ones. The main disadvantage of carrier observables is that the carrier phase is made up of two quantities: the amount of carrier phase since the time of carrier acquisition and the integer ambiguity. Only the former can be known with high accuracy if no temporary interruption or cycle slips have occurred, while the latter, the integer ambiguity, corresponds to the integer number of wavelengths that cover the distance between satellite and user receiver at the beginning of carrier-phase acquisition. This value can be computed, but its determination is time consuming. Moreover, in the case of reacquisition of the carrier phase after a temporary loss of signal tracking, the process of integer ambiguity determination must be rerun, so causing possible problems for real-time application. The carrier-phase model can be derived as follows.

By defining

- $\Phi_r(t)$—phase of the carrier generated in the receiver
- $\Phi_s(t, d)$—satellite signal carrier phase at a distance $d$ from the satellite
- $t_1$—time instant of carrier-phase acquisition by the receiver

it is possible to write

$$\Phi_r(t) = (\omega_0 + \omega_{\text{drift}})(t - t_1) + \Phi_{r1} \tag{2.22}$$

$$\Phi_s(t, d) = (\omega_0 + \omega_{\text{sat.dr}})\left(t - \frac{d}{c}\right) + \Phi_{s0} \tag{2.23}$$

where $\omega_0$ (rad) is the carrier nominal frequency, $\omega_{\text{drift}}$ is the receiver clock frequency drift, $c$ is the speed of light, $\omega_{\text{sat.dr}}$ is the satellite clock frequency drift, $\Phi_{s0}$ and $\Phi_{r0}$ are the initial satellite and receiver carrier phases at the reference time $t_0 = 0$, and finally $\Phi_{r1} = (\omega_0 + \omega_{\text{sat.dr}})t_1 + \Phi_{r0}$ is the receiver carrier phase at $t_1$.

Equation 2.23 can be rewritten in a more convenient form:

$$\Phi_s(t, d) = \Phi_s(t, d) - \Phi_s(t_1, d_1) + \Phi_s(t_1, d_1) \tag{2.24}$$

where $\Phi_s(t, d) - \Phi_s(t_1, d_1)$ is the carrier-phase change continuously tracked after the initial acquisition (assuming no cycle slip has occurred), while $\Phi_s(t_1, d_1)$ is the carrier phase at time $t_1$ when the relative receiver–satellite distance is $d_1$:

$$\Phi_s(t_1, d_1) = \Phi_{\text{frac}} + 2\pi N$$

This can be considered as a sum of two parts: a fractional part $\Phi_{\text{frac}}$ and an (unknown) integer number $N$ that represents the number of periods of $2\pi$ contained in $d_1$.

Hence, Equation 2.24 becomes

$$\Phi_s(t, d) = \bar{\Phi}(t, d) + 2\pi N \tag{2.25}$$

where $\bar{\Phi}(t, d) = \Phi_s(t, d) - \Phi_s(t_1, d_1) + \Phi_{\text{frac}}$ is the measurable part of the received carrier phase.

By defining $\Phi(t, d)$ as the difference between the receiver-generated carrier phase and $\bar{\Phi}(t, d)$,

$$\Phi(t, d) = \Phi_r(t) - \bar{\Phi}(t, d) \tag{2.26}$$

with the result that $\Phi(t,d)$ is a quantity measurable by the receiver. Moreover, the following equation holds:

$$\Phi_r(t) - \Phi_s(t,d) = \Phi(t,d) - 2\pi N \qquad (2.27)$$

The first member of Equation 2.27 provides (after some algebra)

$$\Phi_r(t) - \Phi_s(t,d) = \omega_0 \frac{d}{c} + \omega_{drift} t + \Phi_{r0} - \omega_{sat.dr}\left(t - \frac{d}{c}\right) - \Phi_{s0} \qquad (2.28)$$

By defining the receiver clock offset as

$$b_r(t) = \omega_{drift} t + \Phi_{r0}$$

and the satellite clock-caused offset as

$$b_s(t) = \omega_{sat.dr}\left(t - \frac{d}{c}\right) + \Phi_{s0}$$

Equation 2.27 becomes

$$\Phi(t,d) = \omega_0 \frac{d}{c} + 2\pi N + b_r(t) - b_s(t) \qquad (2.29)$$

Because of the extremely low value of the satellite clock drift assured by both the onboard atomic clock and the continuous monitoring by the control segment, the most significant contribution in this offset is caused by the term $\omega_{sat.dr} t + \Phi_{s0}$.

In Equation 2.29 two important quantities appear:

1. $N$ is called the *initial ambiguity* or *integer ambiguity* and is an integer number of $2\pi$ angular periods contained in the satellite-receiver distance at the time of carrier acquisition $(t_1)$.
2. $\Phi(t,d)$ is the carrier-phase observable or measurement. Its relationship to the satellite-receiver distance is expressed by Equation 2.29, while the way it is measured is given by Equation 2.26.

By means of the carrier tracking loops, the receiver provides the phase measurement

$$\Phi_m(t,d) = \Phi(t,d) + \varepsilon_\Phi(t) \qquad (2.30)$$

that differs from $\Phi^{TM}(t)$ because of the presence of the measurement error $\varepsilon_\Phi(t)$. The carrier-phase measurement model from Equation 2.30 can then be written as

$$\Phi_m(t,d) = \omega_0 \frac{d}{c} + 2\pi N + b_r(t) - b_s(t) + \varepsilon_\Phi(t) \qquad (2.31)$$

It is worth noting that pseudorange and carrier-phase measurements are affected by different errors. In particular, $\varepsilon_\Phi(t)$ is two orders of magnitude smaller than pseudorange error.

## 2.5.2  Doppler Frequency Observables

The (approximated) Doppler measurement model can also be obtained from the previous relations. In fact, by taking the derivatives of Equation 2.24 with respect to the time,

$$\dot{\Phi}(t) = \frac{d}{dt}\Phi(t,d) \tag{2.32}$$

The right-hand part of Equation 2.30 can be expanded in two ways: firstly

$$\frac{d}{dt}\Phi(t,d) = \dot{\Phi}_r(t) - \dot{\Phi}_s(t,d) = \Delta\omega \tag{2.33}$$

where $\Delta\omega$ is the frequency difference between the satellite-receiver-generated signal and the received signal. The Doppler shift frequency $\Delta\omega$ can be measured by the receiver frequency tracking loop.

Secondly, from Equation 2.29 it is possible to write

$$\frac{d}{dt}\Phi(t,d) = \omega_0\frac{v(t)}{c} + \omega_{drift} - \omega_{sat.dr}\left(1 - \frac{v(t)}{c}\right) \tag{2.34}$$

where the relation

$$\dot{d} = v(t) \tag{2.35}$$

has been used, with $v(t)$ indicating the satellite–user relative speed.

By equating 2.33 and 2.34,

$$\Delta\omega = \omega_0\frac{v(t)}{c} + \omega_{drift} - \omega_{sat.dr}\left(1 - \frac{v(t)}{c}\right) \tag{2.36}$$

Finally, the Doppler frequency measurement provided by the receiver is

$$\Delta\omega_m = \Delta\omega + \varepsilon_\omega(t) = \omega_0\frac{v(t)}{c} + \omega_{drift} - \omega_{sat.dr}\left(1 - \frac{v(t)}{c}\right) + \varepsilon_\omega(t) \tag{2.37}$$

where

- $\varepsilon_\omega(t)$ is the error in the measurement of $\Delta\omega$
- $\Delta\omega_m$ is the Doppler shift frequency as measured by the receiver

Equation 2.37 shows the connection between Doppler frequency measurements and satellite-receiver speed.

It is worth noting that the Doppler shift frequency $\Delta\omega_m$ can be obtained from the receiver in two different ways:

1. By the direct measurement of the shift frequency from the NCO in the frequency tracking loop
2. From the discrete time derivative of the carrier phase $\phi_m(t)$ as computed by taking two time-spaced samples

The measurement errors in the Doppler shift observables can be different according to which method (1 or 2) is used. Generally, good receivers allow the user a choice in the method to be employed, depending on the application. Method 2 is based on the observation that, analogously to Equation 2.33,

$$\Delta\omega_m = \dot{\Phi}_m \qquad (2.38)$$

which, by the definition of $\Phi_m(t)$, results in

$$\Delta\omega_m = \dot{\Phi}(t) + \dot{\varepsilon}_\Phi(t) = \Delta\omega + \varepsilon_\omega(t) \qquad (2.39)$$

The derivative of the measured phase is computed numerically and requires a selectable latency time. In fact, the computation of the discrete time derivative is obtained from two carrier-phase samples divided by the sample time.

A carrier-phase measurement performed along a time interval $[t_0, t_1]$ is also referred to as an accumulated delta range and is indicated as $ADR(t_0, t_1)$.

The carrier-phase variations with respect to its initial value can also be determined from the Doppler frequency shift measurements obtained by means of a direct measurement according to method 2. In fact, from Equation 2.38,

$$ADR(t_0, t) = \Phi_m(t) - \Phi_m(t_0) = \int_{t_0}^{t} \Delta\omega_m \, dt \qquad (2.40)$$

Of course, usage of the third member of Equation 2.40 for computing the carrier-phase differences is not encouraged due to the large number of errors affecting this method. In fact, the integration process is performed by assuming that the value of the frequency is constant between two sample times. The accuracy of the result depends on the sample time size, the drift errors in the frequency measurement, and the size of the time interval $t - t_0$.

## 2.5.3   Single-Difference Observables

Equation 2.31 represents the basic mathematical model of the carrier phase from which other observables can be derived. Single differences can be obtained by using two receivers that observe the same satellite as follows. By indicating two receivers with "A" and "B" and using "a" for the GNSS satellite observed by those receivers, Equation 2.31 can be written twice,

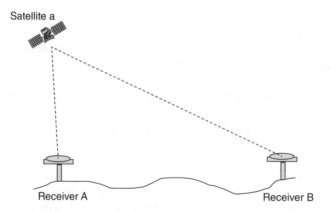

**Figure 2.14**   The single-difference concept

one for each receiver, where the variables involved in the equations have subscripts to identify the receiver and a superscript in order to indicate the satellite. Hence

$$\Phi_A^a(t_1) = \omega_0 \frac{d_A^a(t_1)}{c} + 2\pi N_A^a + b_{rA}(t_1) - b_{sA}^a(t_1) + \varepsilon_{\Phi A}^a(t_1) \tag{2.41}$$

and

$$\Phi_B^a(t_2) = \omega_0 \frac{d_B^a(t_2)}{c} + 2\pi N_B^a + b_{rB}(t_2) - b_{sB}^a(t_2) + \varepsilon_{\Phi B}^a(t_2) \tag{2.42}$$

Figure 2.14 shows a schema for single-difference geometry.
Taking the difference of the two equations results in

$$\Phi_A^a(t_1) - \Phi_B^a(t_2) = \Phi_{AB}^a(t_1,t_2)$$
$$= \omega_0 \frac{d_{AB}^a(t_1,t_2)}{c} + 2\pi N_{AB}^a(t_1,t_2) + b_{rAB}(t_1,t_2) - b_{sAB}^a(t_1,t_2) + \varepsilon_{\Phi B}^a(t_1,t_2) \tag{2.43}$$

where the subscript AB indicates that the term is obtained as the differences among the corresponding terms in the original equations. Should the observation time be the same for both the receivers (i.e., $t_1 = t_2 = t$), then Equation 2.43 leads to the single-difference observable equation

$$\Phi_{AB}^a(t) = \omega_0 \frac{d_{AB}^a(t)}{c} + 2\pi N_{AB}^a + b_{rAB}(t) + \varepsilon_{\Phi AB}^a(t) \tag{2.44}$$

where $b_{sAB}^a(t)$ has been removed because it is zero. The main feature of single differences is that common errors between the two receivers cancel out. In particular, with single differences the satellite clock offset vanishes together with the ephemeris error. Moreover, the portion of the ionospheric and tropospheric errors common to the two receiver sites disappears. The closer the receivers are to each other, the more the cancellation of common errors is effective.

Single differences are used in order to determine an accurate relative positioning, that is, the position vector (baseline) of one receiver (called rover or mobile) with respect to the other (called base). The base receiver is at a known (georeferenced) position and is usually fixed.

## 2.5.4   Double-Difference Observables

If two receivers "A" and "B" concurrently observe two satellites "a" and "b," the double-difference phase observables $\phi_{AB}^{ab}(t)$ can be computed by taking the difference of two single-difference observables taken at the same time:

$$
\begin{aligned}
\Phi_{AB}^{ab}(t) &= \Phi_{AB}^{a}(t) - \Phi_{AB}^{b}(t) \\
&= \omega_0 \frac{d_{AB}^{ab}(t)}{c} + 2\pi N_{AB}^{ab} + \varepsilon_{\Phi AB}^{ab}(t)
\end{aligned}
\tag{2.45}
$$

In this case $b_{sAB}^{ab}(t) = 0$. Also here, analogously with the previous notation, superscript "ab" indicates that the term is obtained as the difference among the corresponding terms in the two single-difference original equations written for satellites "a" and "b," respectively.

Figure 2.15 shows the double-differencing measurement configuration concept. The main advantage of double differences is that the large errors due to receiver clock offset cancel completely, provided that observations of satellites "a" and "b" are taken at the same time or the receiver clock drifts between the observation epochs are negligible.

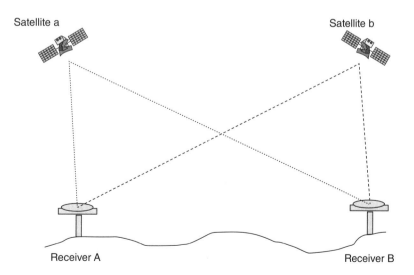

**Figure 2.15**   The double-difference concept

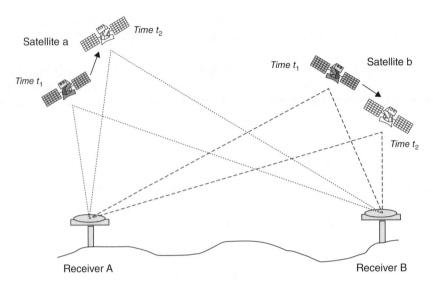

**Figure 2.16**   The triple-difference concept

### 2.5.5   Triple-Difference Observables

The triple-difference concept is illustrated in Figure 2.16. Two observation epochs, $t_1$ and $t_2$, have to be considered in this case, and two double differences have to be taken, one for each epoch. Then, the triple-difference observable can be computed by differencing the double-difference equations obtained from Equation 2.45 in the following way:

$$\Delta\Phi_{AB}^{ab}(t_2,t_1) = \Phi_{AB}^{ab}(t_2) - \Phi_{AB}^{ab}(t_1)$$
$$= \omega_0 \frac{d_{AB}^{ab}(t_2,t_1)}{c} + \varepsilon_{\Phi AB}^{ab}(t_2,t_1) \tag{2.46}$$

where, as usual, the terms on the right are obtained as the differences between the corresponding terms in the double-difference equations calculated in $t_1$ and $t_2$, respectively. The most important feature of triple differences is that the integer ambiguity-related terms cancel: in fact they are time invariant between one epoch and the other. Of course, in order to compute triple differences, it is important that no cycle slips or carrier-phase loss of lock has occurred. Triple differences are considered a preprocessing technique in order to get good approximate positions for the double-difference solutions. In triple differences, cycle slips result as outliers in the computed residuals, and their elimination can be useful to provide cycle slip-free solutions to double-difference observables.

### 2.5.6   Linear Combinations

Linear combinations obtained as the sum or difference of dual- or multifrequency carrier-phase or pseudorange observables improve the final solutions in comparison with those based on original observables. This improvement implies that particular systematic errors can be removed or reduced, resulting in phase ambiguity resolution, improving the reliability, and providing an accuracy increase (Petrovello, 2009).

A number of linear combinations having useful properties can be defined, based on original observables, carrier phases $\Phi_i$, and code pseudoranges $r_i$ (where $i = 1, 2, \ldots$ indicates corresponding measurements in the frequencies $f_i$).

Ionosphere-free combinations $\Phi_{IF}$, $r_{IF}$ remove the first-order (up to 99.9%) ionosphere effect, which depends on the inverse square of the frequency:

$$\Phi_{IF} = \frac{f_1^2 \Phi_1 - f_2^2 \Phi_2}{f_1^2 - f_2^2}; \quad r_{IF} = \frac{f_1^2 r_1 - f_2^2 r_2}{f_1^2 - f_2^2} \tag{2.47}$$

Note that satellite clocks here are defined relative to $r_{IF}$ combination.

Geometry-free combinations $\Phi_{GF}$, $r_{GF}$ remove the geometric part of the measurement, leaving all the frequency-dependent effects: ionosphere refraction, instrumental delays, and windup, as well as multipath and measurement noise. They can be used to estimate the ionosphere electron content, to detect cycle slips in the carrier phase, and also to estimate antenna rotations:

$$\Phi_{GF} = \Phi_1 - \Phi_2; \quad r_{GF} = r_2 - r_1 \tag{2.48}$$

Wide-lane combinations $\Phi_{WL}$, $r_{WL}$ are used to create a signal with a significantly wide wavelength. The longer wavelength is useful for the detection of cycle slips and for ambiguity fixing. Another feature of this combination is the change of sign in the ionosphere term:

$$\Phi_{WL} = \frac{f_1 \Phi_1 - f_2 \Phi_2}{f_1 - f_2}; \quad r_{WL} = \frac{f_1 r_1 - f_2 r_2}{f_1 - f_2} \tag{2.49}$$

Narrow-lane combinations $\Phi_{NL}$, $r_{NL}$ create signals with a narrow wavelength. The signal in this combination has a lower noise than each separate component and can be used to reduce the code noise and to estimate the wide-lane ambiguity:

$$\Phi_{NL} = \frac{f_1 \Phi_1 + f_2 \Phi_2}{f_1 + f_2}; \quad r_{NL} = \frac{f_1 r_1 + f_2 r_2}{f_1 + f_2} \tag{2.50}$$

The Melbourne–Wübbena combination combining $r_{NL}$ and $\Phi_{WL}$ provides an estimation of the wide-lane ambiguity $C_{WL}$ according to the following equation:

$$C_{MW} = \Phi_{WL} - r_{NL} = \lambda_{WL} N_{WL} + b + \varepsilon \tag{2.51}$$

where $N_{WL} = N1 - N2$ is the integer wide-lane ambiguity, $b_{WL}$ accounts for the satellite and receiver instrumental biases, and $\varepsilon$ is the measurement noise, including carrier phase and code multipath. The Melbourne–Wübbena combination has certain advantages: the wide-lane combination has a larger wavelength $\lambda_{WL} = c/(f_1 - f_2)$ than each signal individually, which leads to an enlarging of the ambiguity spacing, and the measurement noise is reduced by the narrow-lane combination of code measurements, so reducing the dispersion of values around the true bias.

The triple-frequency observable combinations in multifrequency GNSS will lead to an additional significant performance improvement in the multifrequency ambiguity resolution technique (Simsky, 2006).

## 2.5.7   Integer Ambiguity Resolution

One of the major problems when dealing with carrier-phase observables is the determination of integer ambiguities. Many methods have been proposed in the literature, and this field is currently under investigation (see Teunissen *et al.* (1997), Vollath *et al.* (1998), Teunissen (1994), Forssel *et al.* (1997), and Donghyun and Langley (2000)). In this section a very short overview of the elaboration processes and ways involved in facing the problem is given; for a deeper and complete discussion, see the indicated references.

The integer ambiguity resolution process is usually split into two main steps:

1. The definition of an appropriate set of possible solutions: of course, the smaller the size of the set, the better and faster will be the subsequent step.
2. The choice of the best solution among those found in the previous step as the preferred integer ambiguity solution. The definition of "best solution" is related to minimum residual criteria.

As to the first step, single-epoch and multiepoch methods exist. Single-epoch methods need measurements collected at one time instant and are therefore particularly suitable for nonstatic applications like moving vehicles. The basic idea in single-epoch methods is that code observables, being ambiguity-free measurements, can be used to assist carrier-phase ambiguity determination. However, due to the high level of noise affecting code measurements, the ambiguity determination does not yield a unique solution, but offers a set of possible candidate solutions, the size of which depends on the level of that noise. Nevertheless, by combining pseudorange $\rho$ and carrier-phase $\phi_m$ observables (i.e., by taking the difference between Equations 2.10 and 2.31), an estimate of the integer ambiguity $N$ can be obtained as a function of $\rho$ and $\phi_m$: the estimated $N$ is characterized by the following standard deviation $\sigma_N$:

$$\sigma_N^2 = \frac{1}{\lambda^2}\sigma_\rho^2 + 2\sigma_\phi^2 \tag{2.52}$$

where $\sigma_\rho$ and $\sigma_\phi$ are the pseudorange and carrier-phase standard deviations, respectively, and $\lambda$ is the wavelength of the carrier. Equation 2.52) shows that the greater is $\lambda$, the better is the estimate of $N$. This observation has suggested how to establish an advantage by using receivers with two frequencies. In fact, the concept of *wide laning* has been introduced: wide lane is a linear combination of the carrier-phase observables obtained from both signals L1 and L2. Consequently, a wide-lane carrier-phase observable $\phi_{12}$ can be obtained with its wavelength $\lambda_{12}$ and its integer ambiguity $N_{12}$ obtained by the corresponding quantities associated with L1 and L2 carrier-phase observables. That is,

$$\phi_{12} = \phi_1 - \phi_2, \quad N_{12} = N_1 - N_2, \quad \frac{1}{\lambda_{12}} = \frac{1}{\lambda_1} - \frac{1}{\lambda_2} \tag{2.53}$$

and $N_{12}$ can be expressed as a function of $\rho$ and $\phi_{12}$. $\lambda_{12}$ is about 85 cm (greater than both $\lambda_1$ and $\lambda_2$), and as a consequence, the estimation error of the integer ambiguity decreases, so reducing also the size of possible solution sets for $N_1$ and $N_2$. If more than two frequencies are available, that is, for the Galileo navigation satellite system or with the new GPS and GLONASS L3 third frequency, wide laning can be further exploited by Three-Carrier Ambiguity Resolution (TCAR) or Multiple CAR (MCAR) algorithms, where various combinations of carrier-phase observables with different frequencies lead to high-performing ambiguity resolution multistep methods.

However, for a single-frequency receiver, wide laning is not available. Hence, multiepoch methods have to be used in order to accomplish the ambiguity resolution by means of observations performed at least in two sufficiently separated time instants. In this case it is possible to exploit the triple-difference observables, as explained in the previous section, where the integer ambiguities cancel out. The major disadvantage of this method is the need to have static receivers during the ambiguity resolution process and therefore is not suitable for navigation applications.

As to the second step, that is the best solution choice among the set of all those possible. In the literature, many methods can be found that work efficiently. Most of them are based on the integer LS estimation problem. These methods are carried out in three steps—the float solution, the integer ambiguity estimation, and the fixed solution. For a description of these techniques, the interested reader can see the cited references and, in particular, Teunissen (1994).

## 2.6   Sources of Error

As explained in previous sections, the GNSS observables are affected by several sources of error, and in order to describe these measurement inaccuracies, the main error terms are considered separately from one another. Moreover, the analysis is carried out with respect to the observables of a single satellite signal since Equation 2.16 relates these errors to the navigation solution by means of a DOP factor depending on the geometry between the user and satellite constellation.

The sources of error described within this section are, in order, atmospheric (ionospheric and tropospheric) effects, SA, and multipath and receiver noise.

Regarding the Earth's atmospheric effects, it is important to note that the spread of GNSS signals through the atmosphere is mainly affected by a propagation speed different from that of electromagnetic waves in a vacuum. The description of the atmosphere is typically split into several parts corresponding to regions with common physical properties. In the context of electromagnetic wave propagation, the atmospheric regions of main interest are the troposphere, tropopause, stratosphere (the overall effect of these regions being referred to as *tropospheric refraction*), and ionosphere (giving rise to *ionospheric refraction*).

As resulting effect, the atmosphere induces on the GNSS signals both an advancement of the carrier phase and a group delay of the codes. These phenomena will be considered in the following sections, and, where necessary, the concepts of phase and group velocities will first be explained. To aid this aim, the following simplified framework is presented:

- The electromagnetic waves propagate in a homogeneous and isotropic medium (i.e., the index of refraction is constant and the propagation speed is independent of the direction).
- The waves are made up only of two sinusoidal components $\varphi_1$, $\varphi_2$ with unitary amplitude.

In this framework the propagation of the first component $\varphi_1$ with respect to time ($t$) and to distance (with coordinate $x$) from the emitting source in a generic direction may be represented as

$$\varphi_1(t,x) = \sin(\omega t - kx) \tag{2.54}$$

where

$$k = \frac{2\pi}{\lambda_\varphi}, \quad \lambda_\varphi = \frac{c_\varphi}{f}, \quad f = \frac{\omega}{2\pi}, \tag{2.55}$$

Here, $\omega$ and $f$ represent the angular frequency and the frequency, respectively, $c_\varphi$ is the wave propagation speed through the medium, $\lambda_\varphi$ is the wavelength, and $k$ is the angular constant (phase propagation along the medium) so that $kx$ corresponds to the starting phase in the point of coordinate $x$. The second component $\varphi_2$ is characterized by an angular frequency and an angular constant which differ by $\Delta\omega$ and $\Delta k$ with respect to those of $\varphi_1$. Hence,

$$\varphi_2(t,x) = \sin\big((\omega + \Delta\omega)t - (k + \Delta k)x\big) \tag{2.56}$$

and as a consequence, the corresponding wave propagation speed is related to the sinusoidal parameters as follows:

$$c_\varphi + \Delta c_\varphi = \frac{\omega + \Delta\omega}{k + \Delta k}. \tag{2.57}$$

In this framework, the sum of the two sinusoidal components gives rise to the overall wave $w$:

$$\begin{aligned}
w &= \varphi_1 + \varphi_2 = \sin(\omega t - kx) + \sin\big((\omega + \Delta\omega)t - (k + \Delta k)x\big) \\
&= 2\sin\left(\frac{\Delta\omega t - \Delta kx}{2} + \frac{\pi}{2}\right)\sin\left(\frac{(2\omega + \Delta\omega)t - (2k + \Delta k)x}{2}\right) \\
&= 2\varphi_g\varphi
\end{aligned} \tag{2.58}$$

where $\varphi_g$ is the so-called group signal and $\varphi$ is the carrier signal.

As to the GPS, $\varphi$ represents either the carrier L1 or L2, while $\varphi_g$ represents the code and data signals. Moreover, the bandwidth of the modulated GPS carriers is sufficiently narrow to consider $\Delta\omega$ and $\Delta k$ as differentials $d\omega$ and $dk$. From Equations 2.56 to 2.58, the group velocity $c_g$ can be defined as follows:

$$c_g = \frac{d\omega}{dk}. \tag{2.59}$$

It is well known (Smith and Weintraub, 1953) that the phase velocity $c_\varphi$ (carrier propagation speed through the atmosphere) and the phase refractive index $n_\varphi$ are related as follows:

$$c_\varphi = \frac{c}{n_\varphi} \qquad (2.60)$$

An analogous relation can be formulated for the group velocity

$$c_g = \frac{c}{n_g} \qquad (2.61)$$

where, as usual, $c$ is the vacuum speed of light and $n_g$ represents the group refractive index that can be related to $n_\varphi$ by means of Equations 2.54 and 2.50 as follows:

$$n_g = c\frac{dk}{d\omega} = \frac{d}{d\omega}\left(\frac{c}{c_\varphi}2\pi f\right) = \frac{d}{d\omega}\left(n_\varphi \omega\right) = n_\varphi + f\frac{dn_\varphi}{df}. \qquad (2.62)$$

When the phase refractive index $n_\varphi$ depends on frequency, the medium is called dispersive. In this case, if from Equation 2.55 the group refractive index $n_g$ is not equal to $n_\varphi$, then the code and phase have different velocities.

## 2.6.1   Ionosphere Effects

The ionosphere is the atmospheric region between approximately 50 and 1000 km over the Earth's surface. The main features of this region are the large numbers of (negative) free electrons and (positive) ions released by the atmospheric gases via ionization resulting from the impact of ultraviolet rays from the sun. In particular, the effect of the ultraviolet radiation is to increase the electron density, whence the electromagnetic wave propagation is significantly influenced by the presence of these free electrons. This is why the exploitation of the relative sun position is very important in modeling the ionosphere effects on GNSS signals. Noting that only carriers with frequencies higher than 30 MHz can penetrate the ionosphere, then GNSS L1 and L2 are appropriate because at their frequencies the ionosphere appears as a dispersive medium and the corresponding phase refractive index $n_\varphi$ is generally described by the following expression (Hargreaves, 1995):

$$n_\varphi = \sqrt{1-2N_I} \approx 1 - N_I, \quad N_I = \frac{\upsilon_2}{f^2}\delta_e \ll 1, \quad \upsilon_2 = 40.3 \qquad (2.63)$$

where $N_I$ is the ionosphere refractivity and $\delta_e$ is the local electron density. Usually $n_\varphi$ is identified by its first-order approximation owing to the smallness of $N_I$. From Equations 2.60 to 2.63, simplified expressions for the group and phase velocities follow:

$$c_g \approx c - \Delta c, \quad c_\varphi \approx c + \Delta c, \quad \Delta c = cN_I = \frac{\upsilon_2 c}{f^2}\delta_e \qquad (2.64)$$

where the group code delay and carrier-phase advancement are highlighted.

The absolute value of the ionosphere range error (which may reach up to 50 m during the day) on the GNSS observables, at a carrier frequency $f$, can be obtained as follows:

$$I_f = \frac{v_2}{f^2} \text{TEC}, \quad \text{TEC} = \int_{\gamma_{iono}} \delta_e ds \qquad (2.65)$$

where *total electron content* (TEC) typically varies from $10^{16}$ to $10^{19}$ el/m$^2$ and is the electron density integrated along the path $\gamma_{iono}$ through the ionosphere.

Users with single-frequency receivers (where only L1 code and data are available) may use an algorithm based on the Klobuchar ionosphere model (ARINC, 2000) in order to halve the ionosphere error on the GNSS observables. The ionosphere model relies on the following parameters:

- Eight coefficients $\alpha_n$, $\beta_n$ (with $n = 0,...,3$) broadcast to users by the navigation message
- The approximated geodetic latitude $\phi_u$ and longitude $\lambda_u$ of the user
- The azimuth $A_i$ and elevation $E_i$ of the $i$th satellite with respect to the local tangent plane
- The GPS time of week $t_{GPS}$

Some of the main quantities, upon which the Klobuchar ionosphere model is based, are depicted in Figure 2.17, where the mean ionosphere height $h_{ion}$ is assumed equal to 350 km above the Earth's surface. Also, the ionosphere point $IP_i$, placed at a height $h_{ion}$ on the line joining the satellite $SV_i$ and the user, is highlighted. The algorithm used to compute the ionosphere effect is summarized in Table 2.5.

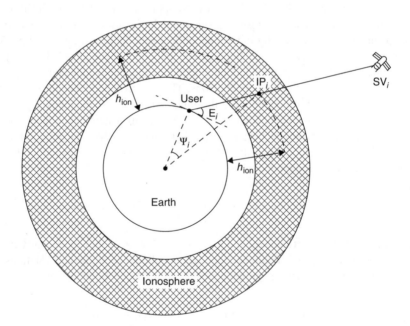

**Figure 2.17**  Ionosphere effects

**Table 2.5**   Klobuchar ionosphere algorithm (angles expressed in semicircles)

| Equation | Description |
| --- | --- |
| $F_i = 1 + 16(0.53 - E_i)^3$ | Obliquity factor (dimensionless) |
| $\Psi_i = \dfrac{0.0137}{0.11 + E_i} - 0.022$ | Earth's central angle between user and $IP_i$ |
| $\phi'_{IP_i} = \phi_u + \Psi_i \cos A_i$ | |
| $\phi_{IP_i} = \begin{cases} \phi'_{IP_i}, & \text{if } \left\|\phi'_{IP_i}\right\| \leq 0.416 \\ 0.416, & \text{if } \phi'_{IP_i} > 0.416 \\ -0.416, & \text{if } \phi'_{IP_i} < -0.416 \end{cases}$ | Geodetic latitude of the $IP_i$ |
| $\lambda_{IP_i} = \lambda_u + \dfrac{\Psi_i \sin A_i}{\cos \phi_{IP_i}}$ | Geodetic longitude of the $IP_i$ |
| $\phi_{m_i} = \phi_{IP_i} + 0.064 \cos\left(\lambda_{IP_i} - 1.617\right)$ | Geomagnetic latitude of the $IP_i$ |
| $t'_i = t_{GPS} + 4.32 \times 10^4 \, \lambda_{IP_i}$ | |
| $t_i = \begin{cases} t'_i, & \text{if } 0 \leq t'_i < 86\ 400 \\ t'_i - 86\ 400, & \text{if } t'_i \geq 86\ 400 \\ t'_i + 86\ 400, & \text{if } t'_i < 0 \end{cases}$ | Local time (s) |
| $AM'_i = \displaystyle\sum_{n=0}^{3} \alpha_n \phi^n_{m_i}$ | |
| $AM_i = \begin{cases} AM'_i, & \text{if } AM'_i \geq 0 \\ 0, & \text{if } AM'_i < 0 \end{cases}$ | Amplitude (s) |
| $PER'_i = \displaystyle\sum_{n=0}^{3} \beta_n \phi^n_{m_i}$ | |
| $PER_i = \begin{cases} PER'_i, & \text{if } PER'_i \geq 72\ 000 \\ 72\ 000, & \text{if } PER'_i < 72\ 000 \end{cases}$ | Period (s) |
| $X_i = \dfrac{2\pi\left(t_i - 50\ 400\right)}{PER_i}$ | Phase (rad) |
| $I_{L1} = \begin{cases} cF_i\left(5 \times 10^{-9} + AM_i\left(1 - \dfrac{X_i^2}{2} + \dfrac{X_i^4}{24}\right)\right), & \text{if } \left\|X_i\right\| < 1.57 \\ cF_i \ 5 \times 10^{-9}, & \text{if } \left\|X_i\right\| \geq 1.57 \end{cases}$ | Ionosphere error on L1 frequency (m) |

It is important to note that users having dual-frequency receivers can essentially eliminate the ionosphere effect on GNSS observables by means of a particular measurement combination. It is sufficient to consider the following relation, obtained from Equation 2.65, between the ionosphere errors at carrier frequencies L1 and L2 (recalling that the ionosphere is a dispersive medium):

$$I_{L2} = \frac{f_1^2}{f_2^2} I_{L1}.$$

Hence a ionosphere-free solution exploiting pseudorange measurements (derived from GNSS code) can be obtained by means of the following new set of $n$ observables:

$$P_{i,\mathrm{IF}} = \frac{1}{1-\left(f_1^2/f_2^2\right)}\left(P_{i,L2} - \frac{f_1^2}{f_2^2}P_{i,L1}\right) = \frac{f_1^2}{f_1^2 - f_2^2}P_{i,L1} - \frac{f_2^2}{f_1^2 - f_2^2}P_{i,L2} \quad i = 1,\ldots,n \quad (2.66)$$

where $\rho_{i,L1}$ and $\rho_{i,L2}$ are, respectively, the pseudorange measurements on L1 and L2.

As to a set of ionosphere-free carrier-phase measurements, an analogous expression to Equation 2.61 can be formulated, but in this case a problem has to be resolved: the initial ambiguities associated with the new observables are not integer owing to the following multipliers:

$$\frac{f_1^2}{f_1^2 - f_2^2} = 2.546, \quad \frac{f_2^2}{f_1^2 - f_2^2} = 1.546.$$

## 2.6.2  Troposphere Effects

Tropospheric refraction is an effect arising in the lowest atmospheric region, already mentioned as the troposphere, which extends above the Earth's surface up to about 50 km. For frequencies below 30 GHz, the troposphere can be assumed to be a nondispersive medium. The corresponding phase refractive index $n_\varphi$ is independent of the frequency, then $n_g = n_\varphi$ from Equation 2.50, and the effect is a delay on both the code/data and carrier phase. In this case

$$n_\varphi = n_g = \sqrt{1-2N_T} \approx 1-N_T, \quad N_T \ll 1, \quad \Delta c \approx cN_T \quad (2.67)$$

where $\Delta c$ is the local decrease in the propagation speed (for the group and phase velocities) and $N_T$ represents the corresponding tropospheric refractivity. Moreover, the tropospheric range error (which may reach up to 30 m when the satellite elevation is very low) is as follows:

$$\mathrm{TR} \approx \int_{\gamma_{\mathrm{tropo}}} N_T ds \quad (2.68)$$

where $\gamma_{\mathrm{tropo}}$ is the path followed by the satellite signal through the troposphere.

Usually, the tropospheric refractivity $N_T$ is split into two parts, $N_{T,hyd}$ and $N_{T,wet}$, corresponding, respectively, to the refractivity of the hydrostatic air component (which results in about 90% of the tropospheric delay) and to the wet air component due to the water vapor, which is very difficult to model. Both the refractivity coefficients depend on some meteorological parameters, modeled by Smith and Weintraub (1953) as follows:

$$N_T = N_{T,hyd} + N_{T,wet}, \quad N_{T,hyd} = k_1 \frac{p}{T} 10^{-6}, \quad N_{T,wet} = \left( k_2 \frac{e}{T} + k_3 \frac{e}{T^2} \right) 10^{-6}, \tag{2.69}$$

$$k_1 = 77.624 \, (K \times mbar^{-1}), \quad k_2 = -12.92 \, (K \times mbar^{-1}), \quad k_3 = 371\,900 \, (K^2 \times mbar^{-1})$$

where $p$ is the total atmospheric pressure (mbar), $e$ is the partial pressure of water vapor (mbar), and $T$ is the temperature (K), while $k_1$, $k_2$, $k_3$ are empirical constants.

Because of the difficulty in determining the actual values of the meteorological parameters along the path $\gamma_{tropo}$ and the consequent problem of integrating Equation 2.63, several approaches are present in the literature. In RTCA (2001a) the tropospheric correction is computed using sea-level meteorological parameters as functions of the receiver latitude and the day of year, after which the error terms of the hydrostatic and wet components are expressed as products of a vertical delay (when the satellite elevation angle is 90°) and a mapping function which is adopted to obtain the actual delay from the vertical one as a function of only the satellite elevation angle. This proposed approach leads to significant tropospheric corrections only for satellite elevation angles greater than 5°.

## 2.6.3   Selective Availability (SA) Effects

SA (when turned on) is the largest error source for unauthorized users making use of the SPS. The accuracy of a navigation solution is degraded by SA in two ways: firstly by virtue of alteration of the broadcast ephemeris parameters and secondly by injecting an error into the satellite clock. The SA effect on the pseudorange measurements can be modeled (RTCA, 2001a) as the sum of:

- A random variable (time constant) with a normal distribution described by a standard deviation of 23 m.
- A second-order Gauss–Markov stochastic process with an autocorrelation time of 28 s and a standard deviation of 23 m. The corresponding power spectral density is

$$s(\omega) = \frac{s_0^2 \omega_0^4}{\omega^4 + \omega_0^4}, \quad s_0 = 353.076 \, \left( m \times s^2 \times rad^{-2} \right), \quad \omega_0 = 0.012 \, \left( rad \times s^{-1} \right). \tag{2.70}$$

It is worth noting that SA is spatially correlated, so it can be removed in a local area with differential GPS applications making use of corrections computed by the ground station.

## 2.6.4   Multipath Effects

Many GPS applications cannot neglect the effects of multipath because of its significant contribution to the overall error and of the difficulties in mitigating it. The term multipath refers to one or more signals that reach the receiver antenna after being reflected by the surfaces encountered between the satellite and the user. Since these reflected signal paths are always longer than that of the direct signal, the multipath effect associated with the observables is a delay.

The most effective way to avoid multipath is via a proper installation. In particular, the receiver antenna has to be placed as far as possible from reflectors so that all reflected signals are received at negative elevation angles: this ensures strong multipath mitigation because of the radiation patterns of typical antennas.

In order to briefly explain the effect of multipath on carrier-phase observables, the following simplified operative condition is considered: the user antenna receives both the direct $s_D$ and a reflected $s_R$ carrier signal:

$$s_D = A\sin(\varphi), \quad s_R = \alpha A\sin(\varphi + \gamma_R) \tag{2.71}$$

where $A$ and $\varphi$ represent, respectively, the amplitude and the phase of $s_D$, while $\alpha$ is the amplitude attenuation factor and $\gamma_R$ is the phase shift of $s_R$. The resulting signal is $s = s_D + s_R$.

## 2.6.5   Receiver Noise

The sources of errors which take place in the GPS receiver are as follows:

- Thermal noise jitter inside the receiver tracking loops—delay lock loop (DLL), frequency lock loop (FLL), and phase lock loop (PLL)
- Receiver resolution that is directly related to the wavelength of the chip (for code measurements) and of the carrier (for phase measurements)

Typical receiver noise is at wavelengths of about 1–2 m for the C/A code, 0.1–0.2 m for the P-code, and 1–2 mm for the carrier-phase measurements.

## 2.7   GNSS Receivers

In the following sections a schematic diagram showing the general architecture of a simple GPS receiver is given and briefly described. Moreover, some important functionalities and software facilities usually implemented in almost all current receivers (such as carrier smoothing) are also described.

## 2.7.1   Receiver Architecture

A GNSS receiver is required to perform two processes sequentially as follows:

1. Acquiring and tracking the GNSS signal to extract from it the data contained in the navigation message together with the phase of the carrier and that of the PRN codes: these quantities are proportional to the pseudoranges between satellites and user.

2. Computing satellite positions and finally the receiver's position and clock offset with respect to the GNSS time.

In the following, the architecture and behavior of a typical civil receiver as shown in the block diagram of Figure 2.18 are briefly described in order to explain the main functions involved in GNSS signal processing. For simplicity, only the general architecture of the single-frequency receiver is considered in these descriptions.

The overall receiver architecture is characterized by two loops for carrier and code tracking, the aim of which is to determine the Doppler frequency shift (due to the satellite–user relative velocity), the phase of the carrier, the time delay of the code, and the data bit of the navigation message. These activities are performed in four steps:

1. Acquisition and tracking of the transmitted carrier frequency and Doppler shift determination
2. Acquisition and tracking of code and time delay determination
3. Acquisition and tracking of carrier phase
4. Synchronization and demodulation of data bits (navigation message)

With reference to Figure 2.18, firstly a low-noise amplifier near the antenna amplifies the very weak GNSS signal. A subsequent frequency down-conversion then takes place in order to transmit via the antenna cable an intermediate frequency (IF) signal which is less sensitive to cable attenuation and more suitable for sampling and digitizing after a filtering process that reduces the out-of-band noise. Sampling and digitization produce in-phase (I) and quadrature (Q) samples, the rate of which is sufficient to preserve the information content of the signal. After digitization, the data stream splits between the channels of the receiver. The subsequent set of blocks is in fact replicated for every channel, usually not less than 12 or 12 per system in combined GNSS receivers, in order to track in parallel the signals of "all-in-view" satellites simultaneously. Each channel consists of an inner and an outer feedback loop (see Figure 2.18), called the code loop (DLL) and the carrier frequency/phase loop (FLL/PLL), respectively.

The purpose of the code loop is to synchronize the received code with an internally generated replica of the same code sequence. Determination of the shift between the two identical codes is a measure of the transit time from satellite to user. The principle is based on the code correlation process where the correlation function $R_{xz}(\tau)$ is defined as

$$R_{xz}(\tau) = \int_{-\infty}^{+\infty} x(t) z(t+\tau) dt \qquad (2.72)$$

and represents a measurement of how much signal $x(t)$ looks like signal $z(t)$. If $x(t)$ is the received signal and $z(t)$ is a shifted, internally generated copy of signal $x(t)$, then $R_{xz}(\tau)$ reaches its peak value for the value $\tau_1$ of $\tau$ such that $z(t+\tau_1) = x(t)$. Here, $\tau_1$ is the amount by which the internal copy has to be shifted to completely correlate with the received signal; $\tau_1$ is proportional to the distance between transmitter and receiver which, by means of a

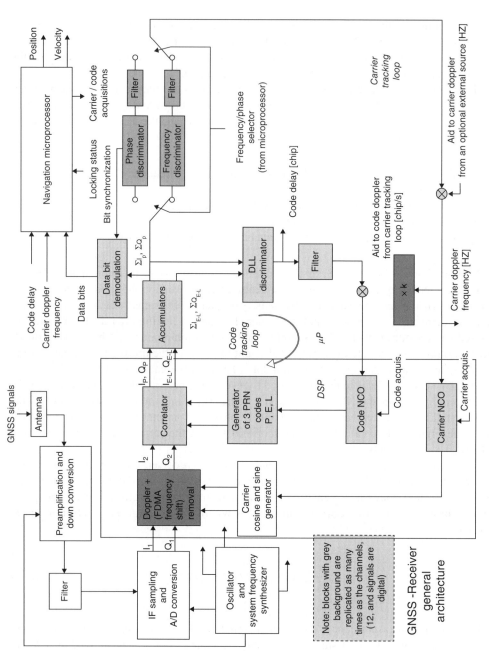

**Figure 2.18** Typical GNSS receiver architecture

speed of light multiplication, becomes the pseudorange. Of course, in order to best exploit the correlation process, the following conditions must hold:

- Cross-correlation—that is the correlation function when $x(t)$ and $z(t)$ are different copies of code corresponding to different satellites—is very low.
- Autocorrelation—that is the correlation function when $x(t)$ and $z(t)$ are the same sequence—must be low for all $\tau \neq \tau_1$.

The code loop performs the correlation process by means of discrete quantities, the I and Q samples, and by multiplications and sums, implemented by correlators and accumulators, respectively. The discriminator DLL block detects the value of the correlation function and, in a feedback loop, steers an NCO for increasing or decreasing the generation frequency of the I and Q replicas in order to perform maximum correlation between received and generated code sequences.

Code loop tracking would be useless if the Doppler effect could not be determined. In fact, the Doppler frequency of the carrier also affects the length of the code chips (which are modulated on the carrier), and the correlation process previously described could not be performed. The FLL has the purpose of tracking the Doppler variations of the carrier frequency, so making possible the locking of the received signal with the internally generated signal. When the Doppler frequency (added to the appropriate constant frequency shift inherent to GLONASS FDMA legacy signals) has been compensated, the FLL becomes a PLL by switching from the frequency discriminator to the phase discriminator (see Figure 2.18). Under these circumstances, the phase of the carrier is tracked, and hence an exact alignment of the code and navigation message bits can be performed. The navigation message can then be demodulated by the navigation microprocessor. This processor, by means of information about the code delay, carrier Doppler frequency, and navigation message, can finally compute the position of the receiver using programmed positioning algorithms.

The features of a receiver are related to the data they provide to the user. Pseudoranges, carrier phases, and ephemerides are called raw data, which not all receivers provide: many of them simply output position and velocity. The availability of raw data can be useful for external processing according to the requirements of the user, to whom the output rate is not greater than 20 Hz. Position and velocity are usually output with a rate of 1–5 Hz.

## 2.7.2  Carrier Smoothing

Carrier smoothing is a well-known technique implemented by high-quality receivers that exploit the joint usage of code pseudorange and carrier-phase observables. The purpose of carrier smoothing is to filter out the high noise level affecting code measurements by means of the low-noise carrier signal. The smoothing process is accomplished by a complementary filter whose architecture is shown in Figure 2.19. The carrier-smoothing filter takes advantages from both the almost noise-free phase observable (that allows for a good estimate of the predicted pseudorange) and the ambiguity-free information represented by the code measurement.

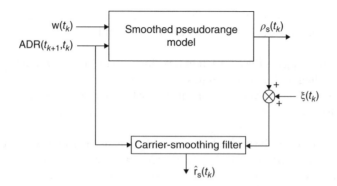

**Figure 2.19**   The carrier smoothing process

It is worth noting that the smoothing process is effective only on high-frequency errors such as receiver noise, while it does not affect biases or slowly varying signals like atmospheric errors that affect both code and carrier signals.

The smoothing filter is based on a mathematical model of the pseudorange as follows.

Consider Equation 2.31 computed at two subsequent time instants, $t_k$ and $t_{k+1}$. The member-by-member difference of the two equations leads to

$$
\begin{aligned}
&\Phi_m\left(t_{k+1}\right)-\Phi_m\left(t_k\right) \\
&= \omega_0 \frac{d\left(t_{k+1}\right)}{c} + b_r\left(t_{k+1}\right) - b_s\left(t_{k+1}\right) + e_\Phi\left(t_{k+1}\right) - \omega_0\frac{d\left(t_k\right)}{c} - b_r\left(t_k\right) + b_s\left(t_k\right) - e_\Phi\left(t_k\right)
\end{aligned}
\tag{2.73}
$$

which, by defining,

$$
b\left(t_k\right) = \frac{c}{\omega_0}\left(b_r\left(t_k\right) - b_s\left(t_k\right)\right)
$$

$$
w\left(t_k\right) = \frac{c}{\omega_0}\left(e_\phi\left(t_k\right) - e_\phi\left(t_{k+1}\right)\right)
$$

where $b(t_k)$ is the clock bias expressed in metres and $\omega_0$ the carrier frequency.

By recalling Equation 2.40, this becomes

$$
d\left(t_{k+1}\right) + b\left(t_{k+1}\right) = d\left(t_k\right) + b\left(t_k\right) + \frac{c}{\omega_0}\mathrm{ADR}\left(t_k,t_{k+1}\right) + w\left(t_k\right)
\tag{2.74}
$$

By remembering the definition of pseudorange $\rho(t_k)$ as given by Equation 2.10 (here subscript $i$ has been omitted and the dependence on time highlighted), it is possible to define the smoothed pseudorange $\rho_s(t_k)$ at time $t_k$ as the range plus clock bias. Hence, Equations 2.10 and 2.69 become

$$\begin{cases} \rho_s\left(t_{k+1}\right) = \rho_s\left(t_k\right) + \dfrac{c}{\omega_0}\,\mathrm{ADR}\left(t_k,\ t_{k+1}\right) + w\left(t_k\right) \\[2mm] \rho\left(t_k\right) = \rho_s\left(t_k\right) + \xi\left(t_k\right) \end{cases} \qquad (2.75)$$

$\rho(t_k)$ being the code pseudorange measurement and $\xi(t_k)$ the code error. The second equation represents the output model, the first is the process model, and $\mathrm{ADR}\left(t_k,t_{k+1}\right)$ is the input signal. A widely used carrier-smoothing filter, designed on the basis of model (2.68), for providing an estimate $\hat{\rho}_s\left(t_k\right)$ of the smoothed pseudorange is

$$\hat{\rho}_s\left(t_k\right) = \alpha\rho\left(t_k\right) + \left(1-\alpha\right)\left(\hat{\rho}_s\left(t_{k-1}\right) + \dfrac{c}{\omega_0}\,\mathrm{ADR}\left(t_k,t_{k+1}\right)\right) \qquad (2.76)$$

where the value of the gain $\alpha$ is chosen to be equal to the sample time divided by 100. The steady state of the filter error can be biased due to the presence of low-frequency errors that cause $w(t_k)$ and $\xi(t_k)$ to have nonzero means. Also, due to ionosphere gradient, code-carrier divergence causes the estimation to be biased (Walter *et al.*, 2004).

## 2.7.3 Attitude Estimation

The GNSS receiver can be exploited not only as a primary source of navigational information but also as a powerful system to estimate the attitude of vehicles themselves. The problem of attitude representation of a vehicle with respect to a reference system may be treated as the problem of representing the orientation of a vehicle body coordinate system (the vehicle being modeled as a rigid body) with respect to a local horizontal frame (a typical requirement) or with respect to an inertial frame in a tridimensional space. Hence, the vehicle attitude can be represented by means of Euler angles (minimum representation), or by a rotation matrix made up of nine direction cosines, or by quaternions of four elements.

Since vehicles are modeled in this framework as rigid bodies, the relative position coordinates of three points integral with the vehicle may be used to form an attitude representation. In this case, differential techniques are used: in particular, there must be many onboard antennas installed in the vehicle and located in positions corresponding to the points selected for defining the attitude. Actually, single GNSS receivers with multiple antennas are frequently adopted in such applications. One of them is called the *master* and is considered to be the reference, whence the others become *slaves*. The vectors between master and slaves are called *baselines*.

In order to obtain sufficient accuracy in attitude estimation, the pseudorange observable cannot be adopted because of the high order of related measurement errors with respect to the (typically short) distance between the onboard antennas. However, the carrier-phase single-difference observable is suitable owing to its subcentimetre accuracy. The main disadvantage related to carrier-phase observables is the necessity of determining the initial integer ambiguities (as described in Section 2.5.1) in order to obtain the projections of the baselines onto the satellite-receiver line of sight.

To facilitate a solution to the problem of attitude determination, some constraints may be used. Theoretically, only two measurements are required for the following reasons:

- The onboard receivers are mutually synchronized by means of a common local time reference; hence the measurements are not affected by differences between receiver clock time bias.
- The antenna placement is fixed and known a priori, so that the additional constraint on the relative distances between antennas makes it possible to decrease the minimum number of required measurements once again.

The basic relation that relates single-difference observables and attitude-related information is (Parkinson and Spilker, 1996)

$$\phi_i^j = \left(\mathbf{b}_i\right)^T \mathbf{s}_j - N_{ij} \tag{2.77}$$

where $\mathbf{b}_i$ is the baseline vector from the master to the $i$th slave antenna, $\mathbf{s}_j$ is the master antenna to $j$th satellite unit vector line of sight, $\phi_i^j$ is the $j$th satellite signal single difference between the master and $i$th slave antenna, and $N_{ij}$ is the integer ambiguity relative to satellite $j$ and baseline $i$. The vectors involved in the previous expression must be expressed in the same reference system. The components of the baseline vector are known in body axes, while the line-of-sight vector is known in an ECEF-related frame. Hence, Equation 2.79 can be written as

$$\phi_i^j = \left(\mathbf{b}_i^{\text{body}}\right)^T \mathbf{A} \mathbf{s}_j^{\text{ECEF}} - N_{ij} \tag{2.78}$$

where $\mathbf{A}$ is the transformation matrix between the ECEF and body frames and $\mathbf{A}$ is a function of the attitude parameters. The superscripts indicate the frames in which the vectors are expressed.

Given a suitable number of available measurements and after the integer ambiguity determination, it is possible to determine the elements of $\mathbf{A}$ and hence the attitude of the vehicle.

Typically the number of measurements is greater than the minimum required. As a consequence, an attitude estimation can be computed with a very high degree of accuracy. Moreover the considerable redundancy in available information improves the integrity.

## 2.7.4   Typical Receivers on the Market

Currently, there are hundreds of GNSS receiver manufacturers and thousands of receiver models on the international market. These models are used for numerous applications, including general navigation (N) and timing (T), defense (D), meteorology (Met), general position reporting (P), real-time differential referencing (R), recreation (Rc), geoinformation systems and surveying (G/S), vehicle vessel tracking (V), and others (O). They may have specific performances to meet user environment requirements such as aero (A), land (L), marine (M), and space (S) applications.

Progress in microelectronics relevant to GNSS receiver units have resulted in few, or even single, chip realizations of the GNSS receiver function. GNSS receiver formats now vary from complete end-user formats (F1 in Table 2.6) to module/chipset formats (F2 in Table 2.6) for OEM applications, including GNSS "system-in-package" (SiP) and "system-on-chip"

**Table 2.6** Key features of some satellite receivers on the market

| Product | Garmin GPSMAP 496 | Javad GGD-112T | Leica GRX1200 | Novatel OEMV-2 | Thales SkyNav GG12 | Trimble TA-12S |
|---|---|---|---|---|---|---|
| GNSS signals | GPS L1, WAAS | GPS L1 and L2, GLONASS L1 and L2, WAAS | GPS L1 and L2, GLONASS L1 and L2 | GPS L1 and L2, GLONASS L1 and L2, WAAS | GPS L1, GLONASS L1 | GPS L1 and L2 + PPS, WAAS |
| GPS accuracy (position, velocity) | 15 m, 0.05 m/s | 5 mm (static), N/A | 10 m (navigation stand-alone mode) and 5 mm (static), N/A | 1.8 m (single point L1), 0.03 m/s | 3.3 m, 0.05 m/s | 16 m, 0.2 m/s |
| DGPS (LAAS) | RTCM SC-104 | RTCM SC-104, RTK | DGPS/RTCM SC-104, RTK | DGPS | RTCA DO-217 | N/A |
| Update rate (Hz) | 5 | 20 | 20 | 20 | 5 | 1 |
| Dynamic range | 6 g | N/A | N/A | 0–515 m/s | 10 g, 0–514 m/s | 0–400 m/s |
| Acquisition time (cold, warm) | 45 s, 15 s | 60 s, 10 s | N/A | 50 s, 40 s | 90 s, 45 s | 1 min (time to first fix) |
| Additional features | Airport maps | RAIM | N/A | OEM board | RTCA DO-208 compliant | RAIM RTCA DO-229B, FAA TSO certified for SPS operations, anti-spoofing, selective availability (authorized users) |
| Typical use | Onboard navigator | Ground station equipment | Survey and geodetic applications | Navigation and ground station equipment | Navigation | Military applications |

(SoC) realizations. Also, major improvements in digital processing rates, accompanied by moderate power consumption, have made it possible to digitize and process spread-spectrum signals in units close to antenna outputs and hence to bring computer-based software (SW) GNSS receivers to the market. The GNSS market is outlined annually in publications such as the "GPS World" Receiver Survey.

Table 2.6 summarizes the main key features of GNSS receivers for aerospace applications currently available on the market during 2013. The product list shown is simply indicative and does not consider many other important manufacturers operating in the satellite navigation receiver area.

## 2.8  Augmentation Systems

Over recent years, specific systems have been developed to increase the performance of GNSS positioning in terms of accuracy and signal integrity (Parkinson and Spilker, 1996). These *Augmentation Systems* (AS) are mainly of two types: *Ground-Based AS* (GBAS) and *Space-Based AS* (SBAS), both of which rely on the concept of *Differential GNSS*, as will be described in the next section. Also known as a *Local Area Augmentation System* (LAAS), a GBAS serves a user area that is local (some tens of kilometres), has to be built up and maintained by the relevant users (airports, harbors, etc.), and has potentially the best performance. Conversely, a SBAS covers continental areas at no cost to users but potentially exhibits a lower performance than GBAS in terms of accuracy and integrity.

### 2.8.1  Differential Techniques

Differential techniques (RTCA, 2001b, 2004) involve the possibility of a mobile GNSS receiver eliminating those GNSS signal errors that are shared with another receiver located at a well-known fixed and georeferenced position. That is, differential techniques require a ground station where one or more fixed reference GNSS receivers are located and georeferenced. Using these known positions, it is possible to compute the exact ground receiver-satellite ranges. Hence, a comparison between the computed ranges and the measured pseudoranges provides the total pseudorange errors added to the GNSS signals at the georeferenced locations. Among these errors, some are common (i.e., atmospheric, satellite clock, and ephemeris errors) in the sense that, within in a certain area, they affect both the reference receivers and all the mobile receivers that are tracking the same satellites. The other errors are noncommon (i.e., receiver noise, multipath error) and affect the reference and mobile receivers differently. A filtering process for cancelling out noncommon errors from the overall computed pseudorange error in the reference receivers provides the so-called differential correction, and this can be applied by all mobile receivers to their measured pseudoranges in order to reduce total errors and to compute position solutions with increased accuracy. Of course, an error is actually common for both reference and mobile receivers depending on their relative separation. In fact, the correlation of atmospheric errors for two receivers decreases with distance.

By taking the components of the error term $\xi$ in the pseudorange model of Equation 2.10, both for reference and mobile receivers tracking the same satellite, the following results appear:

$$\rho_r = d_r + b_r + \eta + v_r - ck \qquad (2.79)$$

$$\rho_m = d_m + b_m + \eta + v_m - ck \qquad (2.80)$$

where subscripts r and m stand for reference and mobile, respectively, and

$\rho$ = pseudorange
$d$ = receiver-satellite range
$b$ = receiver clock bias
$\eta$ = common atmospheric (iono + tropo)errors
$v$ = non-common errors (receiver noise + multipath)
$ck$ = satellite clock error

Every receiver that tracks a satellite can obtain an estimate $\hat{ck}$ of the satellite clock error from the navigation message, that is,

$$ck = \hat{ck} + \Delta ck \qquad (2.81)$$

where $\Delta ck$ is the residual undetermined error. Hence, Equations 2.80 and 2.81 can be written by putting the known quantities in the left side and the unknown ones in the right side:

$$r_r + \hat{ck} = d_r + b_r + h + v_r - \Delta ck \qquad (2.82)$$

$$r_m + \hat{ck} = d_m + b_m + h + v_m - \Delta ck \qquad (2.83)$$

Moreover, an estimate $\hat{b}_r$ of the reference receiver clock bias can be computed using the georeferenced position together with the reference receiver measurements:

$$b_r = \hat{b}_r + \Delta b_r \qquad (2.84)$$

where $\Delta b_r$ is the corresponding estimated residual error. From Equations 2.77 and 2.79

$$h = r_r - d_r - \hat{b}_r + \hat{ck} - v_r - \Delta b_r + \Delta ck = -\Delta r - v_r - \Delta b_r + \Delta ck \qquad (2.85)$$

where

$$\Delta r = -\left(r_r - d_r - \hat{b}_r + \hat{ck}\right) \qquad (2.86)$$

is computed by the reference receiver and is called the *differential correction*.

Finally, the mobile receiver corrected measurement model becomes

$$r_m + \hat{ck} + \Delta r = d_m + b_m + v_m - v_r - \Delta b_r \qquad (2.87)$$

where the influence of errors is strongly reduced with respect to the case without differential corrections. Usually, in order to reduce the effects of the noise components in $v_m$ and $v_r$ on the corrected measurement, the differential corrections are processed by a carrier-smoothing filter with the same time constant both in the reference and the mobile receiver before they are actually used by the mobile receiver. Accuracies of about one to several metres are then attained depending on the distance between the reference and the mobile receivers.

Accuracy degradation in differential range corrections is inevitable due to spatial decorrelation. The distance from the reference receiver that allows the differential corrections to be considered valid is limited to several hundred kilometres. To achieve higher positioning accuracies (to the decimetre or centimetre level) in real time, carrier-phase measurements along with pseudoranges and double-differencing techniques (see Section 2.5.4) should be implemented. The raw pseudorange and carrier-phase observables or their corrections will be transmitted from the reference station to mobile receivers (at a 0.5–2 s update rate) to be processed together with the mobile receiver observables using the double-differencing technique. This process is known as *Real-Time Kinematic (RTK)* GNSS positioning. Since spatial decorrelation degrades the accuracy of double-difference observables and the integrity of ambiguity resolution, the distance between a reference station and a mobile receiver would be greater or smaller depending on whether dual- or single-frequency observables were used, but should not exceed tens of kilometres. The ambiguity terms may be solved for as integer or real numbers, and once integer ambiguities have been fixed, centimetre-level accuracies may be obtained. Alternatively, the floating ambiguity solutions limit accuracies to decimetre level.

## 2.8.2    The Precise Point Positioning (PPP) Technique

The main disadvantages of the differential technique are binding to the reference station and local operation area and spatial accuracy degradation. These have stimulated intensive research for an alternative precise positioning technique and have resulted in the Precise Point Positioning (PPP) technique, firstly in wide area and then globally.

PPP originated as a GNSS data processing technique that utilizes dual-frequency observations from single GNSS receivers, combined with precise satellite orbit and clock corrections, to provide precise estimates of 3D position, clock offset, and tropospheric effects. The term "precise satellite orbit and clock corrections" is used to distinguish it from data available via standard GNSS navigation messages. The contemporary technique providing these corrections was developed as an efficient method of processing code and carrier-phase measurements from networks of static GPS reference stations (Zumberge *et al.*, 1998). In recent years, a number of governmental, academic, and commercial PPP services (IGS, Natural Resources Canada (NRCan), Jet Propulsion Laboratory (JPL), and others) have been developed for various applications such as the precise navigation and positioning of manned and unmanned vehicles, precise timing, precise agriculture, photogrammetry, topography, cartography, hydrography, geoinformation systems, construction, the exploiting of natural resources, the oil and gas industry, environment monitoring, and SAR operations. These services supply the user with precise orbit and clock data products that have several quality grades differing in accuracy and latency:

- "Ultrarapid" (predicted)—real-time data with typical accuracy ~10 cm (orbits) and ~5 ns (clocks)
- "Ultrarapid" (estimated)—1–3 h delayed data with typical accuracy <5 cm (orbits) and ~0.2 ns (clocks)
- "Rapid" (estimated)—17 h delayed data with typical accuracy <5 cm (orbits) and ~0.1 ns (clocks)
- "Final" (estimated)—approximately 13 days delayed data with typical accuracy <5 cm (orbits) and <0.1 ns (clocks)

The combination of precise satellite orbit and clock corrections and carrier-phase measurements processed at the users site can provide positioning accuracy from the centimetre to the decimetre level in static and kinematic applications. Depending on the application, this result can be obtained in postmission mode or in real time, provided that dissemination facilities to produce, transmit, receive, and process precise orbit and clock products are available.

A user position solution is based on the processing of ionosphere-free combinations of undifferenced code and phase observations. Unknown parameters to be estimated from the observation model in PPP include 3D position coordinates, phase ambiguity terms, receiver clock offset, and troposphere effect. Several systematic effects that could introduce an observable variation level in the undifferenced observations and are not considered for standard point positioning and double-differenced RTK positioning will be taken into account in the observation model, and corresponding corrections will be provided to mitigate these effects. The satellite signal biases, phase windup correction, satellite antenna offset, site-dislocation effects (due to solid Earth lunar–solar tide and ocean tide loading), solid Earth and pole motion, and relativity effects are among them.

The PPP technique brings several significant advantages compared to differential precise positioning techniques, but it also presents challenges to the realization of its full potential capabilities. Of course, PPP needs only a single receiver and removes the need to establish local reference stations that have limited operating areas as well as the constraint of simultaneous observations on both rover and reference stations peculiar to the differential RTK technique. PPP solutions are referred to a global reference frame as opposed to differential approaches where position solutions are relative to a reference station in the general case. Furthermore, PPP can support other applications in addition to positioning. By estimating clock offset and troposphere effect along with position coordinates, it provides precise time dissemination and water vapor parameter estimation by a single GNSS receiver.

PPP problems include a long initialization time (typically more than 20 min) for the position solution to converge to centimetre accuracy. Also, the loss of signal tracking lock on a minimum number of satellites requires processing reinitialization and additional time until the position solution reaches convergence. Then, ambiguity terms in PPP undifferenced carrier-phase observations are distorted by satellite and receiver initial fractional phase biases and are floating, not integer ones. Potentially the convergence time may be reduced to minutes or even seconds, but only when reliable ways to identify and determine the initial phase biases and fix ambiguity terms to their correct integer values are found. A promising approach being made in this field involves a new model of the ionosphere-free code and carrier-phase observation equations (Binath and Collins, 2012). This model includes the clocklike parameters as estimated for both satellite and receiver in each observable. Additional opportunities to improve position accuracy and reduce convergence time come from the seamless integration of PPP and RTK techniques (Landau, 2012). An integration of PPP and Inertial Navigation System (INS) in real-time kinematic applications (e.g., georeferencing in airborne mapping or precise LEO satellite orbit determination) can reduce convergence time as well as reinitialization time after signal blockages peculiar to such applications.

## 2.8.3   Satellite-Based Augmentation Systems

Several Satellite-Based Augmentation Systems operate, or are going to operate, in order to improve the performances of GNSS over wide areas or globally.

There are four geostationary SBAS with wide-area coverage:

1. The Wide Area Augmentation System (WAAS), developed by the Federal Aviation Administration (FAA), is intended to be primarily used over the American continental area.
2. The European Geostationary Navigation Overlay System (EGNOS), developed by the ESA, operates within Europe.

3. The MTSAT Satellite-Based Augmentation System (MSAS), based on the Japanese Multifunctional Transport Satellite (MTSAT) system, operates for users within a large part of the Asian continental area.
4. The System for Differential Corrections and Monitoring (SDCM), which is the Russian SBAS that will be based on three GEO satellites from the multifunctional space relay system "Luch" (Луч—ray or beam in Russian), with the first satellites providing service over the main part of Russia (Urlichich *et al.*, 2012a).

In particular, the WAAS area is currently served by two Inmarsat III geostationary satellites: Pacific Ocean Region (POR) and Atlantic Ocean Region-West (AOR-W). The EGNOS area is served mainly by two further Inmarsat III satellites and one ESA geostationary satellite—Atlantic Ocean Region-East (AOR-E), Indian Ocean Region (IOR), and ARTEMIS, respectively. When EGNOS is finally released, ARTEMIS will be the main satellite. Moreover EGNOS shares the AOR-W with the WAAS, whence some parts of Europe will be covered by four satellites. The MSAS area is served by the MTSAT-1 and MTSAT-2 Japanese geostationary satellites. The SDCM area will be served by the Luch 5A (95° E), Luch 5B (16° W), and Luch 4 (167° E) GEOs. As a further detail, in Figure 2.20a and b the deployment of the four Inmarsat III and three Luch geostationary satellites is shown.

For convenience, WAAS, EGNOS, MSAS, and SDCM, being very similar and quite compatible, are here termed as WAAS-like systems.

A WAAS-like system consists of a wide-area network of monitoring ground stations that continuously send information to the Wide-Area Master Station (WMS). The WMS analyzes the received data and computes the parameters needed by users in the wide area to estimate the main sources of errors that reduce the accuracy and to determine the integrity of the GNSS satellite measurements. The Ground Earth Station (GES) then uploads to the geostationary satellites the results (including the ephemeris data) obtained by the WMS. Finally, the geostationary satellites broadcast (at the GPS L1 frequency) the integrity and correction data to users in the wide area by means of messages modulated with a PRN C/A code of the same type as the GPS PRN C/A code described in Section 2.4.1.2. In this way a GNSS receiver need be only slightly modified with respect to the standard hardware architecture in order to use WAAS-like system capability. For this reason, if the accuracy is sufficient (i.e., the position error is typically contained within 3 m for the useful exploitation of a WAAS-like system), this solution is usually preferred to the one based on a GBAS and described in Section 2.8.2.

A WAAS-like system augments a GNSS in order to satisfy the RNP in terms of availability, integrity, and accuracy for all phases of flight up to Category I precision approaches by means of the following three services:

1. An additional WAAS-like system ranging
2. WAAS-like system enhancements to the Receiver Autonomous Integrity Monitoring (RAIM) availability and performance
3. WAAS-like differential corrections to the measurements of the GNSS satellites

The WAAS-like system ranging capability is due to the PRN C/A code modulation of the signal broadcast by the geostationary satellites.

A WAAS-like system provides enhancements to both RAIM (see Section 2.10.1.1) and Wide-Area Differential Corrections by means of an appropriate set of messages broadcast to users.

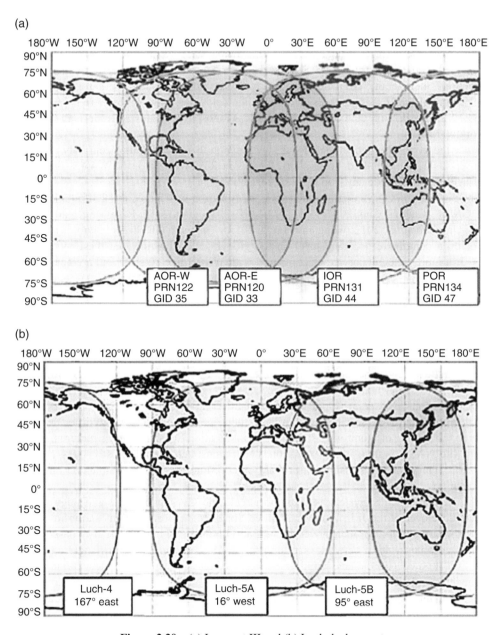

**Figure 2.20**   (a) Inmarsat III and (b) Luch deployment

These messages contain 250 bits and are transmitted with a bit rate of 250 bps, which is five times faster than that of a GPS navigation message. A detailed description of these messages and how to exploit the WAAS-like system capabilities is explained in RTCA (2001a). The integrity message indicates whether to use the GNSS satellite measurements.

Regarding Wide-Area Differential Corrections, there are several messages providing users with a set of parameters for computing the contributions of some of the measurement errors. In particular, the correction data are split into two sets: fast and slow. The fast corrections include compensation for the GPS SA and the short-term satellite clock errors; and the slow corrections include the compensation for both the ephemeris errors and the long-term clock errors of the GNSS satellites. Moreover, the delay due to the ionosphere is provided to users by means of a model composed of vertical ionosphere delays at predefined points on a worldwide grid.

Since the tropospheric delay is a local phenomenon that can be quite accurately computed by users using the standard tropospheric model in RTCA (2001a), the WAAS-like system messages do not include any tropospheric corrections.

The next step in GNSS augmentation system evolution was the Global Differential GNSS (GDGNSS) (Cai and Gao, 2007; Hatch et al., 2012; Landau, 2012; Rizos et al., 2012; Toor, 2012; Urquhart et al., 2012; Wang and Hatch, 2012). The main GDGNSS feature is a decimetre or even centimetre accuracy level on the global scale, showing that these systems can provide a more efficient service as compared with WAAS-like systems and are approaching the accuracy of precise geodetic GBAS that operate in limited local areas.

There are many GDGNSS now offering their capabilities to commercial and governmental customers: the **StarFire** system developed by NavCom Technology Inc. (NavCom) and Ag Management Solutions (AMS), operated by the NavCom; the **OmniSTAR** system supported by the OmniSTAR B.V. from the Fugro Corporation; the **VERIPOS** and **TERRASTAR**™ services provided by Veripos; the **Trimble RTX** service from Trimble; the **Nexteq iPPP** from the Nexteq Navigation company; and, of course, the **NASA Global Differential GPS (GDGPS)** system (Muellerschoen et al., 2000, 2001, 2004; Reichert et al., 2002; Armatys et al., 2003; Bar-Sever et al., 2004; Wu and Bar-Sever, 2005).

All these systems and services are similar in their operating principles but differ from each other in the number of monitoring stations, communication links, message structures, and end users. The NASA system is no exception, but, predestined for the most critical real-time space, air, land, and sea missions, it probably incorporates the most complex and universal structure.

The NASA GDGPS System is actually a real-time GPS monitoring and augmentation system developed at the JPL, Pasadena, United States, to support missions and infrastructure for NASA in addition to other governmental and commercial customers. It is based on the global real-time tracking network of reference GNSS receivers placed at geographical sites with well-known coordinates, including 35 sites equipped with highly accurate atomic frequency standards and four national time laboratories. The reference receivers stream raw measurements—that is, all possible GNSS observables—via highly redundant communication channels to triply redundant, geographically separated, fully independent GDGPS operational centers (GOCs). The GOCs process the GNSS observables to produce real-time differential corrections to the GNSS orbit and clock states along with environmental and ancillary data products. The latency between navigation signal reception by the tracking receiver at the reference site, and the differential correction production and transfer for customers, is about 5 s. The data products of the GOCs are transferred to customers through a variety of communication links including the Internet, frame relay, dedicated landlines, and satellite broadcasting. As a result, the GDGPS System provides 10 cm positioning accuracy and subnanosecond time transfer accuracy anywhere in the world on the ground, in the air, and in space and independently of local infrastructure. Its robust architecture ensures high system reliability that results in continuous service without interruptions. This quality has been accomplished through an extremely high level of end-to-end redundancy without a single point of failure, automatic fault detection, network rerouting, and seamless tailoring between the various components in the system.

## 2.9 Integration of GNSS with Other Sensors

GNSS navigation exhibits the following main advantages and disadvantages:
Advantages:

- High accuracy
- Errors bounded in time
- Worldwide availability

Disadvantages:

- Noisy data
- Low data rate (typically 1 Hz)
- Susceptibility to interference
- Discontinuity of functioning caused by frequent series of signal loss and reacquisition phases
- No attitude information

Briefly, by referring to a dynamic framework, it can be affirmed that GNSS sensors are mainly affected by high-frequency errors, hence exhibiting good low-frequency performance.

Consequently, in order to extend the applications of satellite navigation, data-fusion techniques have been widely exploited for compensating the disadvantages of GNSS by means of other sources of position- and velocity-related information. Integration between GNSS and other sensors with complementary characteristics can be used to mitigate the negative effects of stand-alone sensors and to obtain the best results from a blended position solution. Sensors that can be integrated with a GNSS receiver are inertial sensors (gyros and accelerometers), barometric sensors, pitot (dynamic pressure) and aerodynamic angle sensors, Earth's magnetic field sensors (compasses, magnetometers), Doppler radar, odometers (for ground applications), etc. Of course, the integration architecture depends on which sensor and which measurements have to be blended with GNSS data. However, all the possible integration schemes can be referred back to the complementary filter architecture shown in Figure 2.21, a scheme where good low-frequency performance sensors together with good high-frequency performance sensors can be best exploited.

In this scheme, a quantity y is measured by two different sensors, the output of each being a measurement obtained by adding its particular error to y. Here, $e_1$ is assumed to be a low-band error, while $e_2$ has a high-frequency spectrum (e.g., Sensor 2 could be a GNSS receiver). The difference between the two outputs is therefore the difference between the two sensor errors. This difference is applied to a filter whose low-pass behavior cancels out $e_2$ and outputs an estimate $\hat{e}_1$ of $e_1$. Finally, this estimate is subtracted from the Sensor 1 output, and if the estimate is good, the final output is close to the noise-free value y.

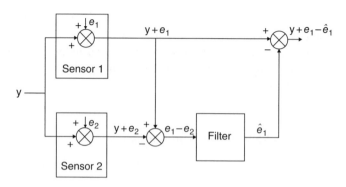

**Figure 2.21**   Complementary filter architecture

Because the integration filter operates on the measurement error and not on the entire measurement, its designed is based on the error models of the GNSS positioning and other sensors. While sensor models are usually highly nonlinear, error models are linear, even if time variant: as a consequence, in this case a linear Kalman filter is usually used to provide optimal performance.

### 2.9.1   GNSS/INS

One of the most relevant and widely used applications of data fusion between satellite navigation data and other sensor information is the integration of a GNSS receiver with an INS (Phillips and Schmidt, 1996; RTCA 2001a). The advantages of an INS are the following:

- It computes the navigation solution by numerically integrating the vehicle acceleration and angular rate measurements obtained from accelerometers and gyroscopes.
- It provides its output at a higher rate than GNSS and maintains the availability of a navigation solution during GNSS receiver outages. In fact, INS is insensitive to jamming and signal or obstacle interference.
- It has a high system bandwidth, so allowing for position computation in high dynamic situations where a GNSS receiver could easily lose signal tracking.
- It provides a full navigation state, that is, position, velocity, and attitude.

However, the INS does have the disadvantage that it suffers from low-frequency errors and in particular it is heavily affected by gyro drifts and accelerometer biases, and as a consequence, position errors grow if not corrected by external means.

Because of their complementary characteristics, GNSS and INS are often used jointly, and many approaches are possible for the achievement of information integration leading to a navigational solution. However, such integration schemes can be divided into two classifications, loosely coupled and tightly coupled. For both types, the INS navigation solution always represents the reference trajectory to which corrections by the complementary filter are added. For simplicity, the following material considers only position or range, but not velocity or attitude.

### 2.9.1.1    Loosely Coupled Integration

This integration scheme is shown in Figures 2.22 and 2.23. In Figure 2.22, the filter processes the difference of the positions separately computed by the Inertial Reference System (IRS, another acronym to indicate a class of laser-gyro INS) and the GNSS receivers. The advantage of this scheme is the simplicity of the integration where the outputs of the GNSS and the INS receivers can be used directly.

An alternative loosely coupled scheme is shown in Figure 2.23. Here, the filter directly processes the pseudoranges and then computes and outputs the position error. In this case a range prediction computation block has to be inserted to transform the INS position into vehicle-to-satellite range values in order to compare homogeneous pseudorange data. This second scheme has the advantage of being independent of the way in which the receiver

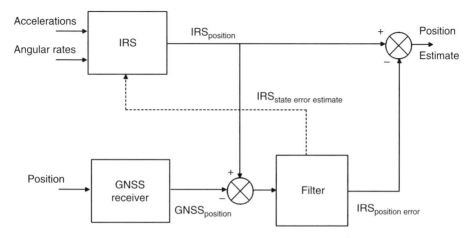

**Figure 2.22**    Loosely coupled integration scheme

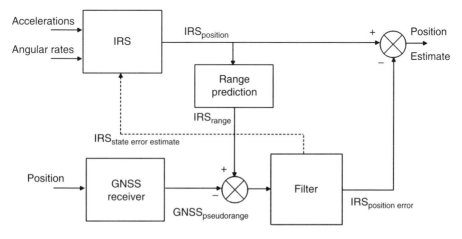

**Figure 2.23**    Alternative loosely coupled integration scheme

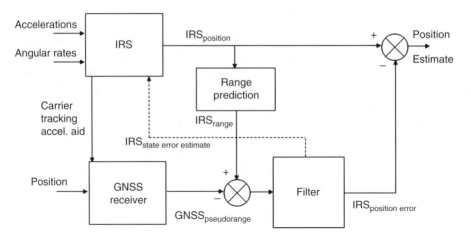

**Figure 2.24**    Tightly coupled integration scheme

computes the position and so improves the overall performance. However, it does have the disadvantage of increased complexity; and it is also necessary to manage the satellite navigation message in order to extract the ephemeris for the satellite position computation.

In the figures, the dashed feedback lines indicate that the IRS error estimates may be fed back to the IRS to improve the system calibration. This can be useful for bounding IRS error growth, but it also risks undetected large GNSS errors causing unacceptable deviations in the reference trajectory.

### 2.9.1.2    Tightly Coupled Integration

Unlike the loosely coupled range integration method, tight coupling uses acceleration information from the INS/IRS to aid the signal tracking of the GNSS receiver, as shown in Figure 2.24. However, during vehicle maneuvers, maintaining signal tracking requires a wide bandwidth carrier tracking loop. However, by incorporating vehicle-to-satellite range acceleration (obtained by projecting the INS/IRS-measured acceleration vector along the line of sight) into the tracking loop, this bandwidth can be significantly reduced, so improving the signal-to-noise ratio. As a consequence, this makes the receiver more immune to interference and jamming. The disadvantage of this approach is the need to have access to the receiver hardware and to the tracking loop registers.

## 2.10    Aerospace Applications

Aerospace applications form very important "bench tests" for satellite navigation systems: in fact they cope with aerospace vehicles with high dynamic performance that navigate in critical situations where safety is a primary requirement. After a brief description of the unique aerospace concept of safety, the problems of integrity are introduced, and a brief treatment of the main aerospace applications of GNSS sensors, including air traffic control (ATC) and air and space navigation, then follows.

## 2.10.1   The Problem of Integrity

The very high accuracy attainable in positioning by means of GNSS suffers the drawback of high disturbance vulnerability (e.g., jamming, unexpected interference levels, and multipath problems) in addition to a sometimes inadequate time to alarm for managing possible failure situations (e.g., sudden satellite clock faults, ephemeris errors, etc.). These factors are critical mainly in aerospace navigation where signal-in-space integrity is a primary concern for safe navigation. In fact, the use of a navigation system in civil aviation depends on its capability of fulfilling the RNP as stated by the ICAO in terms of accuracy, continuity, integrity, and availability. Integrity is the RNP parameter that has the most direct impact on safety because it refers to the reliability and trustworthiness of positional information (ICAO, 1995; RTCA, 2003). More precisely, integrity is the ability of a system to provide a timely warning to the pilot when the system should not be used for navigation. Loss of integrity happens when a system failure remains latent, that is, firstly a failure occurs and secondly no alarm is raised (missed detection (MD)). The loss of integrity probability is therefore given by

$$P_{(\text{loss of integrity})} = P_{(\text{failure})} \times P_{(\text{missed detection})} \tag{2.88}$$

Maximal figures for loss of integrity probability are ICAO requirements: for example, a typical value for the probability of a failure during the approach phase of flight is $10^{-7}$ times per hour. The probability of failure is related to the GNSS satellite mean time between outages (MTBO) and the signal-in-space performance. For GPS, a typical value of $10^{-4}$ times per hour can be assumed. As a consequence, an MD probability of $10^{-3}$ has to be fulfilled by the integrity monitoring system.

A considerable body of literature exists where integrity monitoring systems and fault detection and exclusion (FDE) algorithms for GNSS are proposed and analyzed (Van Graas, 1996a; Parkinson and Axelrad, 1988; Brenner, 1995; Diesel and Luu, 1995). The purpose of FDE algorithms is to detect and exclude a possible fault within a maximum latency time before raising an alarm to the user with a prescribed probability of MD but without exceeding a maximum false alarm rate. Three techniques of integrity monitoring exist: RAIM, Aircraft Autonomous Integrity Monitoring (AAIM), and External Methods.

RAIM and AAIM are essentially based on the availability of redundancy in the information used to compute the navigation solution. Redundancy in GNSS positioning can result from considering a number of satellite signals greater than strictly necessary. This is the case in RAIM methods, where an FDE algorithm is implemented on board the GNSS receiver and makes use of signals provided by at least five GNSS satellites (recalling that only four satellites are strictly necessary for computing the navigation solution). Alternatively, AAIM methods use redundancy obtained by performing data fusion of GNSS signals and measurements provided by other onboard sensors such as INS. External Methods perform integrity monitoring of the navigation solution by transmitting integrity flags or alarms to the GNSS receiver by means of an AS (GBAS or SBAS) where the signal-in-space integrity has been previously checked by the monitoring stations. As an example, Galileo satellites are designed to send their own health status within the navigation message in order to allow any GNSS receiver to monitor the integrity of the signals it receives. However, External Methods cannot send an instantaneous alarm to the user because the Galileo latency time is 1 s. Therefore, the implementation of RAIM or AAIM in a cascade scheme with the eventually available External Methods alarm is always encouraged in order to reduce latency time and to increase the integrity of the navigation solution in critical applications. Due to their importance, a brief description of RAIM methods now follows.

### 2.10.1.1 RAIM

RAIM algorithms can be divided into two main groups, snapshot algorithms and history-based algorithms. However, due to the assumption of linearity in the satellite navigation system measurement model, and under the hypothesis that all errors affect only the pseudor-ange measurement vector, all RAIM algorithms are based on the extensive use of the classical LS or of the Kalman filter estimator. Snapshot algorithms usually rely on LS-based methods that process a set of data sampled at the same time (Van Graas, 1996a), while history-based algorithms make use of multiple Kalman filters fed with different measurement subsets (Brenner, 1995).

RAIM algorithms can also be classified as using range-based or position-based methods respectively, depending on the type of test statistic they adopt for the detection function. Range-based methods work in the pseudorange domain (i.e., the measurement space) and are able to detect the occurrence of a fault in the signal in space according to the aforemen-tioned probabilities. Range-based methods follow the procedures developed for parity-space or residual FDE algorithms. Alternatively, position-based methods use the spread of position solutions compatible with the available measurements in order to detect a possible failure situation. Here, the underlying concept involves the definition of a threshold for the spread of the position subsolutions as computed by removing a satellite signal one at a time from the overall measurement set. In the case of a faulty measurement, the culprit affects all the computed positions but one, so that the position spread serves as a statistical test for detection.

While the equivalence between range- and position-based methods has been demonstrated, a subtle difference holds between them. Given that not all possible faults in a satellite signal necessarily cause a failure in the navigation solution, this difference becomes important, as follows. A range-based algorithm is capable of detecting a fault in GNSS signals, whereas a position-based algorithm can detect a failure in the navigation solution. Hence, the latter seems to conform better with the ICAO requirements in the sense that these are primarily concerned with the impact of a fault on the navigation solution (i.e., the navigation failure and its detection) rather than on the fault itself (and its detection).

Integrity algorithms must also incorporate two important features in order to be imple-mented for air navigation. The first is *exclusion capability*: when a fault is detected, the algorithm must exclude the faulty signal from the navigation solution. Only when the exclusion process fails to find the faulty signal within a specified time-to-alarm period must an alarm be presented to the pilot. The second relevant feature is the computation of the protection levels, both horizontal and vertical (HPL and VPL). This means that whenever no failure is detected, the integrity algorithm has to compute the horizontal and vertical bounds within which the aircraft position must reside, both during normal operation and in the case of an undetected fault. Safe navigation is possible whenever the HPL and VPL are contained within the horizontal and vertical alert limits (HAL and VAL) specified by ICAO requirements for each particular phase of flight.

## 2.10.2   Air Navigation: En Route, Approach, and Landing

Air navigation is one of the most relevant applications of GNSS receivers and where their particular features will be progressively exploited in the near future (Van Graas, 1996b). The usage of a GNSS receiver for air navigation depends heavily on the phase of flight it has to support. The three main kinds of flight phases briefly considered here are *en route*, *terminal area*, and *approach* navigation. RNP for these three phases are in increasing order according to the tightness of the requirements that have to be fulfilled. En route navigation was the first aeronautical application of satellite navigation, and currently many consolidated procedures have been published and now exist for GPS in supporting cruise and terminal area navigation near many important airports around the world. In order that a GNSS receiver may be used for these phases of flight, it must necessarily have RAIM capability. A receiver certified according to Technical Standard Order (TSO) 129a must be able to fulfill the necessary RNP requirements. With these types of receiver, RAIM algorithms are required to further increase their performances by means of data fusion among pseudoranges and pressure information as provided by a barometric altimeter, a device that is always included in basic aircraft instrumentation. This leads to an increase in the availability of integrity and hence to a concomitant improvement in the reliability of the navigation solution.

Instrument approach phases include the most critical tasks for navigation systems in terms of safety, and continuity and integrity requirements demand high capability in those systems. Continuity can be achieved by supporting the satellite receiver with an inertial system and an integration filter that blends together both satellite and inertial data. However, regarding integrity during the approach phase, the RAIM is not sufficient to assure risk values compliant with ICAO normatives (ICAO, 1995). Hence, for these approach phases, augmentations are demanded due to their capability of providing increased integrity in addition to accuracy, and, in particular, either SBAS or GBAS should be considered. SBAS augmentation can support approach phases up to Category I (200 ft decision height (DH)) with a time to alarm of 6 s.

Recently, new approach categories have been introduced by the ICAO to enable the usage of SBAS for approaches that are rather less stringent than Category I: Approach with Vertical Guidance (APV) with different DHs that range from 300 ft down to a minimum of 250 ft (the corresponding FAA equivalent being LPV). Also in these cases, in order to allow the use of GPS receivers for APV approaches to airports with published WAAS procedures, the receivers must be compliant with official requirements as stated in TSO 145a and 146a documents. Category I approach GPS-WAAS procedures have recently been approved, prior to which (2006) only private, special crew, and equipped aircraft were certified for approved Category I approaches. These new procedures require the availability of certified GBAS.

Categories II and III (DH 100 ft and 50 ft, respectively), together with automatic landing, will only be supported by GBAS augmentations. In fact, a maximum time to alarm of 1 s and extremely reduced protection levels (<1 m) are objectives beyond the capabilities of SBAS. At present the use of GBAS augmented satellite navigation for these types of approach is still in the research and development phase.

## 2.10.3   Surveillance and Air Traffic Control (ATC)

GNSS are playing increasing roles not only in air navigation but also in surveillance and ATC applications. Traditionally, surveillance and control of aircraft positions in a controlled airspace have been accomplished by primary radars. Now, the availability of reliable and accurate positioning systems, such as low-cost GNSS receivers with SBAS capabilities, allows most aircraft in either commercial or general aviation not only to know their own positions but also to broadcast them to air traffic controllers. This allow the ATC controllers to accomplish surveillance and monitoring via a radar-like display where the received position data for the aircraft are superimposed on a map of the airspace. This new surveillance technology, which is called Automatic Dependent Surveillance-Broadcast (ADS-B), needs a ground–air digital data link. Moreover, broadcast position data can be received by other aircraft, so giving them knowledge of the surrounding air traffic (situational awareness). In surveillance applications, GNSS receivers are never the only data position sources: rather, they are integrated with other source of information like radars and multilateration sensors. Using data integration, it is possible to obtain a better and more reliable aircraft data position distribution and to cope with any temporary lack of performance by some sensors.

Another emerging application is the surveillance of vehicular traffic on an airport surface. Many different types of vehicle operate at an airport including taxiing aircraft, emergency vehicles, fuel trucks, service carts and luggage vehicles, etc. Surveillance of their positions for planning and controlling their paths and in preventing and monitoring potential collisions is an important task for airport authorities. This can be efficiently accomplished only if the exact position of each vehicle is known. Systems that deal with this surface traffic are called Advanced Surface Movement Guidance and Control System (A-SMGCS) and are currently being developed at most important airports. Positions are determined by GNSS receiver on board the vehicles to be surveyed, and usually GNSS augmentations and integration with airport surface radar improve the reliability and accuracy of this position surveillance.

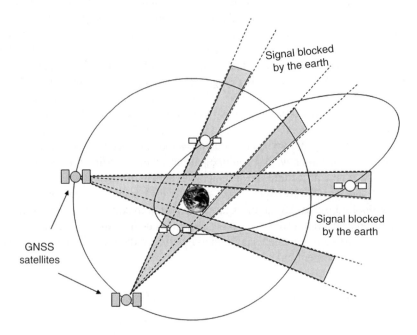

**Figure 2.25**   GNSS signal reception in space

## 2.10.4 Space Vehicle Navigation

There are important applications of GNSS to space vehicle navigation. These include trajectory determination such as position fixes, orbit determination, rendezvous and docking maneuvers, and attitude determination and timing. The advantages of GNSS space applications are reduced sizes and costs of GNSS receivers, precise time information on board, position and velocity vectors directly expressed in ECI and ECEF frames, and real-time trajectory computation. The main disadvantages are the visibility conditions and the high Doppler shift due to high dynamic conditions. For space applications there is also a problem with antenna locations due to the fact that GNSS satellite signals can be received from negative elevations (see Figure 2.25).

# References

ARINC (2000) Navstar GPS Space Segment/Navigation User Interfaces. Interface Control Document, ICD-GPS-200C, ARINC Research Corporation, El Segundo, CA.

Armatys M., Muellerschoen R., Bar-Sever Y., and Meyer R. (2003) *Demonstration of Decimeter-Level Real-time Positioning of an Airborne Platform.* Proceedings of the ION NTM-2003, January 19–24, Anaheim, CA, USA.

Bar-Sever Y., Young L., Stocklin F., Heffernan P., and Rush J. (2004) *The Global Differential GPS System (GDGPS) and the TDRSS Augmentation Service for Satellites (TASS).* Presentation at the ESA Second Workshop on Navigation Equipment, December 8–10, Noordwijk, the Netherlands.

BeiDou (2012) BeiDou Navigation Satellite System Signal in Space Interface Control Document. Open Service Signal B1I (Version 1.0). China Satellite Navigation Office, December 2012, p. 81.

Binath S. and Collins P. (2012) Recent development in precise point positioning, Geomatica, 66 (2), 103–111.

Brenner M. (1995) *Integrated GPS/Inertial Fault Detection Availability.* Proceedings of the Eighth International Technical Meeting ION GPS-95, September 12–15, 1995, Palm Springs, CA, USA.

Cai C. and Gao Y. (2007) Precise point positioning using combined GPS and GLONASS observations, Journal of Global Positioning Systems, **6** (1), 13–22.

Chaffee J. and Abel J. (1994) On the exact solutions of pseudorange equations. IEEE Transactions on Aerospace and Electronic Systems, **30** (4), 1021–1030.

Chen X. (2012) Current State and Further Development of the Compass/BeiDou Navigation Satellite System. Sixth International Satellite Navigation Forum, April.

Diesel J. and Luu S. (1995) *GPS/INS AIME: Calculation of Thresholds and Protection Radius Using Chi-Square Methods.* Proceedings of the Eighth International Technical Meeting ION GPS-95, September 12–15, Palm Springs, CA, USA.

Donghyun K. and Langley R. (2000) *GPS Ambiguity Resolution and Validation: Methodologies, Trends and Issues.* Proceedings of the Seventh GNSS Workshop—International Symposium on GPS/GNSS, November 30–December 2, Seoul, Korea.

Farrell J. and Barth M. (1999) The Global Positioning System and Inertial Navigation. McGraw-Hill, New York.

Forssel B., Martín-Neira M., and Harris R.A. (1997) *Carrier Phase Ambiguity Resolution in GNSS-2.* Proceedings of the Tenth International Technical Meeting ION GPS-94, September 16–19, Kansas City, MO. Institute of Navigation, Satellite Division, Manassas, VA, pp. 1727–1736.

Fujiwara S. (2011) The Quasi-Zenith Satellite System and the Multi-functional Transport Satellite Satellite-based Augmentation System, Sixth Meeting of the International Committee on Global Navigation Satellite Systems (ICG), September 5–9, Tokyo, Japan.

Galileo SIS ICD (September 2010) Galileo Open Service. Signal in Space Interface Control Document (OS SIS ICD), Issue 1, Revision 1.

Ganeshan A.S. (2011) The Indian Satellite-based Navigation System: Description and Implementation Status. Sixth Meeting of the International Committee on Global Navigation Satellite Systems (ICG), September, 5–9, Tokyo, Japan.

GLONASS (2008) GLONASS Interface Control Document. Navigation Radiosignal in Bands L1, L2, Edition 5.1. Russian Institute of Space Device Engineering, Moscow.

Groves P.D. (2008) Principles of GNSS, Inertial, and Multisensor Integrated Navigation Systems. Artech House, Boston, MA.

Hargreaves J.K. (1995) The Solar-Terrestrial Environment: An Introduction to Geospace—the Science of the Terrestrial Upper Atmosphere, Ionosphere and Magnetosphere. Cambridge University Press, Cambridge Atmospheric and Space Science Series, Cambridge, UK.

Hatch R. and Euler H.-J. (1994) *Comparison of Several AROF Kinematic Techniques*. Proceedings of the Seventh International Technical Meeting ION GPS-94, September 20–23, Salt Lake City, UT, USA. Institute of Navigation, Satellite Division, Manassas, VA, pp. 363–370.

Hatch R., Sharpe T., Galyean P. (2012) StarFire: A Global High Accuracy Differential GPS System, Paper 1.6, NavCom Technology Inc., Torrance, CA.

Hofmann-Wellenhof B., Lichtenegger H.K., and Collins J. (1997) Global Positioning System: Theory and Practice, 4th edn. Springer-Verlag, Wien/New York.

Hofmann-Wellenhof B., Lichtenegger H.K., and Wasle, E. (2008). GNSS—Global Navigation Satellite Systems. Springer-Verlag, Wien.

ICAO (1995) Report on the Item 4 of the Special Communications/Operations. ICAO 15th Divisional Meeting on the Application of the (RNP) Concept to Approach, Landing and Departure Operations—SP COM/OPS/95-WP/151, ICAO, Montreal, Canada.

Ipatov V.P. (2005) Spread Spectrum and CDMA: Principles and Applications. Wiley & Sons, Chichester.

Ipatov V.P. and Shebshaevich B.V. (2010) GLONASS CDMA: some proposals on signal formats for future GNSS air interface. Inside GNSS, **5** (5), 46–51.

Ipatov V.P. and Shebshaevich B.V. (2012) Spectrum-compact signals: a suitable option for future GNSS. Inside GNSS, **6** (1), 47–53.

Kaplan E. and Hegarty C. (2006) Understanding GPS: Principles and Applications, 2nd edn. Artech House, London.

Kayton M. and Fried W.R. (1997) Satellite radio navigation, in Avionics Navigation Systems (Van Dierendonck A.J.), 2nd edn. John Wiley and Sons, New York.

Kotelnikov V.A., Dubrovin V.M., Morozov V.A., *et al.* (1958) The use of Doppler effect for determination of Earth artificial satellites orbit parameters. Journal of Communications Technology and Electronics, **7** (7), 873–881.

Kovalevsky J., Mueller I., and Kolaczek B. (1989) Reference Frames in Astronomy and Geophysics. Kluwer Academic Publishers, Dordrecht.

Landau H. (2012) Worldwide Centimeter Accurate GNSS Positioning Using Trimble RTX Technology, PPP-RTK & Open Standards Symposium, March 12–13, Frankfurt am Main, Germany.

Leick A. (1995) GPS Satellite Surveying, 2nd edn. John Wiley and Sons, New York.

Long A.C., Cappellari Jr. J.O., Velez C.E., and Fuchs A.J. (1989) Goddard Trajectory Determination System (GTDS) Mathematical Theory, Revision 1, FDD/552-89/001. NASA, Goddard Space Flight Center, Greenbelt, MD.

Muellerschoen R., Bertiger W., and Lough M. (2000) *Results of an Internet-Based Dual-frequency Global Differential GPS System.* Proceedings of the IAIN World Congress in Association with the U.S. ION 56th Annual Meeting, June 26–28, San Diego, CA, USA.

Muellerschoen, R., Reichert A., Kuang D., Heflin M., Bertiger W., and Bar-Sever Y. (2001) *Orbit Determination with NASA's High Accuracy RealTime Global Differential GPS System.* Proceedings of the ION GPS 2002, September 11–14, Salt Lake City, UT, USA.

Muellerschoen, R., Iijima B., Meyer R., Bar-Sever Y., and Accad E. (2004) *Real-Time Point Positioning Performance Evaluation of Single-Frequency Receivers Using NASA's Global Differential GPS System.* ION GNSS 17th International Technical Meeting of the Satellite Division, September 21–24, Long Beach, CA, USA.

Navstar (2011) Navstar GPS Space Segment/Navigation User Segment Interfaces, IS-GPS-200F, September 21.

NIMA (2000) Department of Defense World Geodetic System 1984 (WGS 84) – Its Definition and Relationships with Local Geodetic Systems, 3rd edn, Technical Report 8350.2. National Imagery and Mapping Agency, Fairfax, VA.

Oleynik E. and Revnivykh S. (2011) GLONASS Status and Modernization. Civil GPS Service Interface Committee, September 19, 2011, p. 20, Portland, OR.

Parkinson B.W. and Axelrad P. (1988) Autonomous GPS integrity monitoring using the pseudorange residual. Navigation: Journal of the Institute of Navigation, **35** (2), 255–271.

Parkinson B.W. and Spilker Jr. J.J. (1996) Global Positioning System: Theory and Applications. Progress in Astronautics and Aeronautics, v. 163–164. America Institute of Aeronautics and Astronautics Inc., Washington, DC.

Petrovello M. (2009) What are linear carrier phase combinations and what are the relevant considerations? Inside GNSS, January/February, 16–19.

Phillips R.E. and Schmidt G.T. (1996) GPS/INS integration. In: System Implications and Innovative Applications of Satellite Navigation. AGARD Lecture Series, 207. Advisory Group for Aerospace Research & Development, Neuilly-sur-Seine, pp. 9.1–9.18.

Reichert A., Meehan T., Munson T. (2002) *Toward Decimeter-Level Real-Time Orbit Determination: A Demonstration Using the SAC-C and CHAMP Spacecraft*. Proceedings of the 15th International Technical Meeting of the Satellite Division of the Institute of Navigation (ION GPS 2002), September 24–27, 2002, Portland, OR, pp. 1996–2003.

Revnivykh S. (2012a) *Global Satellite Navigation Development Trends*. Proceedings of the 19th Saint Petersburg International Conference on Integrated Navigation Systems, May 28–30, State Research Center of the Russian Federation Concern CSRI Elektropribor, St. Petersburg, Russia.

Revnivykh S. (2012b) *GLONASS Status and Modernization*. International GNSS Committee IGC 7, November 4–9, Beijing, China.

Rizos C., Janssen V., Roberts C., and Grinter T. (2012) Precise Point Positioning: is the Era of Differential GNSS Positioning Drawing to an End. FIG Working Week 2012, Knowing to Manage the Territory, Protect the Environment, Evaluate the Cultural Heritage, May 6–10, Rome, Italy.

RTCA (2001a) Minimum Operational Performance Standards for Global Positioning System/Wide Area Augmentation System Airborne Equipment. Technical Report DO-229C. RTCA Inc., Washington, DC.

RTCA (2001b) Minimum Operational Performance Standards for Global Positioning System/Local Area Augmentation System Airborne Equipment. Technical Report DO-253A. RTCA Inc., Washington, DC.

RTCA (2003) Minimum Aviation System Performance Standards: Required Navigation Performance for Area Navigation. Technical Report DO-236B. RTCA Inc., Washington, DC.

RTCA (2004) Minimum Aviation System Performance standards for the Local Area Augmentation System (LAAS). Technical Report DO-245A. RTCA Inc., Washington, DC.

Shebshaevich V.S. (1958) The preliminary estimation of the possibility to use Earth artificial satellites for navigation [PredvaritelnayaotsenkavozmozhnosteyispolzovaniyaiskustvennikhsputnikovZemlidlyatseleynavigatsii//Leningrad]. Proc. Academy-Mozhayskogo (33), 5–38.

Shebshaevich V.S. (1971) The introduction to space navigation theory [Vvedeniye v teoriyu kosmicheskoy navigatsii], Sovetskoye Radio, Moscow, 296 p.

Simsky A. (2006) Triple-frequency combinations in future GNSS. Inside GNSS (July/August), 38–41.

Smith E.K. and Weintraub S. (1953) The constants in the equation for atmospheric refractive index at radio frequencies. Proceedings of the Institute of Radio Engineers, **41**, 1035–1037.

Subirana J.S., Zornoza J.M.J., and Hernández-Pajares M. (2013) Global Navigation Satellite Systems, Volume I: Fundamentals and Algorithms. ESA Communications ESTEC, Noordwijk, the Netherlands, p. 238.

Teunissen P.J.G. (1994) *A New Method for Fast Carrier Phase Ambiguity Estimation*. Proceedings of the IEEE PLANS'94, April 2–15, Las Vegas, NV, USA. IEEE, Manassas, VA, USA, pp. 562–573.

Teunissen P.J., De Jonge P.J., Tiberius C.C.J.M. (1997) Performance of the Lambda Method for Fast GPS Ambiguity Resolution. Navigation: Journal of the Institute of Navigation, **44** (3), 373–383.

Toor P. (2012) Providing GNSS augmentation data: a commercial service provider's perspective. In: *PPP-RTK & Open Standards Symposium*, Frankfurt am Main, Germany, March 12–13.

Urlichich Y., Subbotin, V., Stupak, G., Dvorkin V., Povalyaev A., and Karutin S. (2012a) GLONASS: developing strategies for the future. GPS World (April), 42–49.

Urlichich Y., Subbotin, V., Stupak, G., Dvorkin V., Povalyaev A., Karutin S., and Bakitko R. (2012b) GLONASS Modernization. GPS World November, pp. 1–6.

Urquhart L., Zhang Y., Lee S., and Chan J. (2012) *Nexteq's Integer Ambiguity-Resolved Precise Point Positioning System*. Proceedings of the 25th International Technical Meeting of the Satellite Division of the Institute of Navigation, September 17–21, Nashville, TN, USA. Institute of Navigation, Satellite Division, Manassas, VA, pp. 3046–3054.

Van Graas F (1996a) Signals integrity. In: System Implications and Innovative Applications of Satellite Navigation. AGARD Lecture Series 207. Advisory Group for Aerospace Research & Development, Neuilly-sur-Seine, pp. 7.1–7.12.

Van Graas F (1996b) Requirements on GNSS for civil navigation. In: System Implications and Innovative Applications of Satellite Navigation. AGARD Lecture Series 207. Advisory Group for Aerospace Research & Development, Neuilly-sur-Seine, pp. 6.1–6.8.

Vollath U., Birnbach S., Landau H., Fraile-Ordoñez J.M., and Martín-Neira M. (1998) *Analysis of Three-Carrier Ambiguity Resolution (TCAR) Technique for Precise Relative Positioning in GNSS-2*. Proceedings of the Second

International Technical Meeting ION GPS-98, September 15–18, Nashville, TN, USA. Institute of Navigation, Satellite Division, Manassas, VA, pp. 417–426.

Walter T., Datta-Barua S., Blanch J., and Enge P. (2004) *The Effects of Large Ionosphere Gradients on Single Frequency Airborne Smoothing Filters for WAAS and LAAS*. Proceedings of the ION 2004 National Technical Meeting, January 26–28, San Diego, CA, USA. Institute of Navigation, Satellite Division, Manassas, VA, pp. 103–109.

Wang C. and Hatch R. (2012) StarFire™ GNSS: The Next Generation StarFire Global Satellite Based Augmentation System NavCom Technology Inc., Torrance, CA.

Wu S. and Bar-Sever Y. (2005) *Real-Time Sub-cm Differential Orbit Determination of Two Low-Earth Orbiters with GPS Bias Fixing*. Proceedings of the ION GNSS 2006, September 26–29, Fort Worth, TX, USA. Fort Worth Convention Center, pp. 2515–2522.

Zhu J. (1994) Conversion of Earth-centered Earth-fixed coordinates to geodetic coordinates. IEEE Transactions on Aerospace and Electronic Systems, **30** (3), 957–961.

Zumberge J.F., Heflin M.B., Jefferson D.C., Watkins M.M., and Webb F.H. (1998) Precise point positioning for the efficient and robust analysis of GPS data from large networks. Journal of Geophysical Research, **102** (B3), 5005–5017.

# 3

# Radio Systems for Long-Range Navigation

Anatoly V. Balov and Sergey P. Zarubin
*Institute of Radionavigation and Time, Saint Petersburg, Russia*

## 3.1   Introduction

Radio systems for long-range navigation comprise a subclass of radio-navigation systems based on geodetically referenced and synchronized chains of terrestrial transmitting stations that provide accurate navigation at distances far from those stations. In fact, their development was an attempt to solve the problem of how to minimize the number of terrestrial transmitting stations needed to encompass the global operational area. Later, such attempts resulted in the appearance of Global Navigation Satellite Systems (GNSS).

To reach maximal distances, Low-Frequency (LF) or Very Low-Frequency (VLF) signals are used. Sequences of pulses or pseudo-continuous waveforms at several frequencies are the carrier signal formats. More specifically, these are sequences of Binary Phase Shift Keying (BPSK) changes in the phase angles of the carrier at set time intervals, for example, using two phase points at 0 and 180°. The time of pulse envelope arrival (TOA) and/or signal carrier phases are the observables for navigation solutions.

The resolution of carrier phase ambiguity is based on the concept of a Standard Reading Point (SRP) (see Figure 3.3b) on the envelope of the received signal, or a wide-lane technique combining different carrier phases at several frequencies.

The observables are the time-equivalents of geodetic pseudoranges from transmitting stations to system users. Depending on the situation, a system user can obtain navigation solutions in ranging, pseudo-ranging, or difference-ranging (hyperbolic) modes. The latter was the main mode used at the beginning of long-range navigation procedures, so such systems are sometimes termed "hyperbolic systems." The speed of the signal ground wave propagating through the Earth–ionosphere wave guide from the transmitter to the user is the time-range transformation scale factor. In contrast to the free space propagation speed, it depends on changing waveguide parameters and is therefore not constant. Hence, the observables contain

---

*Aerospace Navigation Systems*, First Edition. Edited by Alexander V. Nebylov and Joseph Watson.
© 2016 John Wiley & Sons, Ltd. Published 2016 by John Wiley & Sons, Ltd.

corresponding systematic and random propagation errors that must be compensated. Methods of compensation define several operational modes as follows:

**Standard**, where corrections are made to compensate for radio wave propagation effects predicted by means of mathematical models. This is the main mode used for predicting such corrections.

**Differential**, where actual corrections by local geodetically referenced differential stations are broadcast to users in real time. The operational area of this mode is limited by the properties of the spatial–temporal correlation of propagation errors.

**Pseudo-differential**, where actual corrections by users as they pass points of known position are used to modify subsequent position solutions along the route. Normally, this operational mode is used when other navigation aids providing significantly higher position accuracies than the standard mode are available only occasionally.

**Adaptive**, where corrections are determined as unknowns simultaneously with position determinations. Such a mode can be operational when many redundant measurements are available.

The first hyperbolic Radio-Navigation System (RNS) was proposed by R.J. Dippy in 1937; it was later implemented during World War II (WWII)-era British Gee VHF short-range system in early 1942, and introduced by the Royal Air Force for use by Bomber Command. This was followed by the Decca Navigator System in 1944 by the Royal Navy.

The history of establishing Loran (Long-Range Navigation) begins with the creation in the United States of the Medium-Frequency (MF) Loran-A in 1941–1942. The initial idea of Loran-A belongs to A. Lumis from Bell Labs, then headed by J. Pierce (Dean, 1996), and the system was awarded the status of "Allied Standard Long-Range Radio-Navigation System" for ship/aircraft navigation during WWII. A similar system using two chains—"Meridian" (1943–1949) and "Normal" (1949–1953)—was developed in the USSR.

Loran-C (project originator R.L. Frank) was developed during 1950–1956 for the navigational support of naval ships and submarines and for military aircraft. Operation of the first chain on the American northeastern coast began in 1957, after realizing a signal format as a phase-coded pulse train. This had an important impact on the system's progress and resulted in the development of mobile Loran-D in 1963 (project originator E. Lipsy).

Later, Loran-C was widely used by civil aviation and overland transportation, and until the early 1990s, the US Coast Guard operated all the Loran-C chains. Then, the United States discontinued overseas chain support and some chains were closed. In 1992, several European countries decided to develop their own Loran-C system known as the Northern Europe Loran Service(NELS), which was phased out in 2005 (Williams, Basker, and Ward, 2008). China, South Korea, Japan, and Russia established the *Far East Radio-Navigation Service* (FERNS) to develop Loran-C chains in the Far East. In 2010, the United States and Canada canceled Loran-C operation in North America. Currently, 20 Loran-C stations forming nine chains are functioning over the territories of Great Britain, Denmark, France, Germany, Norway, Japan, South Korea, and China.

The first CHAYKA (Seagull) chain was developed in the USSR during 1958–1971, the project originators being K.S. Poltorak, S.M. Shargorodsky, and Y.I. Nikitenko. CHAYKA is similar to Loran-C but with insignificant distinctions in waveform and the duration of each pulse in a train (Specification of the Transmitted Loran C Signal, 1994; National Standard of Russian

Federation, 2009). However, the CHAYKA pulse has a sharper front end and a shorter duration because of different Earth surface electrical conductivities within the Loran-C coverage (sea) and that of CHAYKA (dry land, permafrost, and ice) affecting the conditions of ground wave/ sky wave separation. Fifteen CHAYKA stationary transmitting stations forming three chains, the European, Far Eastern, and Northern, currently operate in the Russian Federation.

The creation of the global phase VLF Omega system was largely contributed due to a proposal by J. A. Pierce of the U.S. Naval Ocean Research Center and formulated in 1947 (Swanson, 1983). The system's transmitting stations, located over the territories of Norway, Liberia, Argentina, Australia, Japan, the United States (Dakota and Hawaiian Islands), and Reunion Island, had been put into operation from 1973 through 1982 and phased out in 1997.

The Russian VLF ALPHA was developed from 1956 to 1970 as a three-station system with an operating range of 8 000–10 000 km. These stations are located near Krasnodar, Novosibirsk, and Komsomolsk-on-Amur. This system supports the navigation of ships and aircraft in the regions of Arctic, Atlantic, and Indian Oceans as well as submarine navigation at depths down to 30 m and under ice (Boloshin, 1993). In 1975, an additional station in Revda (the Murmansk region) was inaugurated.

The following sections in the chapter are aimed at representing the VLF and LF long-range navigation systems including their principles of operation, accuracy, coverage, budget of errors, interfering signals, signal processing techniques, and sensor architecture. These subjects are augmented by including various system modernization aspects and their integration with satellite RNS.

## 3.2   Principles of Operation

Relative performances of LF and VLF long-range radio-navigation systems are given in Table 3.1, where the terms used for the characteristics appear in the glossary of terms, *Loran-C User Handbook*—Appendix-C (US DoT, US Coast Guard, 1992).

VLF radio systems employ a phase measurement method; hence, the carrier phases are the observables used for navigation solutions. To resolve phase measurement ambiguity, each station transmits on several frequencies. For transmitting multifrequency signals by a chain of Reference Stations (RS), frequency-timing signal division is used. The transmitting sequence is referred to as the station's emission pattern.

OMEGA operates at four fixed frequencies: $F1 = 10.2\,\text{kHz}$, $F2 = 13.6\,\text{kHz}$, $F3 = 11.33\,\text{kHz}$, and $F4 = 11.05\,\text{kHz}$. During a cycle, each station emitted four signals at the four frequencies with a 10 s emission cycle. The durations of the four signals transmitted by each station varied: at 0.9, 1.0, 1.1, and 1.2 s, and the beginning of the A station's first cycle was at midnight (Greenwich time). The format of signals emitted by the A, B, C, D, E, F, G, and H stations is shown in Figure 3.1 (Swanson, 1983).

ALPHA operates at four fixed frequencies: $F1 = 11.905\,\text{kHz}$, $F2 = 12.649\,\text{kHz}$, $F3 = 14.881\,\text{kHz}$, and $F4 = 12.0907\,\text{kHz}$ (Jacobsen, 2006). The duration of the emitted signals is 0.4 s and the emission cycle is 3.6 s. The format of the signals emitted by the stations is given in Figure 3.2 (Boloshin, 1993).

The single-frequency LF radio system uses the carrier phases of signal pulses as observables. Phase ambiguity resolution is achieved through the measurement of the carrier frequency phase in the time region of the envelope corresponding to the end of the third carrier cycle in the envelope—the so-called *Standard Reading Point* (SRP) or standard zero crossing.

**Table 3.1** Relative performances of Loran-C/CHAYKA and OMEGA/ALPHA

| System | Accuracy | | | Unit availability | Coverage | System availability | Fix rate | Fix dimension | System capacity | Potential ambiguity |
|---|---|---|---|---|---|---|---|---|---|---|
| | Predictable | Repeatable | Relative | | | | | | | |
| Loran-C CHAYKA | At least 0.25 nm (460 m) 1:3 SNR | 60–30 ft. (18–90 m) | 60–30 ft. (18–90 m) | 99+% (transmitting station signal availability greater than 99.9%) | Coastal continental US, RF, and selected overseas areas | 99.7%[a] | 10–20 fixes per minute | Two dimensions | Unlimited of simultaneous uses | Yes, easily resolved |
| OMEGA ALPHA | 2–4 nm (3.7–7.4 km) | 2–4 nm (3.7–7.4 km) | 0.25–0.5 nm (463–926 m) | 99+% | Worldwide continuous | 97%[b] | Six fixes per minute | Two dimensions | Unlimited | Requires knowledge to ±36 nm[c] |

[a] Three-station reliability. Individual station availability normally exceeds 99.9%. Note also that many areas are served by more than one Loran (CHAYKA) chain, so increasing the availability.

[b] Three-station joint signal availability.

[c] Three-frequency receiver.

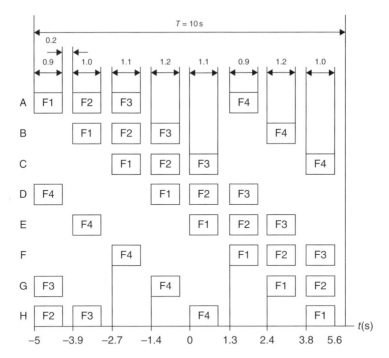

**Figure 3.1**   OMEGA signal format

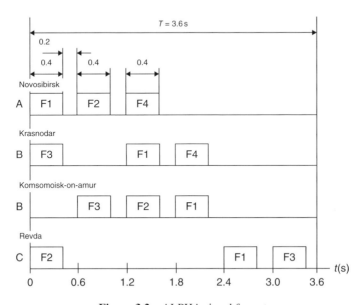

**Figure 3.2**   ALPHA signal format

(a)

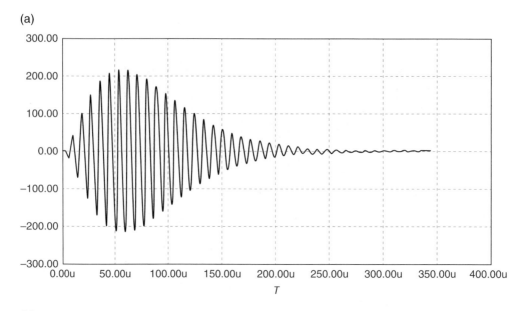

(b)

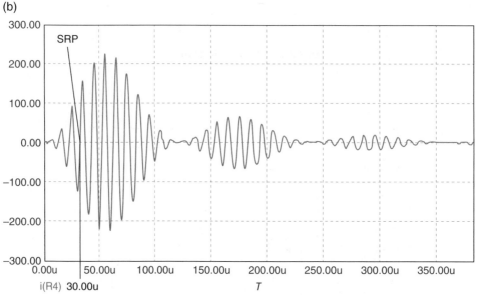

**Figure 3.3**    (a) Loran-C and (b) CHAYKA single pulse general view

Loran-C and CHAYKA stations emit pulse trains at a 100 kHz carrier frequency, and a general view of their respective single pulses in a train is shown in Figure 3.3a and b.

The duration of the Loran-C pulse leading edge from 0.1 to 0.9 of the envelope amplitude is 50 μs and its leading edge duration from 0 to maximum is 65 μs. The pulse duration at the −6 dB level is within 230 μs. The duration of the CHAYKA pulse leading edge from 0.1 to 0.9

of envelope amplitude is $27 \pm 1.5\,\mu s$ and the duration of its leading edge from 0 to maximum is $43.7\,\mu s$. The pulse duration at the $-6\,dB$ level is within $100-120\,\mu s$. The SRP where the phase measurement is made for both systems is $30\,\mu s$ after the pulse start.

Every station operates using a specified *Group Repetition Interval* (GRI), each of which is a multiple of $10\,\mu s$ from $40\,000$ to $99\,990\,\mu s$. In order to increase the average signal power, each LF station emits a pulse train consisting of eight phase-coded pulses per repetition interval. Also, to provide automatic identification of master/secondary signal trains, in addition to improving interference immunity, binary-phase coding of pulses within a train is used with a phase code interval being equal to two GRIs. The position of this binary-phase modulated ninth pulse is denoted by an asterisk (*)in Table 3.2.

The phase codes of all the secondary stations are the same, but the master station's code differs from those of all the secondary stations. Auto- and cross-correlation functions of the phase-coded master/slave signals provide for the elimination of the effect of multiple signal reflections from the ionosphere. The coding laws are given in Table 3.2, where the binary phase-coded pulses with positive/negative polarity are marked as +/–.

The interval between train pulses is $1000\,\mu s$. In Loran-C, the master station can emit the ninth pulse spaced by $2000\,\mu s$ from the eighth pulse. However, all CHAYKA stations additionally emit a ninth pulse used for service communications. The master station can periodically emit a tenth pulse, or "coloring" pulse, showing a coincidence of the emission time of this group with the UTC(SU) time mark. In CHAYKA, the eighth pulse SRP is the moment of coincidence. The coding of the CHAYKA ninth pulse is defined by a data transfer system. Figure 3.4 shows arrangement and function of pulses within a train.

**Table 3.2**   Phase coding laws for CHAYKA, Loran-C master and secondary stations

| Group repetition interval | Master station | Secondary stations |
| --- | --- | --- |
| A | + + – – + – + – + | + + + + + – – + * |
| B | + – – + + + + + * + | + – + – + + – – * |

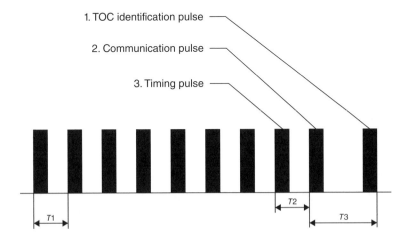

**Figure 3.4**   Pulse train arrangement

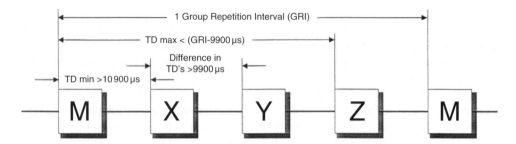

**Figure 3.5** Sequence of master and secondary station pulse trains

Figure 3.5 presents a sequence of trains emitted by chain stations, where the restrictions shown are relevant to the emission delays. Pulses emitted by master station M and secondary (slave) stations X, Y, and Z of a chain are mutually synchronized in time, frequency, and phase. Emissions from secondary stations are delayed relative to the master station's emission by these emission delays. The emission delay values are selected in such a way that the specified sequence of signals arriving from the master and slave station is preserved at any point within the coverage area.

As can be seen from Figure 3.5, the requirements for emission delays are as follows:

- The minimum emission delay, $TD_{min}$, between trains vof master and secondary station should be 10 900 μs.
- The minimum difference between any two delays or time differences should be 9900 μs.
- The maximum emission delay should be equal to the GRI minus 9900 μs.

Transmitting stations comprise the following:

- Control/synchronization equipment
- Transmitter
- Antenna system
- Autonomous power supplies (diesel generators).

Two types of transmitting antennae are used:

1. *Top-Loaded Monopole* (TLM) antennas with heights from 125 to 460 m
2. *Set-Loaded Tower Antennas* (SLTs) of five or more tower masts with heights from 250 to 400 m.

The radiated power of LF transmitters for different stations is 165–1850 kW and of VLF transmitters is 30–50 kW.

## 3.3   Coverage

The coverage of an RNS is that part of the surface within which the user's coordinates could be determined with an error not exceeding the maximum allowable value, and within which the received signal power should exceed a threshold value corresponding to the maximum system range. Thus, the borders of the coverage are determined by such equalities as $D = D_{max}$ and $s_r = s_{r\,max}$, where $s_{r\,max}$ is the maximum allowable Root-Mean-Square (RMS) value of the radial error.

When considering coverage, it is necessary to take into account a system's *geometric factor* G, the factor of accuracy depending on the type of system and the mutual location of the stations and user. The geometric factor can be determined as the ratio of position error to the raw measurement error ($\sigma_p$), and the general expression takes the following form:

$$G = \frac{\sigma_r}{\sigma_p} = \frac{1}{\sin \alpha_M} \sqrt{\frac{1}{g_1^2} + \frac{1}{g_2^2}}$$

Here, $\alpha_M$ is the crossing angle of the lines of position; $g_1$ and $g_2$ are the respective gradients of the measurement field (Bykov and Nikitenko, 1985).

For ranging navigation mode $g_{1,2} = 1$, so

$$G_R = \frac{\sqrt{2}}{\sin \alpha_M}$$

and for difference-ranging mode $g_{1,2} = 2 \sin \psi_{1,2}/2$, so

$$G_{DR} = \frac{1}{2 \sin \alpha_M} \sqrt{\frac{1}{\sin^2 \psi_1 / 2} + \frac{1}{\sin^2 \psi_2 / 2}}$$

where $\psi_1$ and $\psi_2$ are the basic angles of the first and second pairs of stations (angles under which, from the point of reception, the bases can be seen, that is, straight lines connecting pairs of stations) (Frantz *et al.*, 1957; Frank, 1983; Nikitenko, Bykov, and Ustinov, 1992). Figure 3.6 shows the determination of M (M1), the user position in the difference-ranging mode.

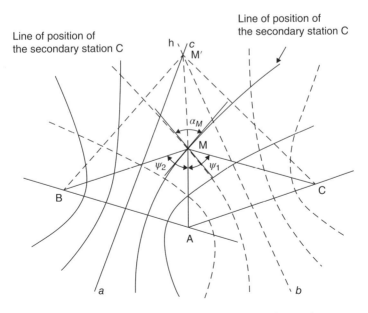

**Figure 3.6**   Positioning in the difference-ranging mode

The expression given before is a criterion for a rational system station disposition (the system's geometry) for servicing a particular district. For this, the stations should be located in such a way that the district of interest is within the region of lowest $G$ values so that the best navigation accuracy is available within this district. Coverage of the OMEGA system was global and the coverages of other long-range navigation systems are shown in Figure 3.7a and b.

For accurate LF radio-navigation, ground waves are used because they are more stable than sky waves reflected from the ionosphere. However, the prediction of propagation velocity and the corresponding corrections is rather complex when the underlying surface along the propagation path is nonuniform. In this case, partial corrections should be determined for individual uniform parts of the path, and then the average velocity for all the individual parts along that path should be calculated. Such factors as irregularities in the underlying surface near the point of signal reception (mountains, hills, coastal line, power transmission line, etc.) also affect the position of the ground wave phase front and, hence, the accuracy of the position solution. These effects can be taken into account only by calibrating a system directly at the reception point area.

It is impossible to separate ground waves and sky waves for VLF radio-navigation. The phases of the received signals have daily and seasonal variations that depend on the Sun's height along the propagation path. Correction prediction would only be possible for a regular component in these changes. It is also impossible to predict the impact of random ionosphere disturbances. However, taking into account any significant correlations in VLF propagation condition changes in a local area (up to several hundred of miles), the introduction of corrections calculated at the monitor station and broadcast to users via a communications link (Grishin, Ipatov, and Kazarinov, 1990) becomes effective.

## 3.4    Interference in VLF and LF Radio-Navigation Systems

Atmospheric noise and narrow-band interference from VLF and LF stations and power supply system harmonics are responsible for the greatest impacts on the functioning of User Equipment (UE). In addition, factors such as industrial interference from power transmission lines and ignition systems, and interference caused by static electricity discharge and radar operation, should be taken into account.

The overall interference level depends on a UE location, antenna type, season, and time-of-day. When analyzing the statistical characteristics of atmospheric noise, it is necessary to separate background and pulse components. The background component is caused by the superposition of signals transmitted by many distant sources and is described by the normal distribution law. The pulse component is caused by local thunderstorms and is approximated by a logarithmic-normal distribution law.

For suppressing smooth interference (inherent noise, background component) as well as the sinusoidal interference from interfering stations, narrow-band linear filtering is advisable, whereas in the case of pulse interference, broadband amplification and limitation of the signal/interference mixture are needed. These contradictory requirements can be met through compromise. One way is to realize UE with gradual band narrowing from several kHz to 10–100 Hz along with simultaneously lowering the limitation threshold to 0. Maximal interference bursts, the most dangerous from the standpoint of input circuitry overloading, are already limited at the input stage within a wide band. For attenuating the effect of electrostatic interference caused by static

(a)

(b)

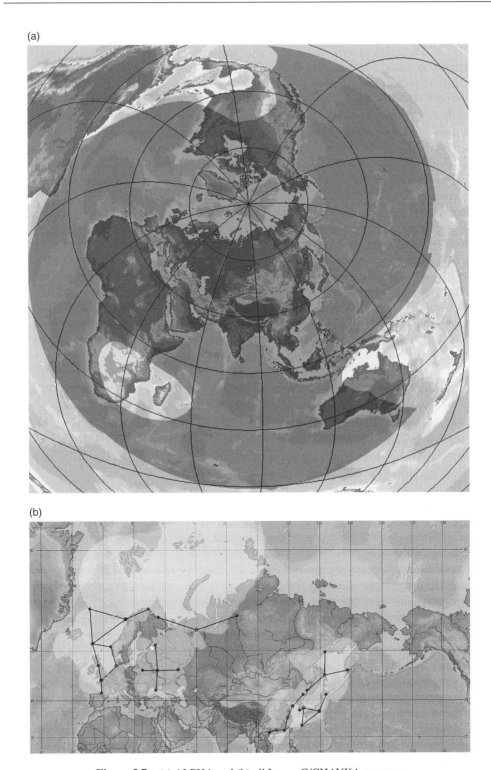

**Figure 3.7**   (a) ALPHA and (b) all Loran-C/CHAYKA coverage

electricity discharges with spectra of hundreds of kilohertz, the signals are received better on H-antennas. This is because, in collapsible whip antennas, the discharge current flows directly through the antenna circuit, whereas in the case of a loop antenna, the interference is generated by an induced current, which has a significantly lower impact (Boloshin *et al.*, 1985).

Loran-C/CHAYKA receivers of all types contain a band-pass filter at the input, which significantly weakens most of interfering stations' signals. The standard band-pass filter of a typical commercial Loran-C/CHAYKA receiver supports the attenuation of out-of-band interference to –50 dB when the interference is within a range of 10–50 kHz relative to the signal carrier frequency. A small number of the remaining interferences can be eliminated using Rejection Filters (RFs). The degree of distortion in the envelope form and phase of the Loran-C signal carrier frequency after passing through these filters depends on the level and frequency of the nonsuppressed remaining interference. While taking into account the possible close proximity of the interfering sources to the Loran-C signal, transmitted power, and spectral interference parameters, it is also necessary to consider all potential interference sources within and nearby the Loran-C coverage when estimating its dimensions and configuration.

The frequency spectrum of the Loran-C/CHAYKA signal consists of a large number of individual spectral lines spaced by the value of the repetition frequency 1/GRI. The degree of the interference impact on the receiver depends on the proximity of the interference frequency to the spectral component of the Loran-C/CHAYKA signal.

According to the minimum operation performance standard for Loran-C/CHAYKA receivers, there are three interference categories.

1. **Synchronous interference**—of which the frequency coincides precisely with a spectral line of the Loran-C/CHAYKA signal. Such interference generates a fixed error value in the range measurement.
2. **Near-synchronous interference**—of which the frequency is close to the signal spectral line (within the pass-band of the receiver's tracking filter). This interference type causes an oscillatory bias in the measured range value.
3. **Asynchronous interference**—which includes all other interferences. Asynchronous interference manifests itself as an increase in the level of the noise at the receiver's input.

The most significant of these are synchronous and near-synchronous interferences, which produce an error when determining phase in the tracking mode. The value of the phase measurement error depends not only on the quality (and quantity) of the RFs but also on the selected GRI of the Loran-C/CHAYKA chain.

The error in the signal phase tracking caused by sinusoidal interference is defined by the following equation:

$$E = T_1 / 2\pi \arcsin\left(\frac{I}{S}\right)$$

Here, $T_1 = 10$ μs—the period of the Loran-C/CHAYKA signal carrier frequency, $E$ is the phase tracking error (μs), and $I/S$ is the Signal-to-Interference Ratio (SIR).

The Loran-C CHAYKA boundary is usually specified by a signal-to-atmospheric noise ratio of –10 dB. In this case, it is assumed that the receiver supports the measurement of TOA

**Table 3.3**   An error budget for LF radio systems

| Source of error | Error value (µs) |
|---|---|
| Secondary stations' synchronization error | No more than 0.1 |
| Error caused by conditions of radio wave propagation | 0.1–0.2 |
| Instrumental error | 0.05–0.1 |
| External noise error | 0.15–0.25 |

differences with an RMS Error (RMSE) less than 100 ns and the position determination with an error of 1/4 NM ($2\sigma$).

The second coverage limitation is when the measurement error exceeds 100 ns under the action of the sinusoidal interference. The SIR is equal to 24 dB for an error of 100 ns but can be decreased by 3 dB if slight sinusoidal variation in the phase tracking error is acceptable (Table 3.3).

It is shown that, depending on the interference frequency, signal phase coding allows an improvement in the SIR of 10–18 dB. In this case, for the estimation of the coverage, the SIR used was 12 dB. When the interferences were close to sinusoidal form, the SIR fell to 9 dB, corresponding to a limited value of the error of 100 ns.

The greater the GRI, the greater the number of interferences affecting the signal. For example, when selecting GRI = 7777 under European conditions, the number of near-synchronous and synchronous interferences that should be taken into account is 29.

Criteria for Loran-C/CHAYKA protection against sinusoidal interference are defined in the ITUR M.589-2 8B/TEMP/5 Document (ITU-R Recommendation) for near-synchronous and asynchronous interference with a detuning range of 0–30 kHz relative to the carrier frequency. Frequency ($f$) of the near-synchronous interference is defined by the following expression:

$$f - \frac{n}{2\mathrm{GRI}} < f_\mathrm{b}$$

Here, GRI is the Loran-C/CHAYKA group repetition interval, $n$ is any integer number, and $f_\mathrm{b}$ is the receiver's band-pass in the tracking mode (0.01 Hz for marine receivers and 0.1 Hz for aviation types).

Using the given estimation procedures, it is possible to define a permissible interference level at the receiver's input by a value of $I \leq 23$ dB (for a typical noise level of 55 dB and a Signal-to-Noise Ratio (SNR) of −10 dB).

In up-to-date Loran-C/CHAYKA receivers, the RFs are realized based on software, thus enabling up to 30 NF (GM 1250, 1995; Beckmann, 1989) without an additional hardware or increase in the receiver's weight, dimensions, power consumption, or cost.

Data for all stations that may constitute a potential threat of interference generation, including frequencies, transmitted power, and coordinates, are loaded into the receiver's RAM. This makes it possible to operatively determine the distances from the user to the most dangerous stations at any coverage points and then verify a number of the interferences to be suppressed. The software NFs can be adjusted manually or automatically. Usually, an effort is made to minimize the number of used NFs to reduce the impact on the desired signal parameters. A large number of filters degrade the SIR and the slope of the signal pulse leading edge. It is not recommended to use automatically adjusted filters in receivers for timing synchronization and measurements requiring high accuracy because they can cause jumps in the pseudorange calculation during adjustment. A maximum number of NFs adjusted to the interference by the operator should not exceed four and that of the automatic NFs should not be more than two.

Note that for suppressing other kinds of interference, such actions can potentially be used as operative replacements of PP RNS signal phase codes and code delays, which were rarely used until the present time due to hardware realization complexity. However, when using up-to-date computing facilities and signal processing techniques, such actions can be realized based on software. Criteria for Loran-C/CHAYKA protection against cross-rate interference (interferences from similar transmitting stations) are defined by Nieuland (Nieuland, A., personal communications).

## 3.5 Error Budget

Error sources and their budgeting for both Loran-C and CHAYKA and for ALPHA and OMEGA systems are considered in Section 3.5.1.

### 3.5.1 Loran-C and CHAYKA Error Budget

The main factors affecting the operation of LF radio systems are as follows (Bykov and Nikitenko, 1985):

- Strength of the ground wave field
- Level of the atmospheric interference at signal reception points
- Errors caused by radio wave propagation conditions.

The range of ground wave reception depends mainly on the emitted power, the interference level at the reception point, and the underlying surface conductivity along the propagation path. The latter also affects the signal strength, usually measured by the SNR (Frank, 1983).

Errors caused by radio wave propagation conditions are compensated using corrections, either calculated or experimentally obtained. The major propagation effects that affect performance are three phase factors: primary, secondary, and additional secondary.

The *Primary Factor* (PF) is due to atmospheric refraction and results in retardation of the wave front with respect to the speed of light. Thus, the average index of refraction results in a 320 ppm effect, or 320 m error, at a 1000 km distance (Hargreaves, Williams, and Bransby, 2012). Variations are usually small (10 m or less) but increase when weather fronts cross over the operational area. This effect could be reduced via the measurement of temperature, pressure, and humidity variations in the operational areas.

The *Secondary Factor* (SF) is caused by the conductivity of the material (terrain or water) under the ground wave, which affects its speed. Seawater has the best conductivity at 5 S/m (Siemens per metre), while rock has a conductivity of two to three orders of magnitude less than that. The corresponding delay can be predicted if the conductivity and its variation are known along the propagation path with the resolution and accuracy required for this application.

The *Additional SF* (ASF) is caused by topographical effects. The ASF and its spatial variations constitute the most significant systematic error, and can degrade the accuracy of LF navigation measurements to hundreds of metres. The ASF can be calculated analytically by integrating the topography and conductivity over the Earth's surface. This requires not only digital terrain models, which are now available with the required accuracy, but also the conductivity, which is generally poorly known at the resolution required (Blazyk *et al.*, 2008). The other method for obtaining the ASF is to map it directly along profiles using a GNSS-LF radio system integrated sensor. The results are good enough along specific profiles, but their extension over an area is problematic due to their rapid spatial de-correlation in mountainous areas (Luo *et al.*, 2006; Jonson *et al.*, 2007).

The combined effect of PF, SF, and ASF is maximal on the Earth's surface and decreases with altitude (Barton, 2008). After a few wavelengths ($1\lambda = 3000\,\text{m}$ at $100\,\text{kHz}$), the ASF effect is very small. For aircraft landing, the ASF variations would have to be mapped in three dimensions.

Among these errors, the ASF exhibits the largest variations because of changing ground properties (the temporal variation) and changing topography (spatial variation). ASF temporal variations comprise long-term variations, caused by seasonal changes of the ground conductivity, and short-term variations due to real-time weather changes. The ASF is a value calibrated at a specified location, which is usually denoted as the calibration point. Within a calibration grid area centered at the calibration point, this ASF value is considered to be constant. On a calibration grid, different positions experience different transmission paths, and thus different transmission delays (Defence Mapping Agency Hydrographic/Topographic Centre (DMANTC), 1989). Delay differences represent ASF spatial variations and are highly topography dependent. In a flat area, the ASF spatial variation is usually much smaller than in a mountainous area. In addition, the smaller the calibration grid area, the smaller the spatial variation from grid point to grid point. When ASF values are used to calibrate LF measurements, the temporal and spatial variations must be taken into consideration, that is, an error bound should be applied simultaneously.

The particular locus-hyperbolic isoline of position corresponds to the measured navigation parameter (Federal Radionavigation Plan, 1999; Jacoby *et al.*, 1999; Lo *et al.*, 2009). The RMSE for determining a line of position depends on the angles under which, from the reception point, the baselines are "seen." These are the base angles $\psi$.

In addition, errors in time difference determinations, $\sigma_d$, must be taken into account. The RMSE for these time difference measurements, which are also named as navigation parameters, comprises the following:

$\sigma_s$—secondary stations' synchronization errors, μs
$\sigma_t$—errors caused by radio wave propagation conditions, μs
$\sigma_i$—the instrumental error in the onboard equipment, μs
$\sigma_n$—the external noise error.

Here, the synchronization error does not exceed $100\,\text{ns}$. The instrumental error of up-to-date onboard equipment depends on the internal noise level, hardware optimization, and the data processing algorithms. It is usually $0.05$–$0.1\,\mu\text{s}$.

The resulting RMSE for measurement of the navigation parameter can be written as $\sigma_d = \sqrt{\sigma_s^2 + \sigma_t^2 + \sigma_i^2 + \sigma_n^2}$ and for an $\text{SNR} \geq 1/3$, $\sigma_d \leq 0.3\,\mu\text{s}$.

An error budget for LF radio systems is given in Table 3.3.

## 3.5.2   ALPHA and OMEGA Error Budget

The error budget for a VLF radio system comprises main components such as those given in the following text (Boloshin et al., 1985).

### 3.5.2.1   Propagation Prediction Errors

Carrier phase observables from the received VLF signal are distorted due to changes in the spherical waveguide (formed by Earth and ionosphere) parameters affecting the phase velocity $v_p$ and the time of radio wave propagation. Due to differences between the actual velocities of radio wave propagation, $\overline{v}_{pA}, \overline{v}_{pB} \ldots$ (where A, B, ... are RS) and the calculated (cartographic) values, $v_c$, the measured values of the signal phases will differ from these calculated values as follows:

$$\phi_{SPF\,A} = \omega_0 \left( \frac{d_A}{\overline{v}_{pA}} - \frac{d_A}{v_c} \right) = \omega_0 d_A \left( \frac{1}{\overline{v}_{pA}} - \frac{1}{v_c} \right),$$

This is referred to as the *Secondary Phase Factor* (SPF). Here $d$ is the geodetic length of the radio path. As a result, the measured values must be adjusted using corrections for radio wave propagation. The ideal correction, $p$, is equal to the SPF with the reverse sign, that is, $p = -\varphi_{SPF}$.

When VLF radio systems are used in a standard navigation mode, corrections for radio wave propagation are predicted based on the SPF mathematical model. The results of both theoretical and experimental studies of wave propagation in the VLF band (i.e., a physical statistical model) are used. The most practical models are parametric global versions, which are models that are linear relative to parameters adjusted via experiments where the following main factors affecting the conditions of radio wave propagation are taken into account:

- Underlying surface conductivity
- Earth's magnetic field
- Latitudinal position of radio path
- Sun's zenith distance
- Seasonal, daily, and frequent variations in propagation medium parameters.

The resulting error reaches a value of 20–30 µs.

### 3.5.2.2   Transmitting Station Synchronization Error

All VLF transmitting stations are equipped with ensembles of frequency standards comprising four cesium standards with a daily frequency instability of no worse than $5 \times 10^{-12}$ (Stein, 1979). In order to prevent a discrepancy in individual stations' times, the phase of each transmitter is adjusted to the average phase of all the remaining system transmitters. This is termed "internal synchronization"; external synchronization is also available to UTC (USNO)

(OMEGA) or UTC (SU) (ALPHA). Overall, errors in VLF stations' external synchronization do not exceed 1–2 µs (Swanson, 1983).

### 3.5.2.3 Synchronization Errors in User's Onboard Reference Oscillators

The user measures a signal phase with an error caused by the onboard time offset relative to the transmitting station's time. The result of measuring the phase of $i$th station's signal at the $j$th frequency in phase cycles can be described using the following relation:

$$\phi_{ij} = \frac{f_j d_i}{v_c} + \phi_{rj} + \psi_{ij}$$

Here, $\phi_{rj} = f_j(t_0 - t_{0r})$, the phase shift is due to the difference between the switch-on instants of the transmitter and reference user's oscillators (i.e., the time offset between user and RS); $\psi_{ij}$ is the random measurement error; and $d_i$ is the geodetic distance from the $i$th RS to the object.

This equation is written for conditions where the measurement ambiguity is resolved and corrections for radio wave propagation are introduced.

When synchronizing UE times from VLF signals, practical results of error assessment show that a synchronization error is achieved not exceeding 1 µs.

### 3.5.2.4 Instrumental Errors in User Equipment and Errors Caused by the Interference Effect

The instrumental error of a UE depends on the internal noise level, hardware optimization, data processing algorithms and various other factors, and is in a range from 0.01 to 0.06 p.c. (phase cycle). This corresponds to position errors from 300 to 1800 m.

In addition, atmospheric noise, narrow-band interference from interfering VLF stations, and industrial interference make negative impacts on measurement quality, mainly on ambiguity resolution reliability, resulting in anomalous positioning errors.

Various techniques, such as narrow-band linear filtering, broadband amplification, hard band-pass signal limiting (10–100 Hz), and magnetic loop antenna deployment instead of electrical whip types, are used to reduce multiple narrow and wide band interference. Hence, due to pulse noise component reduction in the received signal, a minimum level of SNR (from 0.05–0.1 within a 1-kHz band) can be achieved.

Table 3.4 presents an error budget for VLF radio systems.

### 3.5.3 Position Error

When a user's position is determined by the crossing point of two lines of position, the position error will be determined as an $M$ error (Figure 3.7):

$$M = \frac{v}{2\sin\alpha_M} \sqrt{\frac{\sigma_{d1}^2}{\sin^2(\psi_1/2)} + \frac{\sigma_{d2}^2}{\sin^2(\psi_2/2)}} \; (m)$$

**Table 3.4**  Error budget for VLF radio systems

| Source of error | Value of error |
|---|---|
| Propagation prediction error | Depends on models used for radio wave propagation and computation algorithms being corrected using correction tables; achievable values 20–30 μs |
| Transmitting stations synchronization error | 1–2 μs |
| Synchronization error when synchronizing user's onboard reference oscillator time to transmitting station time | No more than 1 μs |
| Instrumental error | From 0.01 to 0.06 p.c. (phase cycle) |

Here, $\psi_1$ and $\psi_2$ are the angles under which the baselines are "seen"; $\alpha_M = (\psi_1 + \psi_2 / 2$ is the crossing angle of the lines of position; $v \approx 300$ m/μs is the radio wave propagation velocity; and $\sigma_{d1}$ and $\sigma_{d2}$ are the resulting RMSE for measurement of the navigation parameters (μs).

Typical figures of predictable position errors are given in Table 3.4.

## 3.6   LF Radio System Modernization

The need for a GNSS backup and augmentation system able to augment its availability, integrity, continuity, and accuracy (disrupted by selective availability S/A at that time) led to the first studies related to Loran-C and CHAYKA modernization. LF radio systems use high-powered transmitters and low-frequency signals (not microwatts and microwaves like GNSS), which are very unlikely to be interfered with or jammed by the same causes that would disrupt GNSS signals. A Loran receiver built into GNSS unit can mitigate the impact of disruptions to GNSS and has the potential to back up GNSS in the case of satellite signal outage, thus allowing users to keep communication and navigation capabilities. Moreover, after modernization the LF transmitting stations could transmit GNSS differential corrections and integrity data in the format of their own navigation signals (Balov, 2010; Basker, 2008; Celano, Peterson, and Schue, 2005; Offermans and Helwig, 2003; Offermans et al., 1997; Peterson et al., 2001; Pisarev et al., 2007; Smirnov, Sorotsky, and Tsarev, 2014; Stout and Schue, 2010; Williams, Basker, and Ward, 2008; Willigen, 1989). EUROFIX—a regional DGNSS—was the first realization of this concept.

In September 2001, the US Government published its "Volpe Report" (Volpe, 2001). The report explained the vulnerability of GPS (and similar GNSS systems) because of possible disruption by intentional or unintentional interference and identified Loran as a potential solution to this important problem. This attracted interest worldwide and provided an impetus to modernize the Loran system in the United States. The US Loran evaluation and modernization program resulted in e-Loran—a new and enhanced version of Loran with significantly improved performance. This has much better accuracy, integrity, and continuity of navigation along with Loran-C's traditional availability maintenance. Different from EUROFIX, its high performance level is provided by e-Loran independently of the presence or absence of GNSS These improvements are realized mainly through the development of a data channel similar to EUROFIX, which is also based on standard navigation additional signal modulation. This data channel allows e-Loran to meet the strict requirements of landing aircraft using non-precision instrument approaches, and to bring ships safely into harbor in low-visibility conditions. At the same time, e-Loran provides the precise time and frequency references needed by the telecommunications systems. One of the data channel messages is used for real-time differential corrections. These corrections are provided via RS that detect variations in the e-Loran signal (similar to the differential GNSS RS), so allowing receivers to compensate for these variations as well as providing information regarding signal integrity. Moreover, e-Loran can do things GNSS cannot, such as acting as a static compass. At sea, a new concept of navigation, enhanced navigation or e-Navigation, is being developed that requires exceptionally reliable inputs of position, navigation, and time data. Uniquely, the combination of GNSS and e-Loran has the potential to meet its needs.

## 3.6.1  EUROFIX—Regional GNSS Differential Subsystem

EUROFIX is an integrated satellite/terrestrial-based system, which combines DGNSS and Loran-C/CHAYKA. The current integration concept (Willigen, 1989; Pisarev *et al.*, 2002) assumes

- the synchronization of terrestrial transmitting station GNSS signals, that is, synchronization of the stations' Times of Emission (TOE) to the GNSS time scale;
- installation of the GNSS RS at the LF transmitting station in order to monitor GNSS integrity, the generation and transfer of the reference data (differential corrections, and warnings of integrity infringement, etc.) in the format of the LF navigation signal; and
- integrated UE that performs separate or joint processing of terrestrial and space-based system signals (pseudorange processing), including the continuous calibration of raw LF observables to compose an onboard LF propagation correction data base over the operational area.

The first full-scale system test was conducted jointly with Megapulse (US) using the US Coast Guard transmitter in Wildwood (1993–1994). Over subsequent years, the EUROFIX concept was realized within the framework of the NELS. Under the framework of the Russian Federal Program "Global Navigation System," the integrated regional differential subsystem for broadcasting GLONASS/GPS corrections was developed, based on three transmitting stations in the CHAYKA European chain.

In addition to GNSS data, EUROFIX makes it possible to use all the data from traditional LF radio-navigation. Two completely different systems, GPS/GLONASS and Loran-C/CHAYKA, combined into EUROFIX, provide better availability, integrity, and continuity of navigation data over all European territory.

EUROFIX benefits are as follows:

- The system's realization is based on existing LF long-range navigation infrastructure.
- It offers large coverage at comparatively low expense.
- It provides improved working ability and availability of the LF data channel under urban and mountainous conditions.
- It has provision for redundancy in case of failure of any GPS/GLONASS or Loran-C/CHAYKA (Offermans, Helwig, and Willigen, 2000).
- High-accuracy positioning from GNSS signals can be used for calibrating Loran-C/CHAYKA as well as for compensating errors caused by LF radio wave propagation.
- Loran-C/CHAYKA can be used for GPS/GLONASS integrity monitoring (given a GNSS receiver).

The EUROFIX structure includes a DGNSS RS combined with a Loran-C or CHAYKA transmitting station. The correcting data at the output are coded to improve the interference immunity of the data link. From the output of the coding device, the signal is applied to the input of the EUROFIX modulator, which performs the message modulation using a technique of balanced three-level Pulse-Position Modulation (PPM) of the carrier frequency with respect to the last six pulses in the navigation signal train, by a value of −1, 0 or +1 μs. The modulation index is selected to be sufficiently low to prevent signal-level losses in the tracking mode. The same number of leading and delayed pulses in the message allows for minimization of the PPM impact on navigation data losses up to a value not exceeding 0.79 dB. The first two train pulses are not modulated, so maintaining a capability for transferring notification signals about the system's working ability using a "blinking" technique for Loran-C/CHAYKA users.

The EUROFIX user should have a Loran-C/CHAYKA receiver containing a demodulator and a message decoder along with a DGNSS receiver supporting the reception and processing of GNSS signals and correcting data via the EUROFIX link.

For supporting the reception of signals from distant stations with the allowable error probability under conditions of atmospheric, sine, and mutual interference from adjacent Loran-C/CHAYKA chains, the correcting data are specially coded using a combination of Forward Error Correction (FEC) of the Reed–Solomon code with cyclic redundancy checking. For standard and commercial receivers, the EUROFIX data are converted to the RTCM SC-104 message type 9 standard. The full EUROFIX description and signal format is given in Offermans and Helwig (2003).

While using corrections from a single EUROFIX RS, the position solution error is in the range of 3–7 m (95%) at distances of up to 1000 km. Such results were experimentally obtained within NELS coverage as well as within those of CHAYKA European and Northern Chains (Balov et al., 2000; Pisarev et al., 2002, 2004; Zholnerov, 2002). With simultaneous processing of data from a chain of three to four stations, the position solution errors decrease to 3.1 and 2.3 m respectively (95%), due to the elimination of spatial and temporal de-correlation.

American specialists proposed various methods for improving the data transfer rate using various kinds of modulation (Peterson et al., 2001, 2002) as follows:

- Five-level PPM
- Incidental Frequency Modulation (IFM)
- Combined PPM/IFM.

However, after a number of theoretical and experimental studies with the US Loran-C variant, it was proposed to transfer the data using PPM on the carrier frequency of individual ninth navigation train pulses.

Russian specialists managed to improve the effective correcting data transfer rate up to 40 baud by reducing the correcting code's volume in a message and increasing the PPM index up to 1.5 μs. In this variant, the relative interference immunity of the data link is lower compared to the characteristics adopted in EUROFIX, but it meets the requirements as regards the probability of adequate correcting data reception from four navigation satellites $P_{ad}(s) \geq 0.9999$ (Balov et al., 2005a and 2005b). Another format improvement was proposed in 2009 (but not yet published), allowing up to five satellite corrections in one message. Taking into account the length of the message, this is 3.2 times more effective than in the native EUROFIX.

## 3.6.2   Enhanced Loran

e-Loran meets a set of worldwide standards and operates wholly independently of GPS, GLONASS, Galileo, or any future GNSS. Each user's e-Loran receiver will be operable in all regions where an e-Loran service is provided. e-Loran receivers work automatically, with minimal user input.

The core e-Loran system comprises modernized control centers, transmitting stations, and monitoring sites.

e-Loran transmissions are synchronized to an identifiable, publicly certified, Coordinated Universal Time (UTC) source by a method wholly independent of GNSS. This allows the e-Loran Service Provider to operate on a time scale that is synchronized with, but operates independently of, GNSS time scales. Synchronizing to a common time source also allows receivers to employ a mixture of e-Loran and satellite signals.

The principal difference between e-Loran and traditional Loran-C is the addition of a data channel on the transmitted signal. This conveys application-specific corrections, warnings, and signal integrity information to the user's receiver. It conveys corrections, warnings, and signal integrity information to the user's receiver via the Loran transmission. The data transmitted may not be needed for all applications but will include at a minimum:

- The identity of the station; an almanac of Loran transmitting and differential monitor sites
- Absolute time based on the UTC scale; leap-second offsets between e-Loran system time and UTC
- Warnings of anomalous radio propagation conditions including early sky waves and warnings of signal failures, aimed at maximizing the integrity of the system
- Messages that allow users to authenticate the e-Loran transmissions and official-use only messages
- Differential e-Loran corrections, to maximize accuracy for maritime and timing users
- Differential GNSS corrections.

This data channel allows e-Loran to meet the following very stringent requirements of aircraft making nonprecision approaches for landing, and of ships approaching and entering harbors under low-visibility conditions:

**Accuracy**: 0.004–0.01 nautical mile (8–20 m)
**Availability**: 0.999–0.9999
**Integrity**: 0.999999 ($1 \times 10^{-7}$)
**Continuity**: 0.999–0.9999 over 150 s.

e-Loran is also capable of providing the exceedingly precise time and frequency references needed by the telecommunications systems that carry voice and Internet communications, and it also meets ITU requirements in G.811 for primary reference clocks.

All e-Loran transmitters use modern solid-state transmitter (SSX) and control technology. They have Uninterruptible Power Supplies (UPS) that ensure that any failure of the incoming power will neither interrupt nor affect the transmitted signal. The time and frequency control systems of the transmitter are designed for e-Loran operation, and they apply phase corrections in a continuous manner. The time reference system uses multiple cesium clocks, or an alternative technology of at least equal quality. When an e-Loran station is detected as being out of tolerance, it is immediately taken off the air to ensure that receivers promptly cease to use its signals. Traditional Loran-C pulse blinking is used to show that a station is under test and should not be used.

e-Loran transmitting stations run unattended, for which reason adequate personnel must be at the control centers and on call to respond rapidly to failures and to maintain the published very high levels of availability and continuity. Scheduled maintenance work is planned carefully to minimize the impact on users of stations being off the air. Users are given adequate notice of interruptions via well-publicized channels of communication. Security of these sites and of any critical communications systems is of a high level, reflecting the importance of the applications for which the transmitted signal is being used. Monitor sites, located in the e-Loran coverage area, are used to provide integrity for the user community. The receivers used at these sites monitor the e-Loran signals and provide real-time information to the control centers regarding signals in space. Users are notified immediately if any abnormalities are detected. Some of the monitor sites are used as RS to generate the data channel messages, and selected sites also have at least one highly accurate clock for synchronization to UTC, so providing time and frequency corrections for timing users. A monitoring network provides real-time maritime differential corrections and supplies warnings for aviation users.

e-Loran user receivers operate in an *all-in-view* mode. That is, they acquire and track the signals of many Loran stations (the same way GNSS receivers acquire and track multiple satellites) and employ them to make the most accurate and reliable position and timing measurements. Another benefit of using the all-in-view mode is that it ensures that the e-Loran receiver is always tracking the correct cycle of each individual signal.

An e-Loran receiver is capable of receiving and decoding the data channel messages and applying this information based on the user-specific application. This information, coupled with published signal propagation corrections, provides the user with a highly accurate PNT solution. Loran errors are $\geq 10\,\mu s$.

The main factor restraining the e-Loran capabilities is the rather slow EUROFIX-similar data link (about 30 bps) and the corresponding data latencies—a complete set of e-Loran correction data transmissions takes about 90 s. This has proved to be the source of significant differential e-Loran errors, and therefore modern e-Loran systems cannot provide accuracies better than 10 m (New Loran at 5 Meters, 2014; Willigen *et al.*, 2014).

### 3.6.3   Enhanced Differential Loran

The basic concept of Enhanced Differential Loran (e-DLoran) is depicted in Figure 3.8 (Willigen *et al.*, 2014), and it implies the following three important improvements resulting in an accuracy at about the 5 m level, and at extremely low cost:

1. A significant reduction in the data transmission latency of up to 1–2 s
2. Simultaneous data transmissions by a large number of RS
3. More accurate ASF data measurements provided by low-cost RS without precise (atomic) clock implementation.

Instead of using a EUROFIX data channel, e-DLoran employs a mobile GSM network. Also, the e-DLoran RS are connected to the Internet, which may be implemented via cable or

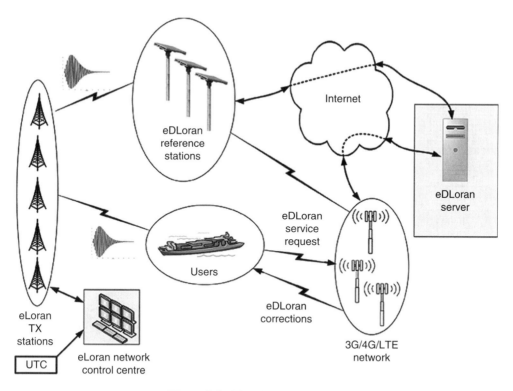

**Figure 3.8**    The e-DLoran concept

**Table 3.5**    dLoran (e-Loran) and e-DLoran comparison

| DLoran | e-DLoran |
| --- | --- |
| Measure all pseudoranges of useful transmitters | Calculate position from measured uncorrected pseudoranges |
| Apply pseudorange corrections from receiver's ASF database | Apply position corrections from receiver's ASF database |
| Apply pseudorange corrections from reference station via Eurofix | Apply last-second position corrections optimized for user's location from server via public mobile telecom network |
| Calculate position and clock offset from corrected pseudoranges | |
| Generate ASF database with special equipment (costly operation) | Refine daily ASF database without special equipment |

a GSM modem. By these means, the first two improvements are attained, while the third is a result of positioning instead of a pseudorange method to generate ASF correction data. Table 3.5 (Willigen *et al.*, 2014) compares the main signal processing steps in the dLoran (e-Loran) and e-DLoran approaches.

The e-DLoran RS (Figure 3.9) comprises eLoran antenna, UMTS/GPRS antenna, Loran receiver (including UMTS/GPRS module and microcomputer), solar cells, buffer battery, and metal mast (4–5 m for fixed and 2 m for test RS).

(a)                                          (b)

**Figure 3.9**   e-DLoran reference station: (a) test and (b) fixed

e-DLoran has exhibited proven high performances in both static and dynamic tests. During a Rotterdam approach tests the results were proved +/– 5 m in the mode GPS-RTK.

## 3.7   User Equipment

Since the OMEGA (1997) and North American and Canadian Loran-C (2010) operation cancellation, a few manufacturers still support long-range navigation UE production, and long-range navigation system receivers (mainly Loran-C/Chayka) are presented in the market as Original Equipment Manufacturer (OEM)-products, complete autonomous devices, and integrated systems.

The relatively low-frequency band occupied by long-range navigation radio systems and the wide choices of integral multibit Analog-to-Digital Converters (ADC) and Digital Signal Processors (DSPs) make possible a *Software-Defined Radio* (SDR) approach to receiver architecture. This means that the major parts of the signal processing procedure, including digital signal pass-band filtering, acquisition and tracking, phase ambiguity resolution, data decoding, interference suppression, and Automatic Gain Control (AGC), may be performed at the software level along with the navigation and timing solutions. This technique provides opportunities to adjust these processing procedures for changing the reception environment as well as the specific navigation signal waveforms transmitted by different long-range-radio-navigation systems.

Typical receiver architecture (Figure 3.10) comprises an antenna with preamplifier, linear receiver (active band filter and amplifier with regulated gain), reference oscillator, synthesizer,

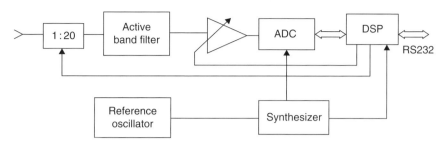

**Figure 3.10**   Receiver architecture

an ADC, and a DSP. VLF and LF signal processing architectures are the same, though the bandwidth of the active band filter is 85–115) kHz for LF and 10–17 kHz for VLF signals. The 16-bit ADC has a sampling rate more than 400 kHz for LF and more than 40 kHz for VLF signals.

When operating from a loop antenna, a dual-channel receiver is used to provide operation from several transmitting chains as well as interference mitigation.

Two antenna types are used: electrical (collapsible whip) and magnetic (loop). The collapsible whip antennas have relatively large effective heights and circular direction patterns. Such antenna heights make possible the use of simple preamplifiers with low gains of 10–20 dB. The function of the main preamplifier is to match the high output resistance of the collapsible whip antenna (several megohms) to the low resistance of the coaxial cable (50 or 75 Ohms) as well as to the preliminary signal band filter. As distinct from the electrical antenna where a discharge current weakened by the isolator flows directly to the antenna circuit, in a loop antenna interference is generated by the induced current, and its effect in both the LF and VLF bands is significantly lower. The main disadvantages of loop antennas are as follows:

- Low effective heights (around the millimetre level), resulting in the need for high gain, low-noise input signal amplifiers.
- The directional pattern of the loop antenna is shaped like a figure of eight, so requiring two orthogonal magnetic antennas and a dual-channel receiver to provide for reception of signals from all directions.
- Circular directional pattern generation implies broadband phase shift over the 90° between the orthogonal magnetic antenna signals, though the phases of the received signals, dependent on the reception direction, must be corrected to produce this circular pattern (see Figure 3.11). This leads to additional measurement errors caused by the non-coordination of antenna gains.

The directional pattern of an omnidirectional loop antenna and its phase characteristic depends on the course angle to the transmitting station. Since the phases of the received signals contain additional shifts equal to the course angle to the radio station, the mismatching of signal phase and envelope takes place at the summator output, and the phase ambiguity resolution through the envelope's delay measurement needs additional compensation.

Figure 3.12 shows H-field antennas for various applications.

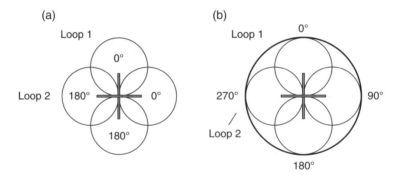

**Figure 3.11**  Directional pattern and pulse characteristics for two-component orthogonal magnetic antennas

**Figure 3.12**  Aviation H-field antenna (bottom) and magnet antenna integrated with GPS patch antenna (top) from Locus Inc., USA

Contemporary OEM-board receivers are linear, DSP-based, operate in "all-in-view mode" and are capable of tracking signals of strength 30–120 dBμV/m and dynamic range 90 dB from about 40 transmitter stations out of 12 chains (Doty *et al.*, 2003).

The interference mitigation network within an OEM-board receiver includes up to 40 automatically adjusted notch filters. Also, e-Loran data channel decoding and message formatting is provided to enable the DLoran mode of operation and DGNSS corrections for an integrated or external GNSS receiver. OEM boards are small-sized with moderate power consumption. Due to the similarity of LF and VLF band signal processing, some models integrate both bands in a single unit (Figure 3.13). A typical measurement output would include time of signal arrival, Time Difference (TD), position, true-North heading (even when the receiver is stationary), SNR, Envelope–Carrier Discrepancy (ECD), and Eurofix data.

A complete Loran-C/Chayka autonomous receiver (Figure 3.14) contains all the necessary component options inside the housing: an LF antenna, OEM board, power supply, application

(a)                                                    (b)

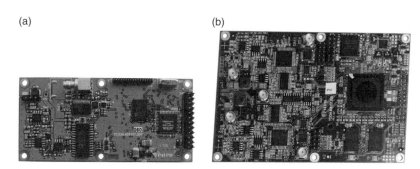

**Figure 3.13**   LF, VLF OEM-board receivers from RIRT: (a) e-Loran-C/Chayka (4.5×9.5×1.5 cm, 1.5 W) and (b) e-Loran-C/Chayka/Alpha (8×12×2 cm, 6 W)

(a)                                                    (b)

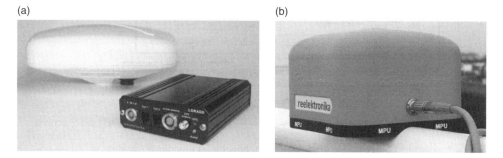

**Figure 3.14**   Loran-C/Chayka complete autonomous receivers from Reelektronika (Holland): (a) e-Loran-C/Chayka (11×8.5×3 cm, 4 W) and (b) e-DLoran-C/Chayka (14×14×10 cm)

processor, and GPRS modem, so providing the receiver with mechanical, electrical, and data interfaces with the environment and the user. The receiver outputs independent e-Loran positions, enhanced by the use of ASF maps where available. Optionally, it can also combine its e-Loran measurements with additional raw measurements from an externally connected GNSS receiver, and output an integrated position solution, taking full advantage of the strong points of both navigation systems.

To take full advantage of the GNSS and Loran-C/Chayka (all Loran) strong points, it is necessary to optimize the system measurement processing of both in order to improve the availability, integrity, and continuity of navigation with the required accuracies. The general idea of GNSS/all Loran integration is presented in the functional diagram of Figure 3.15 (Doty *et al.*, 2003). This approach implies multiple solution generation and cross-checking:

**Raw GNSS-Receiver**—raw position solution from the GNSS receiver before time correction.
**Raw All Loran Receiver**—raw all Loran receiver's position solution.
**Time-Adjusted GNSS Receiver**—time-delayed version of the GNSS receiver's position solution. (The GNSS position is computed by transferring the nearest valid GNSS receiver's position/velocity to the Loran TOV and extrapolating to match the all Loran time delay.)

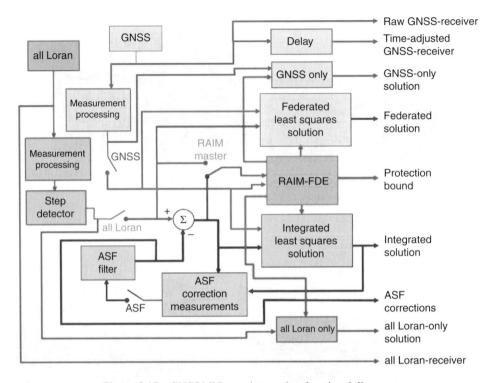

**Figure 3.15**  GNSS/all Loran integration functional diagram

**GNSS-Only Solution**—position solution utilizing only the GNSS satellites determined to be valid by the GNSS/all Loran RAIM-FDE function. (This solution utilizes time-adjusted GNSS pseudorange measurements.)

**All Loran-Only Solution**—computed position solution based on the Loran delta-range measurements determined to be valid by the RAIM-FDE algorithm.

**Federated Solution**—least-squares position solution utilizing time-adjusted GNSS measurements and uncorrected Loran delta-range measurements determined to be valid by the RAIM-FDE algorithm.

**Integrated Solution**—a least-squares position solution utilizing time-adjusted GNSS measurements and Loran delta-range measurements after correction by the ASF values produced by the Kalman filter estimator. (Only GNSS and Loran measurements determined to be valid by the RAIM-FDE algorithm are utilized.)

**Protection Bound**—horizontal protection integrity limit on position estimate as determined by the RAIM-FDE algorithm.

**ASF Corrections**—output of the ASF error estimator provides corrections to all Loran delta-range measurements.

All Loran-only and GNSS-only solutions are provided to show how the GNSS or all Loran data contributes to the combined solutions. If an anomaly is observed in the Integrated or

Federated Solutions, the GNSS or all Loran-only solutions based on exactly the same set of satellites or stations used in the combined solutions may be examined to help determine the cause of the anomaly.

Both the Integrated and Federated Solutions produce least-squares position estimates with weighting based on estimates of the characteristic noise and uncertainty of the signals. Because GNSS has lower fault-free noise and range error characteristics than the Loran signals, the GNSS is more heavily weighted in the solution. Therefore, when GNSS has good availability (with a large number of satellites in a good geometry), the Integrated, Federated, and GNSS-only solutions are very similar. The Integrated Solution tends to be within about 1 m RMS of the GNSS solution. The Federated Solution difference is slightly larger due to the contribution of the uncorrected ASF all Loran errors, but it still tends to be within a few metres of the GNSS solution. A much more dramatic difference in performance characteristics may be observed when GNSS is lost. Because the Integrated Solution utilizes Loran data corrected for ASF errors, the average position accuracy remains very good. There is a slight increase in position noise but the Integrated Solution tends to remain within a few metres of the true position for some time after loss of GNSS. The Federated Solution utilizes uncorrected all Loran measurements, so after loss of GNSS, its position estimate will instantly revert to the all Loran position that may be in error by 100 m or more due to the lack of ASF corrections.

The system is capable of detecting erroneous data in the GNSS or all Loran data and excluding it from the position solutions due to several levels of integrity monitoring and fault exclusion. The first level is realized in the signal processing of the GNSS and all Loran receivers, which monitor the parameters of each signal received and flag any satellite or station with anomalous characteristics. The second level is provided by *step detectors* that monitor the data and exclude any measurements with unexpectedly large changes or steps. The Loran ASF estimation filter also includes both step detectors and *maximum value detectors* to identify stations with excessively large or rapidly changing errors in their range measurements. Stations with unacceptable ASF errors are excluded from the Integrated, Federated, and Loran-only solutions. To detect less obvious faults, the RAIM-FDE fault detection and exclusion algorithm is implemented: four dimensional (three position axes plus time) RAIM-FDE when GNSS is available, and two-dimensional when not.

By combining GNSS and all Loran data, fault detection may be improved over the performance of either system individually. The Federated Solution does not utilize GNSS data to correct all Loran errors and, therefore, maintains independence of any GNSS fault. However, without some type of ASF correction, the large uncertainty in the all Loran data makes detecting small position errors in GNSSS data unreliable. The Federated Solution was ineffective in detecting the error, so allowing a radial position error of 3.8 km to build up before the satellite error was detected and the measurement excluded. Although performance can be improved by decreasing the uncertainty estimate for the Loran data, without some type of technique to estimate the Loran ASF errors (i.e., ASF maps or differential corrections), the integrity bound of the Federated Solution is large. Because the Integrated Solution estimates and removes the Loran ASF errors, Loran data uncertainty is greatly reduced, so providing a capability for detecting much smaller GNSS errors. The Integrated Solution utilizes both GNSS and al -Loran data to estimate ASF corrections. Therefore, the corrected all Loran data is not independent of potential GNSS faults. Fortunately, ASF values tend to change slowly, so the

**Figure 3.16**  e-Loran/Chayka/GNSS integrated receiver. Single housing option from Reelektronika (8.3 cm high × 20.3 cm dia)

estimation filter can be made very slow, which prevents GNSS faults from rapidly corrupting the Loran data. The Integrated Solution can therefore detect rapidly changing GNSS errors as well as moderately slow errors such as range ramps of a few metres per second.

Examples of e-Loran/Chayka/GNSS integrated receivers are shown in Figure 3.16.

# References

Balov A. V. (2010). Radio Navigation, Present and Future, *Gyroscopy and Navigation*, 1 (3), 224–235.

Balov A., Abramov L., Hitrun G., Johannessen E., and Marshall D. (2000). Broadcast of Eurofix/Chayka/Loran-Service from a Single Site. In Proceedings of the International Symposium on Integration LORAN-C/Eurofix and EGNOS/Galileo, Bonn, Germany, March 22–23, 2000.

Balov A. V., Zholnerov V. S., Malyukov S. N., and Choglokov A. E. (2005a). Data Transfer Using Navigation Signals from Long-Range Radionavigation Systems. In Proceedings of the All Russian Conference on Fundamental and Applied Positioning and Timing Support, vol. 13. Institute of Applied Astronomy RAS, St. Petersburg, Russia, April 11–15, 2005. Nauka, Moscow, pp. 322–332 (in Russian).

Balov A., Zholnerov V., Malyukov S., Tsarev V., and Choglokov A. (2005b). Data Transmission Format Via a Datalink Using PPM Modulation of the Six Last Pulses in The Pulse Phase RNS Signal. *The Journal of Navigation News*, annual, No 1–4 (3), 22–29.

Barton C. (2008). Loran (Legacy, Modernized and Enhanced) and Research Efforts at Ohio University. ION Southern California Section Meeting, Torrance, CA, USA, December 16, 2008.

Basker S. (April 2008). e-Loran Securing Positioning, Navigation and Timing for Europe's Future. European e-Loran Forum. Document of the UK General Lighthouse Authorities.

Beckmann M. (1989). Interference Detection and Suppression in Loran-C Receivers. *IEE Journal on Radar and Signal Processing*, 136 (5), 109–117.

Blazyk J., Barton C., Adler F., and Narins M. (2008). The Loran Propagation Model: Development, Analysis, Test and Validation. In Proceedings of the International Loran Association 37, London, England, October 28–30, 2008, pp. 1–14.

Boloshin S. B. (1993). Russian VLF RNS and Some Problems on Its Integration with OMEGA. *Journal of the Russian Institute of the Radonavigation and Time*, 1 (2), 18–21.

Boloshin S. B., Semenov G. A., Guzman A. S., Golovushkin G. V., and Olyanyuk P. V. (1985). *VLF Radionavigation Systems*. Radio and Communications, Moscow, p. 264 (in Russian).

Bykov V. I. and Nikitenko Y. I. (1985). *Pulsed-Phase Radionavigation Systems for Ship Navigation*. Transport, Moscow (in Russian).

Celano T., Peterson B., and Schue C. (2005). Low Cost Digitally Enhanced Loran for Tactical Applications (LC DELTA). Available at: www.ursanav.com/doc (accessed on October 30, 2005).

Dean W. (1996). History of Loran-C. In ILA Proceedings of the 25th Annual Technical Conference, San Diego, CA, USA, November 1996. ILA, The Institute of Navigation. Available at: www.ion.org (accessed on January 27, 2016), pp. 89–127.

Defence Mapping Agency Hydrographic/Topographic Centre (DMAHTC). (1989). Loran Correction Tables, Stock: LCPUB 2211200C.

Doty J. H., Hwang P. Y., Roth G. L., and Narins M. J. (2003). Integrated GPS/Loran Navigation Sensor for Aviation Applications, ION GPS/GNSS, Portland, OR, USA, September 9–12, 2003. ION. Available at: www.ion.org (accessed on January 27, 2016).

Federal Radionavigation Plan (1999). US DOD, US DoT.

Frank R. (1983). Current Developments in Loran-C. *Proceedings of the IEEE*, 71 (10), 1127–1139.

Franz W. P., Dean W. N., and Frank L. R. (1957). A Precision Multi-Purpose Radio Navigation System. IRE (Institute of Radio Engineers). Correctional Recommendation or Navigational Convention Record, V.5, part 8, pp. 79–98.

GM 1250 (June 19, 1995). *Monitor User Guide*. Geometrix, Tromsoe.

Grishin Y. P., Ipatov V. P., Kazarinov Y. M. (1990). *Radio Systems*. Vysshaya Shkola, Moscow, pp. 1–496.

Hargreaves C., Williams P., and Bransby M. (2012). ASF Quality Assurance for eLoran. *IEEE/ION PLANS: Position, Location and Navigation Conference 2012*, April 23–26, 2010, Myrtle Beach, SC, USA, pp. 1169–1174. Available at: www.ion.org (accessed on April 11, 2016).

Jacobsen T. (2006). The Russian VLF Navaid System, ALPHA, RSDN – 20. Available at: www.vlf.it/alphatrond/alpha.htm (accessed on April 11, 2016).

Jacoby J., Schick P., Rischwalski E., and Zamzov K. (1999). Advantages of a Combined GPS/Loran-C Precision Timing Receiver. In Proceedings of the 28th Technical Symposium, ILA.

Jonson G., Swaszek P., Harnett R., and Nichols C. (2007). Navigating Harbors at High Accuracy without GPS: eLoran in the United States. Time. Nav' 07, Switzerland, D/Loran-2007, May 29–31, 2007, pp. 1400–1049.

Lo S., Leathem M., Offermans G., Gunther G., Hamilton B., Peterson B., and Enge P. (2009). Defining Primary, Secondary, Additional Secondary Factors for RTCM Minimum Specifications (MPS). In 38th Annual Convention and Technical Symposium of the International Loran Association 2009 (ILA – 38), Portland, ME, USA, October 13, 2009, p. 18. Available at: www.ion.org (accessed on January 27, 2016).

Luo N., Mao G., Lachapelle G., and Cannon E. (2006). ASF Effect Analysis Using an Integrated GPS/e LORAN Position System. Presented at the Institute of Navigation, NTM 2006 Conference, Monterey, Canada, January 18–20, 2006.

National Standard of Russian Federation. (2009). Radionavigation System "Chayka". Signals of Transmitted Stations Technical Requirements. GOST P53168 – 2008. Standardinform, Moscow.

New Loran at 5 Metres, GPS World, № 6, June 6, 2014.

Nikitenko Y. I., Bykov V. I., and Ustinov Y. M. (1992). *Maritime Radionavigation Systems*. Transport, Moscow, pp. 1–335.

Offermans G. and Helwig A. (2003). *Integrated Navigation System Eurofix – Vision, Concept, Design, Implementation*. Technische Universiteit Delft, Delft, pp. 1–298.

Offermans G. W. A., Helwig A. W. S., van Essen R. F., and van Willigen D. (1997). Integration Aspects of DGNSS and Loran-C for Land Applications. In 53rd Annual Meeting of the Institute of Navigation, Albuquerque, NM, USA, June 30–July 10, 1997.

Offermans G. W. A., Helwig A. W. S., and Willigen D. (2000). Eurofix System Overview: Differential GNSS and Integrity Service Through Loran C. In International Symposium on Integration of Loran-C/Eurofix and EGNOS/Galileo, Bonn, Germany, March 22–23, 2000. DGON, Bonn, pp. 281–296.

Peterson B., Dykstra K., Swaszek P., and Boyer J. (2001). High Speed LORAN-C Data Communications—June 2001 Update. In ION 57th Annual Meeting/CIGTF 20th Biennial Guidance Test Symposium, Albuquerque, NM, USA, June 11–13, 2001.

Peterson B., Dykstra K., Swaszek P., Boyer J., Carroll K., Narins M., and Johannessen P. (2002). WAAS Messages via LORAN Data Communications—Technical Progress towards Going Operational. ION NTM 2002, San Diego, CA, USA, January 28–30, 2002.

Peterson B., Celano T. R., and Schue, III. (October 2006). Loran Data Channel Communication Using 9th Pulse Modulation, Version 1.3. US CG Loran Support Unit, Wildwood Crest, NJ, pp. 1–16.

Pisarev S., Balov A., Zholnerov, V., Zarubin S., Borovsky V., Kichigin V., and Neuymin B. (2002). CHAYKA Current Status and Problems to be Solved for its Integration with LORAN-C, GNSS, EGNOS, WAAS Using EUROFIX Technology. In Proceedings of the International Symposium on Integration LORAN-C, GNSS, EGNOS, WAAS and EUROFIX, Munich, Germany, June 3–10, 2002. DVD, file: 30, pp. 1–10.

Pisarev S., Balov A., Zholnerov V., Malyukov S., and Shebshaevich B. (2004). Analysis of Characteristics as Regards the Datalink Using Various Modulation Techniques for PP RNS Signal. *The Journal of "Navigation News"*, (2), 17–25.

Pisarev S., Shebshaevich B., Balov A., Zarubin S., Efremov P., and Gevorkyan A. (2007). Russian Long Range Navigation System, Modernization and Development. In Proceedings of the Second All Russian Conference on Fundamental and Applied PNT Support (PNT 2007), St. Petersburg, Russia, April, 2–5 2007.

Radio Technical Commission for Marine Services (RTCM). (1977). Minimum Performance Standards, Marine Loran-C Receiving Equipment. RTCM Services, Valley Forge, PA, p. 86.

Smirnov V., Sorotsky V., and Tsarev V. (2014). New Concept for Development of Competitive Transmitters for Advanced LW Navigation Systems and Initial Steps to Practical Implementation. *The Journal of Navigation News*, 1, 17–21.

Stein K. (1979). Omega Station Stress Timing Stability. *Aviation Week and Space Technology*, 110 (4), 53–56.

Stout C. and Schue C. (2010). Designing, Developing, and Deploying a Small Footprint Low Frequency System for PNT, and Data Services. *Proceedings of the IEEE/ION Position, Location and Navigation Symposium (PLANS 2010)*. Indian Wells, CA, USA, May 4–6, 2010, pp. 952–956.

Swanson E. R. (1983). Omega. *Proceedings of the IEEE*, 71 (10), 1140–1155.

US DoT, U.S. Coast Guard. (1994). Specification of the Transmitted LORAN-C Signal. COMDTINS M16562, 4A. US DoT, U.S. Coast Guard, Washington, DC, p. 60.

US DoT, US Coast Guard. (1992). Glossary of terms. In *LORAN-C User Handbook*, COMDTPUB P16562.6. US DoT, US Coast Guard, Washington, DC, pp. C1–C20.

Volpe J. A. (August 2001). Vulnerability Assessment of the Transportation Infrastructure relying on the Global Positioning System. Final Report. National Transportation Systems Centre, Cambridge, MA. Available at: www.volpe.dot.gov (accessed on April 11, 2016).

Williams P., Basker R., and Ward N. (2008). e-Navigation and the Case eLoran. *The Journal of Navigation*, 61, 473–484.

Willigen D. (1989). Eurofix. *The Journal of Navigation*, 42 (3), 375–381.

Willigen D., Breeuwer E. J., Offermans G. W. A., and Helwig A. W. S. (June 1996). "Eurofix". Information Paper. TVS Memorandum.

Willigen D., Kellenbach R., Dekker C., and van Burren W. (July 2014). "e-DLoran: the Next-Gen Loran." GPS World, pp. 36–43. Available at: http://gpsworld.com/cameron-named-gps-world-publisher-12759/ (accessed on April 11, 2016).

Zholnerov V. (2002). Experimental Investigation of EUROFIX Technology in Chayka. In 11th Session of the Council of the Far East Radionavigation Service, Xian, China, October 14–18, 2002.

# 4

# Radio Systems for Short-Range Navigation

J. Paul Sims[1] and Joseph Watson[2]
*[1] East Tennessee State University, Johnson City, TN, USA*
*[2] Swansea University (retd.), Swansea, UK*

## 4.1   Overview of Short-Range Navigational Aids

The Federal Aviation Administration (FAA) and the Federal Communications Commission (FCC) are the two bodies that regulate navigational aids in the United States. The FCC is only involved in those systems that emit/sense radio frequencies. The principal source of information for pilots on those navigational aids that are approved for flight is the *Aeronautical Information Manual* (AIM), and this is available free online through the FAA. With few exceptions, this is the same list of navigational aids used in other regions of the world.

For the purpose of this chapter, the following short-range navigational aids will be described:

1. Nondirectional radio beacons
2. VHF omni-directional range equipment
3. Tactical air navigation instrument landing systems
4. The VORTAC station
5. The RSBN system.

Over the years, there have been many short-range systems that are no longer operational and these are not covered in this text.

*Aerospace Navigation Systems*, First Edition. Edited by Alexander V. Nebylov and Joseph Watson.
© 2016 John Wiley & Sons, Ltd. Published 2016 by John Wiley & Sons, Ltd.

## 4.2   Nondirectional Radio Beacon and the "Automatic Direction Finder"

The Nondirectional Radio Beacon (NDB) is a ground station shown on navigational charts, and the term Automatic Direction Finder (ADF) is also used to describe the relevant system. NDB transmission does not contain directional information but simply provides a bearing to the transmitting station.

Aeronautical NDBs broadcast on relatively low frequencies (200–415 kHz), and most ADF receivers will also work over the standard AM broadcast band of 550–1600 kHz. All NDBs have a two- or three-letter identifier broadcast in Morse code. Some NDBs also broadcast audio reports that can provide station weather information and possibly other content.

All ADF receiver systems have both loop and sense antennae. The loop antenna is a flat plate antenna located on the bottom of the aircraft, while the sense antenna is a simple wire or foil-type antenna mounted in a fairing. There are also options that allow the loop and sense antennae to be mounted in the same pod. The loop antenna consists of two perpendicular windings on a square ferrite core so that the H-field induces a voltage into the two windings of the loop. Since these windings are on a closed loop, the phase angle of the voltages varies as the antenna is rotated relative to the transmitter. Within this rotation there are two points where the voltages exactly cancel out. These are the NDB null points, one of which defines the bearing to the NDB and the other to a bearing 180° away. The sense antenna is used to determine which null point defines the bearing to the transmitter. It does this by using the electric portion of the electromagnetic field to produce a voltage that is always in phase with the transmitter. The receiver measures the combined voltage of the two windings in the loop antenna and compares it to the voltage received by the sense antenna so that the ADF is able to determine the direction to the beacon. The minimum "working" E-field for the NDB at the aircraft is 70 µV/m as defined by the International Telecommunications Union (ITU). Below this field strength the receiver becomes intermittent or unable to detect a bearing to the station (Collinson, 1997; Rockwell Collins, 1997).

If the loop antenna is rotated to achieve a minimum voltage output, or null, this is indicated by a pointer on the Radio Magnetic Indicator (RMI) dial. Depending on the age of the system, the ADF either mechanically or electronically aligns this null to the ADF station. The older mechanical system allows for the loop antenna itself to be rotated, whereas in the later electronic systems the loop antenna is fixed and the same result is obtained by rotating a goniometer. If the goniometer search coil is not exactly at null, a loop voltage is generated, which is applied to a biphase motor that rotates the goniometer until null is achieved. Since the phase of the loop antenna signal either leads or lags that of the sense antenna, depending upon which side of the null the rotor is positioned, the goniometer is rotated in the correct direction to achieve null. The output from the goniometer is then used to drive the pointer on the RMI display in the cockpit. More simply, while in operation the ADF receiver compares the loop antenna voltage to the sense antenna voltage. When both loop and sense antenna signals are in phase prior to the null, they will be additive. If both signals are out of phase before the null, they will be subtractive. This allows the ADF receiver to determine the difference between the two nulls. The manufacturer will set one of the nulls as the "To" null, based on the antenna installation geometry and aircraft mounting. If, for example, both the loop and sense antenna voltages are in phase and add to each other while tracking to the

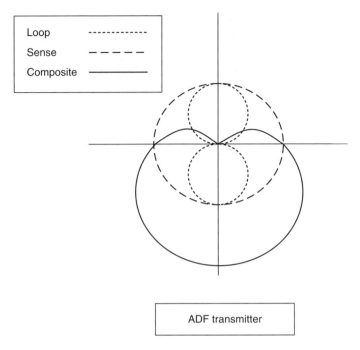

Loop — - - - - - - - - - -

Sense — — — — — —

Composite —————————

ADF transmitter

**Figure 4.1**   The ADF antenna pattern

null position, the loop antenna will be facing toward the transmitter, and this will be set as "To" in the receiver circuit. This Composite ADF antenna pattern has the form of a cardioid pattern as shown in Figure 4.1 (ICAO, 2000).

## 4.2.1   Operation and Controls

ADF receivers have several operating modes that can be selected. The most common form of RMI consists of a compass card and a single pointer (Figure 4.2 upper), and in the ADF mode as selected by a control panel switch (Figure 4.2 lower), both the loop and sense antennae are used and the RMI pointer is activated. If the compass card is rotated using the heading knob at the bottom left of the RMI so that the (magnetic) heading of the aircraft appears at the top in Figure 4.2 left, then the pointer, which always points to the selected NDB, will show the (magnetic) track to that NDB. Older versions used a fixed compass card, which necessitated some mental arithmetic to operate, whereas later ones incorporated a compass card driven from a gyro compass so that the aircraft heading was automatically placed at the top of the display.

In the Antenna (ANT) mode, the loop antenna is disabled and only the sense antenna is used. The pointer will then not move even if the loop antenna control is turned left or right. This mode is used simply to identify a station.

NDB radio frequencies are relatively low and range is usually limited, as shown in the list of typical ADF specifications (CAA, 2000; Rockwell Collins, 1997).

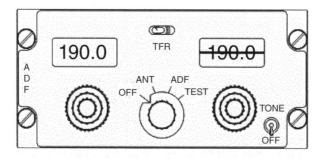

**Figure 4.2**   Typical ADF display and control panel

*Classes and Maximum Range*: Compass Locator 15 nm, Medium Homing 25 nm, Homing
   50 nm, High Homing 75 nm (Typical 70 μV/m)
*Frequency Range*: ADF 190–1750 kHz, NDB 190–535 kHz
Continuous Carrier Modulation (NDB only) 400 or 1020 Hz
*Angular Accuracy*: Depending on type and range to transmitter ±3 to ±10°

   NDB radio waves can be reflected off the ionosphere, and they can also be affected by
shorelines (the "coastal effect" can result in errors up to 30°), mountains, and cliffs. Lightning
and atmospheric static also emit radio frequencies that are close to NDB frequencies and can
affect ADF reception resulting in a wandering ADF pointer and noise on the ADF channel
even while in range. Precipitation can cause similar problems (CAA, 2000).

   However, under appropriate conditions, NDB systems can be used for instrument approaches
and allow aircraft to land under 600 ft ceilings or less, and only 1-mile visibility. Compared to
the nominal 1000 ft ceiling and 3-mile visibility requirement of a "Visual Flight Rules (VFR)
flight," this brought about a significant improvement in the provision of guidance during poor
weather and mountainous-area landings.

   Figure 4.3 shows typical NDB and Simplified Directional Facility (SDF) approach plates,
and these legacy illustrations may be compared with VHF Omni-Directional Radio Range
(VOR) and Ground-Based Augmentation System (GBAS) plates later in the chapter.

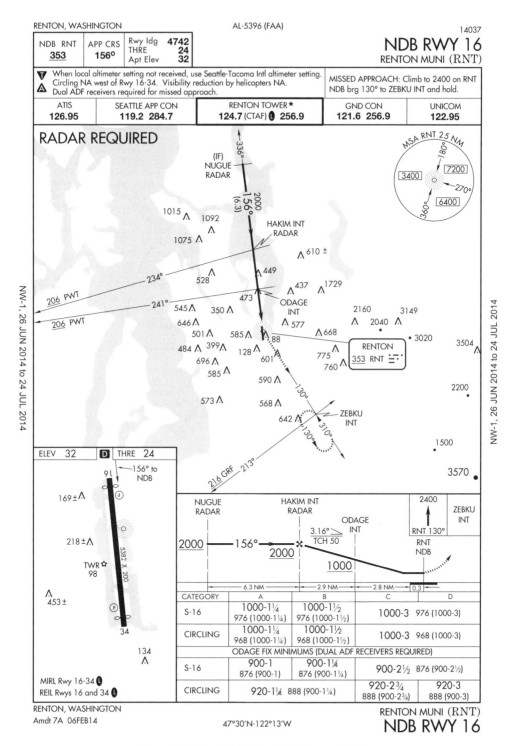

**Figure 4.3**   Typical ADF and SDF approach plates

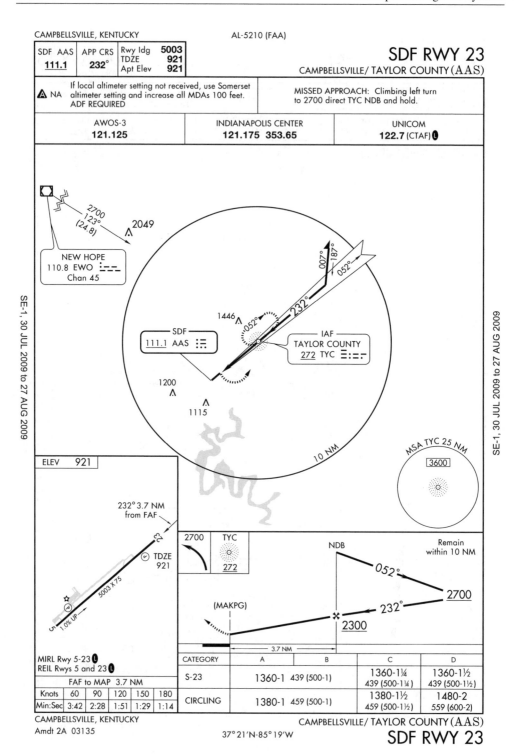

**Figure 4.3**   (Continued)

**Table 4.1**   Typical specifications for a commercial Rockwell Collins ADF

| Power requirements | | |
|---|---|---|
| ADF-60A | ANT-60A | ANT-60B |
| Power | 28 V DC 0.6 A, including ANT | Powered by ADF-60A | Powered by ADF-60A |

| Frequency | |
|---|---|
| Control | ARINC two out of five or CSDB with use of a CAD-62 adapter unit |
| Range | 190.0–1749.5 kHz |
| Tuning increments | 0.5 Hz |
| Capture range | ±250 Hz typical |

| Other specifications | |
|---|---|
| *Intermediate frequencies* | |
| First IF | 15 MHz |
| Second IF | 3.6 MHz |
| Bandwidth | 6 dB—2.8 kHz (nom) 80 dB – not more than 14 kHz |
| Sensitivity | Modulated 30% at 400 Hz |
| ANT | 70 μA V/m max at 6 dB $(s+n)/n$ |
| ADF | 100 μA V/m max at 6 dB $(s+n)/n$ |
| ADF accuracy | ±3° over the operating temperature range, with input signals of 70–0.5 μA V/m driving two ARINC standard ADF indicators |
| Spurious response | At least 80 dB below desired response |
| Audio gain | 100 mW (nominal) into 600 Ω load with a 1000 μA V/m input signal, 30% modulation at 400 Hz |
| CW identifier | 1000 Hz tone |
| *Output loading* | |
| DC sin/cos | 8 V DC max output, 5 kΩ min |
| Synchro bearing | Capable of driving up to five loads |
| output | Additional loads may be driven with the following increase in bearing error: three loads, 0.3°; four loads, 0.7°; five loads, 1.2° |

Reproduced with permission from Rockwell Collins, 1997.

The GPS system is replacing many current navigational aids, but the FAA still recognizes NDB approaches. Effective from July 16, 1998, pilots have been allowed to substitute IFR-certified GPS receivers for Distance Measuring Equipment (DME) and ADF avionics for all operations except NDB approaches without a GPS overlay. The GPS can be used in lieu of DME and ADF on all localizer-type approaches as well as VOR/DME approaches, including when charted NDB or DME transmitters are temporarily out of service. The exception to this is shown in Figure 4.3—the SDF approach requires an ADF—a GPS cannot be substituted. Full specifications for a currently available ADF are given in Table 4.1.

## 4.3  VHF Omni-Directional Radio Range

The ground-based VOR transmitter operates in the VHF band between 108.0 and 117.95 MHz. Adopted by ICAO as early as 1960, the VOR has been the main short-range navigational aid for several decades but is now being displaced by GPS methods. However, VOR systems are still in use and will be supported by the FAA until 2020 and perhaps beyond.[1] While the ADF/NDB, discussed in the previous section, transmits a nondirectional signal, the signal transmitted by the VOR contains directional information. The principle of operation is bearing measurement by phase comparison. Simply put, the transmitter on the ground produces a set of signals that make it possible for the receiver to determine its position in relation to that ground station by comparing the phases of these signals to an omni-directional signal. From the pilot's viewpoint the VOR produces a number of tracks all originating at the transmitter. These tracks are called radials and are numbered from 1 to 360, expressed in degrees. The 360° radial is the track leaving the VOR station and pointing toward Magnetic North, and the remaining 359 points of the compass are also represented by signals generated by the VOR transmitter.

### 4.3.1  Basic VOR Principles

The ground-based VOR transmitter transmits in the radio-frequency VHF band between 108.0 and 117.95 MHz. The discrete frequencies used by VOR stations are: 108.0, 108.05, 108.2, 108.25, 108.4, and 108.45 spaced in this manner up to 111.8 and 111.85 MHz. This provides a total of 40 VOR channels, and the same frequency band between 108 and 112 MHz also allows for 40 more frequencies allocated to ILS localizer transmitters (see Section 4.5.2). From 112 to 117.85 MHz the spacing is 0.05 Mhz, providing for another 120 possible VOR channels.

All VOR stations transmit a three-letter Morse identification code (1020 Hz amplitude modulation) that is repeated six times a minute, the only exception to this being if the VOR is paired with a DME—see Section 4.4), or if a Broadcast VOR station is being received. A Broadcast VOR is used to carry secondary information via a voice channel. If the code received is TST, the station is under "Test" or calibration and cannot be used for navigation. The typical transmission power ranges from 200 W for enroute VORs giving approximately 200 NM useful range for VOR navigation. Airfield beacons typically transmit at 50 W giving less than 100 NM of useful range and are called Terminal VORs (TVORs).

---

[1] A recent Federal Register Notice (FRN) published by the FAA describes a proposed federal rule for a long-awaited transition of the conventional navigation infrastructure of the National Airspace System (NAS) to Performance-Based Navigation (PBN) for NextGen. Benefits of this transition are increased capacity and efficiency, reduction in aircraft noise and emissions, and enhancement of safety all enabled or are dependent upon the implementation of RNAV and RNP routes, arrivals, departures, instrument approaches, and other procedures. While the FAA has alluded to a transition to PBN in the past, this FRN is the first time a formal transition plan has been defined.

The FAA's proposed plan would transition from defining airways, routes, and procedures using existing VOR and other legacy navigation aids toward an NAS based on RNAV and RNP procedures. The plan goes further to define the gradual removal of VOR facilities in the United States. Currently, over 80% of the 967 VORs in the NAS are past their economic service life, costing the FAA more than $220M per year to operate. The phase-out of VOR is planned to be completed by January 1, 2020. Capabilities will be enabled through the Global Position System (GPS) and the Wide Area Augmentation System (WAAS). The FAA is also considering programmatic changes under the Airport Improvement Program (AIP) that would favor WAAS approaches at airports, rather than Instrument Landing Systems (ILSs). Existing ILSs would provide an alternative approach and landing capability in support of recovery and dispatch of aircraft during GPS outages (Universal Navigation Bulletin, Volume 5 Issue 2 April 1, 2012).

The ground equipment is set up on a surveyed site and consists of a transmitter driving an aerial system, one part producing a reference signal (REF), the other producing a variable signal (VAR). The REF signal is an omni-directional continuous wave transmission at the fundamental frequency for the VOR station that is frequency modulated by a 9960 Hz subcarrier that is in turn frequency modulated at 30 Hz. Since this is an omni-directional transmission, its polar diagram is a circle. The 30 Hz component of this signal is used as a reference for measuring the phase difference relative to a VAR.

The VAR is transmitted from an aerial that is effectively a loop producing a figure-of-eight polar diagram that is electronically rotated at 30 revolutions per second. When the VAR and REF signals are mixed, the resulting polar diagram becomes a cardioid called a *limacon* that rotates at 30 revolutions per second. The rotation of this limacon results in amplitude modulation at 30 Hz, but unlike the ADF cardioid, it does not have a null position (Moir and Seabridge, 2003; Helfrick, 2007).

The VOR receiver splits the two signals into their two original components and the phase of the 30 Hz modulations of the fixed REF signal and the rotating VAR signal are compared in a phase comparator. The phase difference between these two signals is directly proportional to the angular position of the aircraft with reference to the VOR transmitter. Magnetic North is the normal reference for the radials, so when 0° phase difference is detected, the receiver is on the 360° radial from the station.

Figure 4.4 gives examples of the signal comparisons that are used by the airborne receiver to create driving signals for the Course Deviation Indicator (CDI) display needle. The description above is valid for the conventional VOR, usually called a CVOR. However, CVORs suffer from reflections from objects in the vicinity of the VOR site that can cause errors that in principle could be reduced if the horizontal antenna dimensions were increased. However, they can also be reduced by a new ground-based transmission system that produces the same VOR navigation display in the cockpit called DOPPLER VOR (DVOR), and currently CVOR transmitters are being replaced by DVOR transmitters. Aircraft VOR receivers behave in the same manner regardless of whether a CVOR or a DVOR signal is being received.

## 4.3.2   The Doppler VOR

Operation of the DVOR is also based on the phase difference between two 30 Hz signals modulating a 9960 Hz subcarrier wave, these forming *reference phase* and *variable phase* signals.

The reference phase signal is obtained by amplitude modulating the subcarrier with a 30 Hz sinusoidal waveform, the whole composite signal at the relevant VOR frequency being radiated omni-directionally in the horizontal plane by a central antenna. The radiation pattern forms a circle and produces at the aircraft receiver a 30 Hz signal with a phase independent of azimuth.

The variable phase signal is obtained from a 9960 Hz frequency-modulated subcarrier which amplitude modulates the radio-frequency carrier so that the upper and lower sideband signals are displaced 9960 Hz above and below that carrier, respectively. So, when added in phase with the carrier, they will produce a resultant signal that is amplitude modulated at 9960 Hz. This subcarrier is frequency modulated at a 30 Hz rate.

The sideband signals are sequentially distributed to, and radiated from, 48 sideband antennae arranged in a circle to simulate two diametrically opposed antennae rotating counterclockwise around the antenna ring at 30 revolutions per second, with one antenna radiating the upper sideband signal and the other the lower sideband signal. Since the effective length of the path of travel between the rotating sideband sources and the distant point of reception

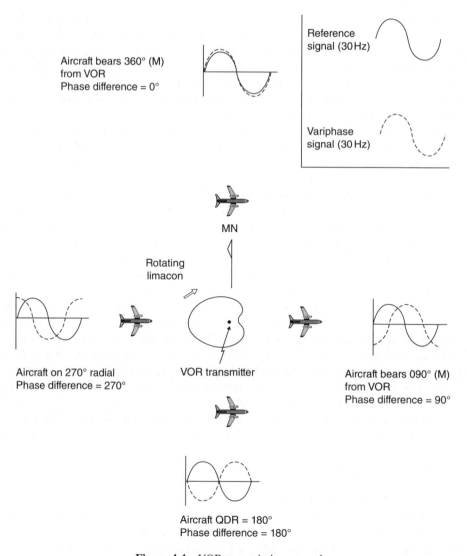

**Figure 4.4**   VOR transmission example

varies at a 30 Hz rate, the observed frequency of the sideband signals also varies at a 30 Hz rate and the subcarrier then appears frequency modulated at 30 Hz.

The frequency deviation is proportional to the diameter of the sideband antenna ring expressed in wavelengths at the operating frequency. Setting the diameter to 44.0 ft (13.4 m) produces a peak frequency deviation of 480 Hz at a frequency of 118.85 MHz, 454 Hz at 108 MHz, and 497 Hz at 118 MHz. The corresponding deviation ratio therefore varies from 15.13 at 108 MHz to 16.57 at 118 MHz. The deviation frequency is determined by the following formula:

$$f_d = \omega \cdot \lambda \cdot \pi \tag{4.1}$$

Here, $f_d$ is the deviation frequency in Hertz, $\omega$ is the angular velocity of the signal (30 Hz), $\lambda$ is the diameter of the ring in wavelengths, and $\pi = 3.14$. The deviation ratio is determined by the following formula:

$$r_d = \frac{f_d}{30} \tag{4.2}$$

The aircraft receiver uses the 30 Hz signal which it extracts from the 9960 Hz FM subcarrier. The phase of this second 30 Hz signal varies linearly with changes in the azimuth bearing of the receiving point: for each degree of azimuth change, the phase of the 30 Hz variable phase signal also changes by 1°. The sequential energizing of the sideband antennas and the 30 Hz amplitude modulation of the carrier are time-related in such a way that the reference and the variable phase 30 Hz signals are in phase at 0° magnetic from the DVOR ground station. As the receiving point is moved clockwise around the station, the variable phase signal (a 30 Hz FM waveform) begins to lead the reference signal (a 30 Hz AM waveform). For example, an observer West of the DVOR sees the 30 Hz FM signal leading the 30 Hz AM signal by 2700 Hz. The aircraft receiver determines the phase difference between the two 30 Hz signals and converts this to a bearing in degrees relative to the station, that is, to the number of degrees by which the 30 Hz AM signal lags the 30 Hz FM signal (Moir and Seabridge, 2003; FAA, 2013).

Summarizing, the DVOR antenna system thus simulates a rotating arm with a transmitting antenna at each end, radiating the upper sideband signal from one end and the lower sideband signal from the other end. This is achieved electronically by using 48 antennas spaced equally around the perimeter of a circle 44 ft (13.4 m) in diameter, with an antenna in the center of the circle radiating the reference carrier. This configuration and transmission method presents to the aircraft receiver a signal, which is decoded to produce the same guidance information regardless of transmission from a DVOR or CVOR.

### 4.3.2.1  The VOR/DVOR Display

Using the output from the onboard receiver, the calculated phase differences from the demodulated signal are displayed on the Course Deviation Indicator (CDI). This is depicted in Figure 4.5 and shows a compass rose and a CDI needle that indicates the aircraft position to the left or right of a selected course, along with a TO/FROM and a NAV validity "flag." This display can be realized by either a mechanical movement display or a digital representation. This type of deviation presentation is so well understood that it is used in many cases regardless of the type of navigation source. A CDI can be driven by GPS, VOR, and Tactical Air Navigation (TACAN )and represents a gold standard for cockpit displays.

To determine whether a bearing is TO/FROM a station, a 90° phase shift is applied to the variable phase signal, which is again compared with the reference signal by another phase detector. The 90° phase shift leading or lagging the reference signal produces a TO/FROM flag drive signal that actuates a galvanometer movement to provide a TO/FROM indication on the CDI. A small metal "flag" moves behind an opening on the face of the instrument indicating either TO/FROM.

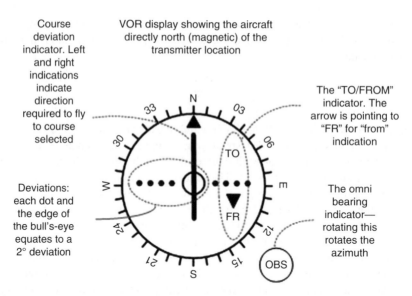

Course deviation indicator. Left and right indications indicate direction required to fly to course selected

VOR display showing the aircraft directly north (magnetic) of the transmitter location

The "TO/FROM" indicator. The arrow is pointing to "FR" for "from" indication

Deviations: each dot and the edge of the bull's-eye equates to a 2° deviation

The omni bearing indicator— rotating this rotates the azimuth

**Figure 4.5** Typical course deviation indicator. Not shown is a red NAV flag for indicating when the navigation display information is not useable for navigation (Source: Langley Flying School)

The CDI needle is driven left or right depending on the amount of deviation from the course selected by the Omni Bearing Selector (OBS). Each dot in the display represents 2° off course in the left or right direction.

There are two areas that represent ambiguous displays for the VOR receiver, these being directly over the transmitter and directly left or right of the transmitter while flying past the station on a selected heading. The ICAO defines this flyover *cone of confusion* as extending a maximum of 50° over the transmitter. When flying abeam and perpendicular to the transmitter, the display will provide incorrect CDI indications over a 20° sector on either side of it. These areas are referred to as *Sectors (or Zones) of Ambiguity*.

Typical errors for the VOR are as follows:

| | |
|---|---|
| Transmitter error | ±1° |
| Airborne equipment error | ±2° |
| Station interference | ±1° |
| Site error | Insignificant with Doppler VOR |
| Propagation error | Decreases to near zero if the VOR is within range |
| Total error | ±2 to 5° |

The sensitivity for VOR receivers is typically 3 μV, and the selectivity is from 6 dB ± 15 kHz to 60 dB at ±41 kHz. Receivers are currently made by Collins, Garmin, Honeywell, and other suppliers. Many receivers integrate both VOR and ILS circuitry and some displays combine both VOR and ILS data. The selection of which functionality is preferred depends on the selected frequency.

For comparison with the NDB and SDF approach plates of Figure 4.3, a typical VOR equivalent is shown in Figure 4.6.

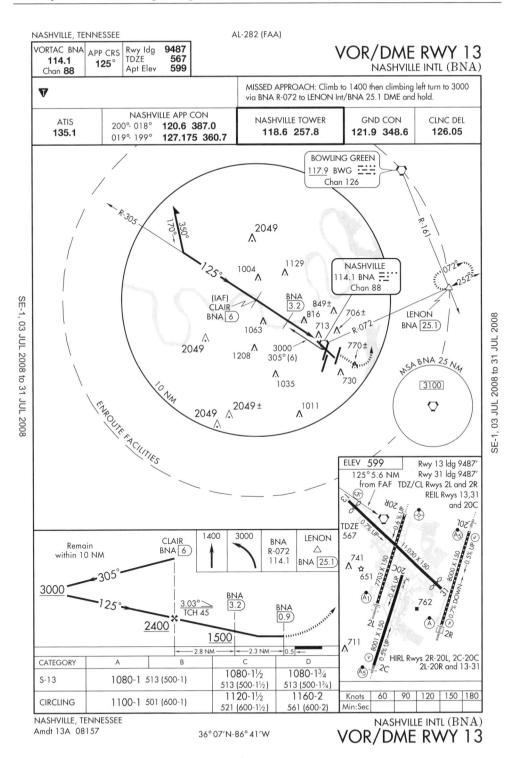

**Figure 4.6**   A typical VOR approach plate

## 4.4    DME and TACAN Systems

### 4.4.1    DME Equipment

DME is a type of radar system used to determine the slant range distance of an aircraft that contains a DME interrogator, to a DME transponder at a ground station. To accomplish range calculations, shaped RF double pulses are transmitted by the aircraft to a ground station and, after a defined period called the *reply delay*, this ground station returns the pulses. The receiver in the aircraft uses the round trip time of the double pulses to determine the distance to the ground station. The method is defined in International Civil Aviation Organization (ICAO) Annex 10 to the Convention on International Civil Aviation and also FAA Technical Standard Order T66a (Moir and Seabridge, 2003; Helfrick, 2007; Ostermeirt, 2009; FAA, 2013).

Most DME ground stations are co-located along with a VOR system to allow an aircraft to determine its position relative to this station in terms of both distance and bearing. The DME channels range from 1025 to 1150 MHz for the aircraft transmitter and 962 to 1213 MHz for the DEM ground stations. The frequency difference between the transmitted and received signal is always 63 MHz; and the spacing between the various DME channels is always 1 MHz. Each channel has two codings ($X$ and $Y$) that differ with regard to their pulse spacings. The ground equipment is known as an $X$ beacon (TACAN/DME). In order to increase the number of channels from 126 to 252 within the same frequency allocation, modified beacons became known as $Y$ beacons. $Y$ beacon receivers operate on the same frequency as $X$ beacon receivers of the same channel number, but the transmitted frequencies are different. An $X$ beacon transmitter frequency is 63 MHz lower than the associated receiver frequency and forms channel numbers from 1 to 63, whereas the $Y$ beacon transmitter is 63 MHz higher and forms channel numbers from 64 to 126. At this point it should be noted that there is no difference between the range pulse system used in the civilian DME and the range pulse system used in the military TACAN systems. Also, civilian aircraft can use a TACAN for range data to the station using standard DME equipment but are not able to use the bearing-to-station portion of the TACAN signal (US Department of Defense, 1998).

### 4.4.1.1    DME Pulse Details

As noted earlier, the aircraft DME interrogator emits a sequence of pulses that are received at the ground station and, after a defined delay time, are returned at a different frequency, the offset between the sent and received signals being always 63 MHz. The receiver in the aircraft filters out its own pulse sequence from all the received pulses and in this way determines the time difference between the transmitted and received pulses. This makes possible a determination of the slant distance to the station by calculating the round trip time. This distance is indicated in nautical miles (where 1 NM = 1852.02 m) and a signal round trip time of 12.36 µs is used as an internal reference. Note that as either the aircraft altitude increases or its distance from the station decreases, the slant angle and hence the measured slant distance from that station will diverge more and more from the horizontal distance.

By also taking the flight altitude Above Ground Level (AGL) into account along with the azimuth angle to the VOR ground station into consideration, it is possible to determine the precise position of the aircraft. Within the aircraft DME interrogator equipment, a distinction is made between *search mode* and *track mode*. In search mode, the interrogator attempts to set up a connection to a ground station and to synchronize to this ground station. In this mode, the pulse repetition rate can be increased up to 150 pulse pairs per second (pp/s). When the

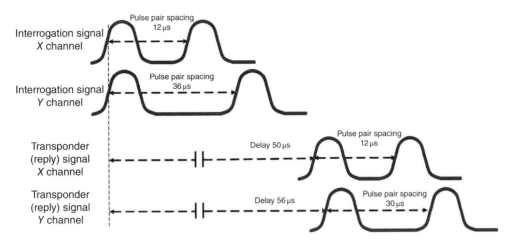

**Figure 4.7**  Time envelope for DME interrogation and reply (Courtesy of Rohde and Schwarz, München, Germany)

interrogator has synchronized to a ground station, it changes to track mode and performs its distance measurements at regular intervals. The pulse repetition rate in track mode is maximally 16 pp/s, and the transmission power of the aircraft interrogator can range from 100 to 250 W. Figure 4.7 shows the time envelope for a DME interrogation and reply.

Pulse and space delay times are different for the *X* and *Y* channels, this being simply a function of signal channel identification: the *X* channel has a pulse spacing of 12 μs, while the *Y* channel has a pulse spacing of 36 μs. DME ground stations are divided into two types: DME *en route transponders* that have a 1 kW pulse power output and provide range information up to 200 NM from the station; DME *terminal transponders* that have a 100 W pulse power output and provide range information up to 60 NM from the station. Both types of stations emit identical signals for range detection (Ostermeirt, 2009).

To fully understand DME operation, two unique modes of DME ground station operation must be considered—the *squitter pulse* and *automatic gain* modes. The DME ground station uses automatic gain and squitter (which is a noise-generated output) control methods to maintain a constant pulse output. If too few interrogations are being received from aircraft, the gain and squitter of the receiver increases and subsequently adds noise-generated pulses to the pulse train, so keeping the number of pulse pairs at a constant 2700 pp/s. If more interrogating aircraft come into range, the gain and squitter decreases and a reduced number of noise-generated pulses are imposed on the transmission. As noted earlier, beacons transmit a constant number of pulses per second so that the greater the number of interrogation pulses received, the fewer are the noise pulses needed by the ground station to maintain 2700 pp/s. As more aircraft enter the operational zone of the DME ground station, the gain will be reduced until noise pulses with small amplitudes fail to satisfy the requirements of the response ("ground-station-sends-a-reply") circuit. As the tracking rate of the interrogating pulses is 24–30 per second, the trigger circuit will be saturated by the interrogations of more than 100 aircraft. If this happens, the receiver gain will be reduced until replies are generated only by the strongest 100 interrogations, so reducing the useable range envelope for the DME ground station. Because of this, there is a maximum number of aircraft in a given location that the DME station can service.

Typical accuracies for DME equipment are about ±0.1 NM, and while the minimum accuracy in the FAA regulations is given as 1/2 mile or 3% of the distance at 199 NM, most receivers exceed this specification.[2] Knowing the distance to a station and the change in distance over time will also provide time and ground speed to that station.

### 4.4.2  Tactical Air Navigation

This is the military version of DME. The DME portion works in the same way as does as the civilian version, but the TACAN signal also provides bearing information between the aircraft and the ground station. Hence, the accuracy of azimuth direction determination is higher than in the civilian VOR method. To allow a TACAN receiver on board an aircraft to determine the direction to a TACAN ground station, that station transmits 900 specially coded pulse pairs per second in addition to the DME pulses. All pulses are transmitted by a rotating antenna, thereby creating a specially formed cardioid radiation pattern and applying two-tone (15 and 135 Hz) amplitude modulation signals to the envelope of the DME pulses received from a TACAN aircraft interrogator. The TACAN receiver determines the azimuthal direction by measuring the phase relation between the amplitude modulation and the 900 specially coded TACAN pulses. Since the amplitude modulation is generated by a rotating antenna, the pulse peak amplitude at the transponder output (or antenna input) is constant, as is the case for a DME transponder.

The calculation of bearing information is accomplished by comparing the 15 Hz modulation signal with a 15 Hz reference burst signal received from the ground facility. The phase relationship between the 15 Hz modulation signal and the 15 Hz reference burst signal depends on the location of the aircraft in the cardioid pattern. The 15 Hz reference burst signals are transmitted when the maximum signal in the cardioid pattern aims due East. This group of 12 pulse pairs is commonly referred to as the North or main reference burst. An additional phase comparison using the 135 Hz modulation further reduces the error and improves accuracy. Figure 4.8 shows the TACAN signal pattern (US Department of Defense, 1998; Moir, 2006).

The bearing accuracy of the TACAN system is nominally ±0.5° to a station, but like the VOR systems considered earlier in the chapter, it also has a 70° cone above the ground station in which this bearing function is inoperative. One particularly interesting component of the TACAN is the *air-to-air mode*. This is part of the military system and is not available to civilian aircraft. In this mode two aircraft can set up the TACAN equipment as though they were each transmitting to a ground station—that is, each aircraft replies as though it were a ground station. This allows a group of up to 33 aircraft to determine their ranges to any given aircraft. This is ideally suited for tanker operations where a group of aircraft can obtain range data to a tanker. Figure 4.9 shows both a typical military TACAN and a civilian DME control/display module.

### 4.4.3  The VORTAC Station

The VORTAC station is in common use at civilian airports and consists simply of a VOR and a TACAN co-located at an airport or remote station. From the *Aeronautical Information Manual* (AIM), "A VORTAC is a facility consisting of two components, VOR and TACAN,

---

[2] FAA (2012).

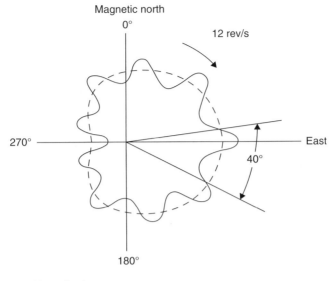

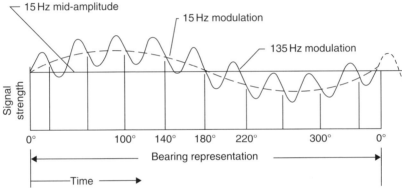

**Figure 4.8**   TACAN bearing signal waveforms

which provides three individual services: VOR azimuth, TACAN azimuth, and TACAN distance (DME) at one site. Although consisting of more than one component, incorporating more than one operating frequency, and using more than one antenna system, a VORTAC is considered to be a unified navigational aid. Both components of a VORTAC are envisioned as operating simultaneously and providing the three services at all times. Transmitted signals of VOR and TACAN are each identified by three-letter code transmission and are interlocked so that pilots using VOR azimuth with TACAN distance can be assured that both signals being received are definitely from the same ground station. The frequency channels of the VOR and the TACAN at each VORTAC facility are "paired" in accordance with a national plan to simplify airborne operation" (FAA, 2002, 2013).

Figure 4.10 shows a typical VORTAC installation wherein the top portion is for the bearing facility of the TACAN system. The smaller antenna allows for TACAN systems to be mobile or even ship-mounted where it would be impossible to mount a standard VOR system.

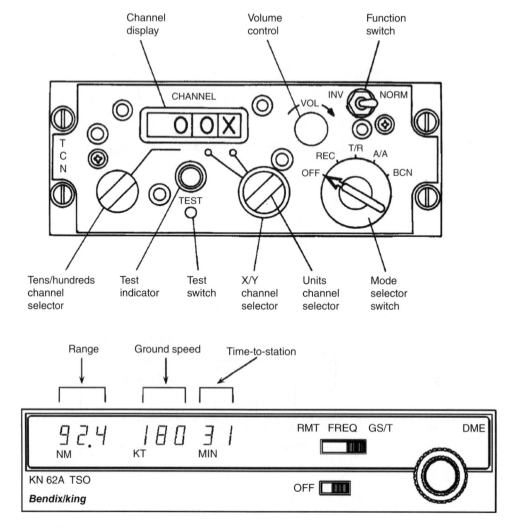

**Figure 4.9** Military TACAN and civilian DME control/display modules (On the KN662 control, note that the mode selected is GS/T which provides range, time, and distance to the station.)

### 4.4.4   The Radiotechnical Short-Range Navigation System

The non-ICAO Radiotechnical Short-Range Navigation (RSBN—Радиотехническая Система ближней Навигации) system originated in Russia (Baburov and Ponomarenko, 2005) but is now used largely for military purposes there and also where Russian aircraft are employed, notably in India and China.

It is essentially a short-range navigation system that involves ground-to-air azimuth transmissions along with air-to-ground identification requests. The DME and glide-path subsidiary systems also exist including the PRMG (Приводная Радиомаячная Группа) landing system that automatically takes the aircraft from about 80 km to the decision height for the runway,

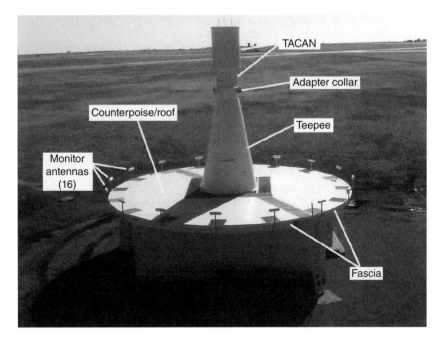

**Figure 4.10** A VORTAC installation

where the pilot takes over the final landing. Unlike the VOR system, it uses a mechanically rotating azimuth antenna as will be described later and is potentially more accurate, which is why it continues to be developed as an RSBN-2, 4, 6, 7, etc. series.

## 4.4.5 Principles of Operation and Construction of the RSBN System

In the RSBN system, the method of azimuth measurement involves the time-of-arrival of a signal from a ground-based rotating transmitting antenna with respect to an omni-directional reference signal that defines the north meridian at that transmitter. The combination of RSBN and DME equipment constitutes a polar coordinate positioning system that can provide an accuracy of $\pm 0.25°$ and $\pm 200\,\text{m}$.

The azimuth antenna is a rotating truncated parabolic reflector with three slot irradiators that provide a vertical coverage of $45°$. It transmits an unmodulated radiation pattern in the horizontal plane that consists of two lobes (Figure 4.11) rotating at $\Omega_A = 100$ revolutions per minute, corresponding to a sweeping frequency of $f_A = 1.6\overline{6}\,\text{Hz}$, that is, $600°\,\text{s}^{-1}$.

The airborne receiver, therefore, extracts an *Azimuth Signal* (AS) having the shape of a double pulse whose recurrence frequency is $f_{AS} = f_A = 1.6\overline{6}$ Hz. The minimum point between these two lobes is used as the azimuth measuring point as it sweeps past the airborne reception antenna. As can be seen from Figure 4.11, at any azimuth this signal is delayed for a time $t_A = A/\Omega_A$ relative to a time of origin $t_0$ where the radiation pattern minimum coincides with the northern direction N of the meridian passing through the rotating ground antenna. This time interval $t_A$ is therefore a measure of the direction from the aircraft to the ground station, which

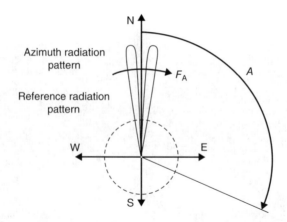

**Figure 4.11**   Radiation patterns of the RSBN azimuth and reference antennae

in turn implies that $t_0$ must also be determined by the aircraft receiver. This is achieved as follows.

The omni-directional ground antenna is co-located with the rotating azimuth antenna and radiates a Reference Signal (RS) consisting of two sequences of pulses, one at 35 pps and the other at 36 pps. When the azimuth radiation pattern minimum (between its two lobes) passes through the north meridian direction N, and pulses from both the RS sequences also coincide with this, $t_0$ is defined. It is the coincidence of the two RS signals that is used by the aircraft to define $t_0$, after which the azimuth pulse arrives at $t_A$. This delay is used to determine the azimuth of the aircraft from the ground station. (Obviously, if the aircraft were due north of the ground station, then $t_A$ would be zero.)

A coarse azimuth timing is provided by counting the number of pulses from the 36 pps signal between $t_0$ and $t_A$; and a fine timing addition to this measure is provided by an accurate timer that measures the interval between the last pulse and the minimum of the AS signal. The total time $t_A$ is the addition of these two measures.

In theory, this method could provide an accuracy down to ±0.02°; but in practice, the observed accuracy is limited by the processing in the subsequent electronics and the display device to about ±0.25°.

Interrogation and responder equipment makes it possible to initiate this procedure from either the aircraft or the ground station, and appropriate antennae are provided for this purpose at both.

# References

Baburov V I and Ponomarenko B V (2005), *Principles of Integrated Onboard Avionics*. Agency "RDK-Print", St. Petersburg (in Russian).

Civil Aviation Authority (2000), Description of NDB and ADF Operation and Definition of Protection Requirements, Report Number 8AP/88/08/04. Civil Aviation Authority, London.

Collinson R P G (1997), *Introduction to Avionics*, Chapman & Hall, London.

Federal Aviation Administration (FAA), (2002), Order 8260.3B Change 19, United States Standard for Terminal Instrument Procedures (Terps).

FAA (2012), *Aeronautical Information Manual*, dated July 7–26, 2012. FAA, Washington, DC, Change 3.

Helfrick A D (2007), *Principles of Avionics*, 4th Ed., Avionics Communications, Leesburg, VA.

Moir I and Seabridge A (2003), *Civil Avionics Systems*, Professional Engineering Publishing, Bury St. Edmunds.

Moir I and Seabridge A (2006), *Military Avionics Systems*, John Wiley and Sons Ltd,Chichester.

Ostermeirt J (2009) Rhode & Schwartz, Test of DME/TACAN Transponders, Application Note 03.2009-1.10.

Rockwell Collins (1997, April 11), ADF-60A Installation Manual, Manual Number 523-0766186-00A118.

U.S. Department of Defense (1998), Standard Tactical Air Navigation (TACAN), MIL-STD-291C, U.S. Department of Defense (September 1986), Navigational Set AN/URN-25, MIL-N-29510.

# 5

# Radio Technical Landing Systems

J. Paul Sims
*Department of Engineering, Technology, and Surveying, East Tennessee State University, Johnson City, TN, USA*

## 5.1 Instrument Landing Systems

The Instrument Landing System (ILS) is a final navigation aid used for approach and landing and consists of three components, a *Localizer* (LOC), a *Glide Path* (GP), and a series of *marker beacons* that include an *outer marker*, a *middle marker* and, sometimes, an *inner marker*. Each group generates radio signals independently and continuously. The LOC supplies left–right navigational information, the GP supplies up–down navigational information, and the marker beacons supply distance-to-threshold information. Normally, all three facilities can be controlled and monitored from equipment in the control tower. The overall layout of the system is shown in Figure 5.1 and is taken from the FAA *Airways Information Manual* (FAA, 2013).

### 5.1.1 The Marker Beacons

The marker beacons radiate fan-shaped patterns of RF energy vertically upward, and Figure 5.2 shows an installation using all three, though in new installations, the inner marker is usually omitted. All marker beacons transmit on a fixed frequency of 75 MHz, but there is no interference between adjacent beacons because of the narrow extent of the radiation patterns along the GP. The marker beacon transmissions are amplitude modulated with dots and/or dashes at given tones. (One or two locator beacons—low-powered Nondirectional Beacons (NDBs)—are often positioned on the same sites as the outer marker and middle marker, but if only one locator is used, it is usually colocated with the outer marker.) Marker beacons are placed at distances of 75 m from the threshold for the inner marker, 1050 m (or more) for the middle marker and 7200 m to a maximum of 11.2 km for the outer marker. Each transmits a particular pulse code vertically upward at a carrier frequency of 75 MHz, as noted earlier. The modulated tones are at frequencies of 3000 Hz (inner marker), 1300 Hz (middle marker), and 400 Hz (outer marker). The aircraft flies through the *transmission cones* in the

*Aerospace Navigation Systems*, First Edition. Edited by Alexander V. Nebylov and Joseph Watson.
© 2016 John Wiley & Sons, Ltd. Published 2016 by John Wiley & Sons, Ltd.

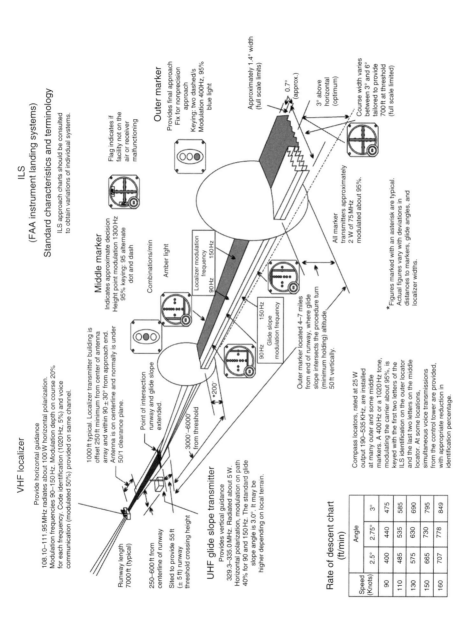

**Figure 5.1**  The standard FAA Instrument Landing System

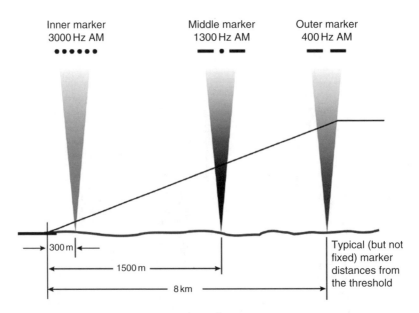

| Inner marker | Middle marker | Outer marker |
| 3000 Hz AM | 1300 Hz AM | 400 Hz AM |

**Figure 5.2**  A typical marker beacon layout (distances from beacon to threshold will vary depending upon approach requirements for holding and alignment)

approach path, and the pilot receives an audible indication of the pulse code and the identity signal. The marker outputs are adjusted to ensure the following beam widths, measured along the GP and LOC axes:

- Inner marker: $150 \pm 50$ m
- Middle marker: $300 \pm 100$ m
- Outer marker: $600 \pm 200$ m

A Distance Measuring Equipment (DME) system (see Section 4.3.1) can also be installed instead of marker beacons because this system provides a continuous distance readout between the aircraft and the runway touchdown point.

## 5.1.2   Approach Guidance—Ground Installations

The next two components, the LOC and the GP operate on the same principles but at different frequencies, both being based on measurements of the difference in *Depth of Modulation* (DDM) between two signals with frequencies of 90 and 150 Hz.[1] These are the navigation frequencies used to detect the correct approach course (DDM=0) and the specified GP angle (DDM=0). The LOC operates over the frequency range 108–112 MHz and generates a vertical guidance plane that permits the pilot to select a left/right approach course from a distance of up to 25 NM with only 25 W of radiated power. The antenna radiation pattern transmits exactly the same amplitude for the two modulation frequencies of 90 and 150 Hz in the guidance plane.

---

[1] DDM is calculated by subtracting the percentage of modulation depth of the smaller signal from the percentage of modulation depth of the larger signal and then dividing by 100.

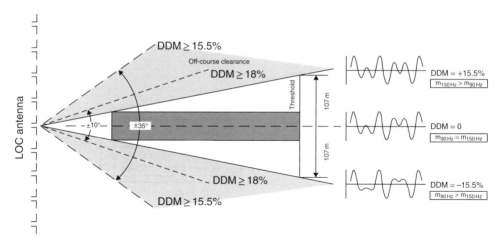

**Figure 5.3**   Typical localizer layout (note 170 m widths at threshold) (image courtesy of EASA)

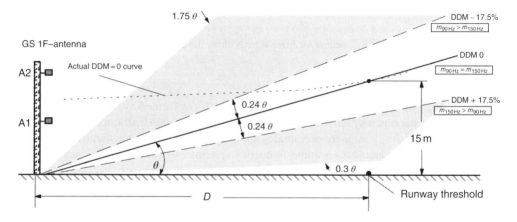

**Figure 5.4**   Glide slope layout (note 15 m crossing point at threshold) (image courtesy of EASA)

If the pilot deviates to the left of the guidance plane, the 90 Hz modulation signal will predominate, causing the cockpit indicator to show a "fly right" indication. If the pilot deviates to the right, the 150 Hz modulation signal will predominate, causing the cockpit indicator to show a "fly left" indication. Figure 5.3 shows the LOC values for a typical installation (ICAO, 2000).

The GP operates over the frequency range 328–336 MHz and generates the proper GP, which is elevated above the runway by the selected glide angle. The radiated power of a typical installation is 5 W with a reception range of 5 NM. The antenna radiation pattern results from an interaction with the earth's surface, and it contains a 150 Hz modulation below the GP and 90 Hz modulation above the GP. In the GP itself the amplitudes of the two modulation signals are equal. The beam defining the correct landing approach path is formed by the intersection of the vertical course guidance plane and the horizontal GP plane. Figure 5.4 shows a typical GP layout and signal structure. Note that at the *runway threshold* the height of the GP is 15 m (Moir and Seabridge, 2006).

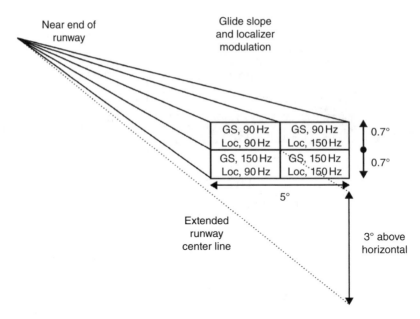

**Figure 5.5**   Glide slope and localizer signals along the approach (Battlett, 2008)

Figure 5.5 shows the rectangular form of the signal pathway formed by the LOC and the GP signals. Ground stations may have differing antenna configurations depending on local terrain and obstacles. For the LOC there are single- and dual-frequency systems and for the glide slope there are three systems (two single-frequency systems and one dual-frequency system), both providing the same apparent guidance to the aircraft.

Along with course information, the LOC's RF signal includes an airport identity signal in Morse code. This code includes three or four characters with a frequency of 1020 Hz. It is also possible to use an auxiliary feature to externally modulate the LOC RF signal with a voice signal (Automatic Terminal Information Service (ATIS) and to or from a tower). This feature provides a tone frequency range of 300–3000 Hz and a modulation depth of up to 40% (Battlett, 2008; Thales, 2010).

Consider now the relationship between the aircraft and the ILS. Section 5.4 covers optical guidance systems, and it is worthwhile noting here that where there are no obstacle clearance problems at a given airfield, both the ILS glide slope and the visual glide slope (either a Precision Approach Path Indicator (PAPI) or a Visual Approach Slope Indicator (VASI)) will be set at about 3°. Hence, there should be a direct correlation between the information provided by both systems. However, on a long-bodied aircraft, such as a Boeing 747 or A300 Airbus, the wheels of the aircraft will be several metres below the pilot's eyes, and it is important that he visualizes a parallel but higher slope to ensure adequate wheel clearance at the runway threshold on landing. The three-bar VASI was developed specifically to ensure that pilots of long-bodied aircraft use only the second and third wing bars of the VASI and ignore the first (lower) bar. When on the correct visual glide slope, the top bar will appear red and the middle bar will appear white, so ensuring proper landing gear clearance.

**Figure 5.6**   Typical ILS HSI display showing the vertical localizer and horizontal glide slope needles

## 5.1.3   Approach Guidance—Aircraft Equipment

In the aircraft, the *Horizontal Situation Indicator* (HSI) provides ILS guidance for advanced displays, and its basic functionality is analogous to the common ILS/VOR two-bar display. Referring to Figure 5.6, when used in the VOR mode, the vertical needle has a sensitivity of 2° per dot of deviation from the LOC; and when used in the ILS mode, it is far more responsive at one dot for each half-degree of deviation.

For the GP, full-scale deflection of the horizontal needle will occur when the aircraft is off guidance by approximately 0.75° above or below the GP. Hence, for a standard five-dot display, one dot of needle displacement represents approximately 0.15° of deviation above or below the GP.

## 5.1.4   CAT II and III Landing

ILS services provide for standardized signals but, depending on the *approach category* (CAT), the ground portion must include certain types of transmitters. For example, the LOC must consist of a dual transmitter/dual monitor design for CAT III approaches. Monitoring systems must ensure that the ILS signal is coherent and within operating specifications.

Category approach numbers simply describe the lowest approach visual minimums.[2] To make such approaches, there are three requirements: specific ground equipment as noted earlier; in some cases, specific aircraft equipment coupled to the ILS that can automatically fly the aircraft by adjusting both attitude and thrust; and a specially trained crew.

---

[2] • Cat I—decision height of 200 ft (60 m) and a runway visual range of 1800 ft (550 m)
  • Cat II—decision height of 100 ft (30 m) and a runway visual range of 1200 ft (350 m)
  • Cat IIIa—decision height of 50 ft (15 m) and a runway visual range of 700 ft (200 m)
  • Cat IIIb—decision height of 50 ft (15 m) and a runway visual range of 150 ft (50 m)
  • Cat IIIc—decision height of 0 ft, (0 m) and a runway visual range of less than 150 ft (50 m).

|            |     | ICAO | FAA | JAA |
|------------|-----|------|-----|-----|
| CAT II     | DH  | 100 ft ≤ DH < 200 ft | 100 ft ≤ DH < 200 ft | 100 ft ≤ DH < 200 ft |
|            | RVR | 350 m ≤ RVR<br>1200 ft ≤ RVR | 350 m ≤ RVR < 800 m<br>1200 ft ≤ RVR < 2400 ft | 300 m ≤ RVR<br>1000 ft ≤ RVR |
| CAT III A  | DH  | No DH or DH < 100 ft (1) | No DH or DH < 100 ft (1) | DH < 100 ft (1) |
|            | RVR | 200 m ≤ RVR<br>700 ft ≤ RVR | 200 m ≤ RVR<br>700 ft ≤ RVR | 200 m ≤ RVR<br>700 ft ≤ RVR |
| CAT III B  | DH  | No DH or DH < 50 ft | No DH or DH < 50 ft | No DH or DH < 50 ft |
|            | RVR | 50 m ≤ RVR < 200 m<br>150 ft ≤ RVR < 700 ft | 50 m ≤ RVR < 200 m<br>150 ft ≤ RVR < 700 ft | 75 m ≤ RVR < 200 m<br>250 ft ≤ RVR < 700 ft |
| CAT III C  | DH  | No DH | No DH | |
|            | RVR | No RVR limitation | No RVR limitation | |

**Figure 5.7**   Differences in Runway Visual Range (RVR) and Decision Height (DH) for different regions

Common factors affecting ILS range and accuracy include beam bending caused by local terrain, though its effect is reduced to near zero by the fact that as the aircraft approaches the threshold, the LOC accuracy increases. Another issue is that of propagation error, again due to terrain. Finally, large objects near the approach path may cause multipath problems. For airports with CAT II and III approaches, these errors are continually monitored. As a precaution, the FAA maintains a fleet of aircraft that daily fly into different approaches across the United States to establish that ILS systems are working correctly to provide safe and secure approach corridors. All ILS approach installations require protected zonesat the front of the LOC and GP transmitters called the *sensitive areas*. These are established to keep away vehicles, other aircraft, and any object that might penetrate the transmission zone of the ILS signals. Because of signal integrity issues, these sensitive areas occupy very large areas on the airport. Sensitive areas for CAT III approaches have the largest footprints, and as an example, the sensitive area at one airport for the CAT III approach forms a rectangle, which extends 2750 m along the runway and 90 m across the runway from the LOC while an aircraft is on this approach.

The pilot uses an *ILS approach plate* to assist navigation into an airport, this being a single-page document containing all the information necessary to provide a safe approach. Figure 5.7 shows the difference in definitions of CAT II and III operations between the International Civil Aviation Organization (ICAO), Joint Aviation Authority (JAA), and Federal Aviation Administration (FAA) that pilots on international flights must deal with during approaches. The equipment is standard between the FAA and the JAA (ICAO, 2000; Moir and Seabridge, 2003; Helfrick, 2007).

During a CAT II or III approach, the pilots must depend on the automatic systems that are coupled to the ILS transmissions and are flying the aircraft, and must constantly monitor

**Table 5.1**  Deviation limits for CAT II and III approaches above which crews must make a determination for the continuation of an approach

| Parameters | If deviation exceeds | | Call required |
|---|---|---|---|
| IAS | +10 kt | | "SPEED" |
| | −5 kt | | |
| Rate of descent | −1000 ft/min | | "SINKRATE" |
| Pitch attitude | 10° nose up | | "PITCH" |
| | 0° (A330/340), −2.5° (A320/321) | | |
| **Bank angle** | 7 | | **"BANK"** |
| Localizer | Excess deviation warning | 1/4 DOT (PFD) | "LOCALIZER" |
| Glide slope | | 1 DOT (PFD) | "GLIDE SLOPE" |

From Airbus Category II and III Operations, Airbus Technical Document STL 472.3494/95 October 2011.

the approach for variations that exceed prescribed limits. Table 5.1 provides an example of variations that would lead to a crew callout requiring a decision on how to proceed. The three options that follow these excess limits are as follows: continue the approach, revert to higher minima, or go around. However, as an example of the robust design of autopilot and coupled ILS systems in a in a CAT III fully redundant dual approach, even the single failure of one of the redundant autopilots or one engine failure below the *Alert Height*[3] on certain Airbus aircraft still allows for the safe continuation of the landing (Airbus, 2001).

At the time of writing, significant changes involving GPS systems are in progress for both short- and long-range navigation, but it is nevertheless expected that ILS systems will be operational well into the future. In 2011 the FAA estimated that ILS systems would remain operational for the next 20 years and possibly beyond, but will be phased out and replaced with space-based precision approach systems.[4] As an example, American Senate Report 111-069, dated 2010, allocated 18 million dollars to support installation of new ILS systems across the United States.

The modern *Ground-Based Augmentation System* (GBAS) is a GPS-based precision approach system designed to replace the current ILS and provide for CAT II and III approaches, and this will now be described.

## 5.2  Microwave Landing Systems—Current Status

Other than for a few operational civilian approaches and some military applications in the United States, Microwave Landing Systems (MLS) have been or will be replaced with GPS-based approach systems. However, the military systems are still in wide general use; for example, the AN/ARA-63 Tactical Instrument Landing System is used by the US Navy and also by other countries. This is MLS-based and is installed on many aircraft such as the F/A-18E/F/G and E-2D aircraft, and the US Navy has continued to order these systems up to the time of writing.

---

[3] An *Alert Height* is a height above the runway, based on the characteristics of the airplane and its fail-operational automatic landing system, above which a CAT III approach would be discontinued and a missed approach initiated if a failure occurred in one of the redundant parts of the automatic landing system, or in the relevant ground equipment.
[4] FAA Global Navigation Satellite System Update, ICG-6, September 5, 2011.

MLS offers many advantages over standard ILS. There are no site difficulties involving signal blockage or other terrain problems, and there are few interference issues with MLS signals. Perhaps the most important advantage of the MLS system is its ability to support approach entry at any point within the beam coverage. As an example, approach entry can be made up to ±60° off the runway center line and varying glide slopes are possible from 0.9 to 30° with a capture range of up to 20 NM. Because of the wide bandwidth used by MLS, current designs have the capacity to host 200 channels compared with the ILS maximum of 40.

### 5.2.1   MLS Basic Concepts

The MLS is technically very robust and provides for approach azimuth, back azimuth, approach elevation, and range and data communications. Runway-based transmitters radiate information, which is interpreted by aircraft receivers that calculate angles in both azimuth and elevation by measuring the time intervals between successive passes of a narrow fan-shaped beam. Ranging is derived from an accurate DME installation, so rendering marker beacons are unnecessary. Two unique parts of the system are the back azimuth guidance, which provides the aircraft with lateral navigation data following a missed approach. The data communications part of the system can provide multiple sets of data such as weather or wind shear warnings.

### 5.2.2   MLS Functionality

Functionally, the system works by providing a set of signals that allow aircraft position determination three dimensionally from the airport landing point, and the display to the pilot is the same (with some variations) as the normal ILS display. Elevation data is provided by an elevation station that transmits signals on the same frequency as the azimuth station. This elevation station provides a wide range of GP angles so that an appropriate approach angle for a specific aircraft can be selected by the pilot. The elevation signal coverage extends through the azimuth coverage, and so provides precision GP guidance at all points where azimuth guidance is available. Range guidance is provided by a DME/P station (see previous section) with a range of up to 22 NM.

The azimuth signal is a narrow vertical fan-shaped beam that sweeps back and forth across a coverage area, with the center of the beam normally aligned with the runway centerline, as shown in Figure 5.8. From the approach side, the beam starts at the left and sweeps at constant angular velocity to the right. This is known as the TO scan. After a short period, known as the guard time, the beam sweeps back to the starting point, and this is known as the FRO scan (derived from "to and from"). So, within a complete cycle of the TO and FRO scan, two pulses will be received by the aircraft, and by accurately determining the time interval between these two pulses a location can be derived, which is proportional to the angular location (azimuth) of the aircraft. The repeat rate of this scan is 13.5 scan cycles per second. The maximum time interval between the passage of the TO and FRO beams is measured from the extreme left edge of coverage, and the minimum time from the extreme right edge of coverage. The actual measured time interval represents angular position and therefore displacement from the center line (FAA, 2002, 2013; Battlett, 2008; McShea, 2010).

The elevation scanning method is much the same as the azimuth scanning method, and the elevation beam sweeps up and down at a rate of 40.5 scans per second. The aircraft receiver combines the elevation information with the azimuth and range (from the DME/P) to determine

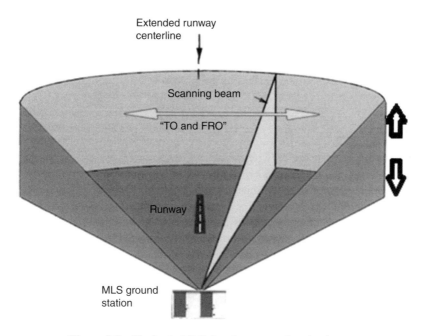

**Figure 5.8**   The basic MLS signal sweep and station layout

a three-dimensional (3D) position that can be used by the onboard equipment to compute steering commands in relation to curved approaches, varying glide slopes and segmented approaches. The typical MLS frequencies are spaced 300 KHz apart from 5031.0 MHz to 5090.7 MHz. All the functions except the DME take place on a single channel, and time multiplexing and signal preamble is used to segment azimuth and elevation scans. Figure 5.8 shows a typical MLS sweep pattern both across and up-and-down, so giving 2D data with the range data supplied by the DME/P.

## 5.3   Ground-Based Augmentation System

The Ground-Based Augmentation System (GBAS) is actually a satellite-based precision landing system that has been adopted by the ICAO as a replacement for the current ILS. This new system involves a fixed ground station that transmits position-correction data for satellite-based GPS signals to provide aircraft with very precise positioning guidance during the final stages of an approach, followed by very accurate 3D positioning during the landing phase of flight. Just one GBAS station can guide up to 26 highly precise approach flight paths simultaneously.

The FAA provides the following relating to GBAS:

A Ground Based Augmentation System (GBAS) augments the existing Global Positioning System (GPS) utilized in U.S. airspace by providing corrections to aircraft in the vicinity of an airport in order to improve the accuracy of, and provide integrity for, these aircrafts' GPS navigational position. The goal of GBAS implementation is to provide an alternative to the Instrument Landing

System (ILS) supporting the full range of approach and landing operations. Current non-federal (non-Fed) GBAS installations provide Category I (CAT-I) precision approach service. The Federal Aviation Administration (FAA) work program is now focused on validating standards for a GBAS Approach Service Type-D (GAST-D) (CAT-III minima) service. The program currently projects a GAST-D GBAS system can be available in 2016.[5]

### 5.3.1   Current Status

At the time of writing, there are only two working GBAS systems in the United States, one in Newark, New Jersey and the other in Houston, Texas. The only other approved approach is in Bremen, Germany. Additional stations about to come on line within the coming year are in Malaga, Spain and in Sydney, Australia. The only experimental CAT III station in operation is at Frankfurt airport, Germany, and this came on line on May 31, 2013, and is currently under test by the European regulatory authorities. It is expected that the majority of airports will have certified GBAS systems by 2025.[6]

### 5.3.2   Technical Features

Many of the advantages of GBAS relate to the number of ground components necessary to support multiple runways. For example, just one GBAS station can support 26 approach paths at multiple runway ends, so reducing the total number of legacy systems otherwise needed at an airport. Communication bandwidth is less problematic too—GBAS requires only one Very High Frequency (VHF) channel assignment for up to 48 individual approach procedures. Furthermore, the GBAS ground station and the supporting GPS reference receivers can be sited at almost any location on an airport, which allows the GBAS to serve runways that ILS could not support.

The current FAA standard calls for GBAS systems to provide 1 m accuracy, which along with concomitant reliability is sufficient for CAT II/III operations. (The Honeywell GBAS system currently in operation provides for 4 m accuracy, which is sufficient for a CAT I approach.) The basic GBAS system comprises a central horizontally polarized VHF antenna and four GPS reference antennas that provide a grid reference to allow for a very accurate position location. This position is used to correct errors in the satellite-based GPS signal. This corrected data is then transmitted to the aircraft via the central ground station and is processed by the aircraft avionics to establish a very precise 3D position. The avionics can then provide Autoland systems and detailed crew guidance for any runway used by the airport. Typically, GBAS coverage has a range of 30 miles from the airport ground station (Honeywell, 2001).

The basic operational concept is based on the fact that over a given area (of about 50 miles radius) the atmospheric and ionospheric corrections will be the same. Therefore, the ground stations, which are precisely surveyed when installed, provide fixed positional references. The differences taken from all the ground stations are determined and a computed correction

---

[5] http://www.faa.gov/about/office_org/headquarters_offices/ato/service_units/techops/navservices/
[6] http://www.faa.gov/about/office_org/headquarters_offices/ato/service_units/techops/navservices/gnss/laas/

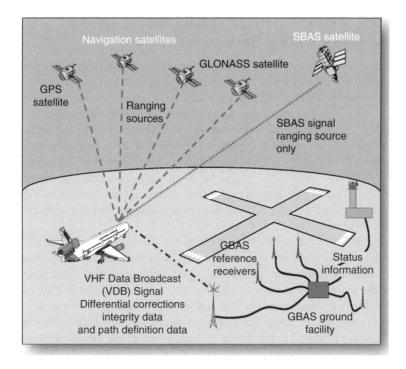

**Figure 5.9**  Typical GBAS system layout. http://ieeexplore.ieee.org/ieee_pilot/articles/96jproc12/jproc-TMurphy-2006101/article.html

is sent to the aircraft. The current ICAO standard calls for the following data to be encoded into the VHF database message[7]:

- Differential Code Corrections and Integrity data. (This includes differential correction and integrity-related data for each satellite tracked by the ground system.)
- Reference Point and GBAS station data.
- Final Approach Path Description(s). (Approach Path Definitions include Final Approach Segment definitions for each runway end or approach served by the ground segment.)

This information is then used by the aircraft avionics to calculate approach and guidance data for the crew and the Autoland system. There are many advantages to this, both for the aircraft and the airport. As an example, for the airport there are no longer any ILS critical areas as discussed in the ILS section, and this can also lead to better noise abatement because aircraft can fly flexible approach geometries. For aircraft and airlines, it may result in reduced holding times and fuel costs. As with MLS, variable GPs are possible as well as displaced thresholds. Currently, Honeywell and Rockwell Collins offer GBAS avionics, which are certified for installation on Boeing 737-NG, 747-8, 787 aircraft and also on the Airbus A320, A330/340, and A380. Current airlines flying GBAS include United Airlines, Qantas, and Air Berlin. Figure 5.9 shows a typical GBAS system layout.[8]

---

[7] http://ieeexplore.ieee.org/ieee_pilot/articles/96jproc12/jproc-TMurphy-2006101/article.html
[8] http://honeywellsmartpath.com/system_features.php

## 5.4 Lighting Systems—Airport Visual Landing Aids and Other Short-Range Optical Navigation Systems

As with most things relating to aircraft and airports, there are specific regulations relating to airport visual landing aids. The first component that a pilot identifies for an airport in visual meteorological conditions is normally the airport beacon. There are several types of these outlined in the FAA AC 150/5345-12E circular—"Specifications for airport and heliport beacons," dated 11/17/05. Table 5.2 gives most of the important information related to the various beacon types.

A typical airport beacon produces significantly more light than the minimum requirements shown in the fourth column. As an example the Manairco Company of Mansfield, Ohio, produces airport beacons with visible ranges of 30–40 miles and luminous intensities up to 400,000 candelas. The airport beacon is the first visible airport navigation aid a pilot will encounter, and it provides not only a visual bearing to the airport but also indicates the type of airport based via the characteristics of the visual signal.

As the pilot approaches the airport, the next visual aid will be the runway lights. These provide an outline of the runway itself and in some cases the magnetic heading is also outlined. During the approach phase of flight, there are several visual guidance systems that help the pilot to follow the proper path. These are of considerable assistance to the pilot and are mandatory for Instrument Flight Rule decision height and transition to visual guidance in IFR approaches where the airport may not be visible due to weather until a very low altitude has been reached.

After the beacon a pilot will normally see the approach lighting system. Approach Lighting Systems (ALS) are a configuration of high-intensity or medium-intensity sequenced flashing signal lights designed to guide the pilot from the approach zone to the runway threshold. Approach lights can also provide additional visual guidance for nighttime approaches under visual flight rules.

The ALS consists of a variety of ALS approved for use in the United States and other countries. These include a high-intensity array with sequenced flashing lights (ALSF-2) for use on

**Table 5.2** Airport beacon descriptions

| Beacon type | Application | Signal provided | Minimum intensity (cd) (elevation angle, 3–7°) |
|---|---|---|---|
| L-801A | Medium-intensity airport beacon | Alternate white and green | 50 000 |
| L-801H | Medium-intensity heliport beacon | Alternate white, green, and yellow | 25 000 |
| L-801S | Medium-intensity seaplane base beacon | Alternate white and yellow | 50 000 |
| L-802A | High-intensity airport beacon | Alternate white and green | 75 000 |
| L-802M | High-intensity military airport beacon | Alternate white, white, and green | 95 000 |
| L-802H | High-intensity heliport beacon | Alternate white, green, and yellow | 37 500 |
| L-802S | High-intensity seaplane base beacon | Alternate white and yellow | 75 000 |

**Figure 5.10**  Typical ALS layout (note that the green light line indicates the runway threshold)

*Aerospace Navigation Systems*, First Edition. Edited by Alexander V. Nebylov and Joseph Watson.
© 2016 John Wiley & Sons, Ltd. Published 2016 by John Wiley & Sons, Ltd.

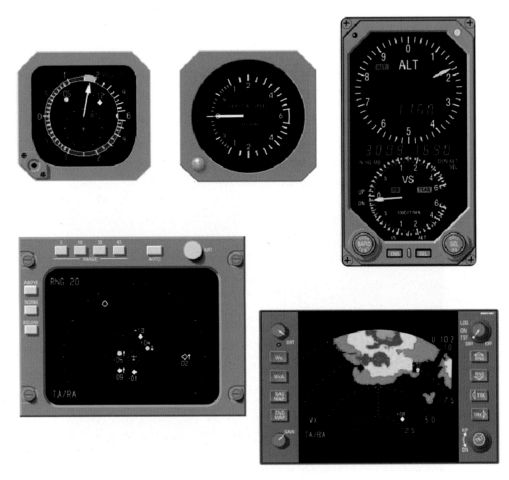

**Figure 9.15** Typical TCAS displays

**Figure 5.10**  Typical ALS layout (note that the green light line indicates the runway threshold) (*See insert for color representation of the figure.*)

CAT II and III precision-instrument approaches, high-intensity sequenced flashing lights (ALSF-1), and three medium-intensity approach lighting arrays (MALSR, MALS, and MALSF). From the 2012 *Aeronautical Information Manual* (AIM) "ALS are a configuration of signal lights starting at the landing threshold and extending into the approach area at a distance of 2400–3000 ft for precision instrument runways and 1400–1500 ft for nonprecision instrument runways. Some systems include sequenced flashing lights, which appear to the pilot as a ball of light traveling towards the runway at high speed (twice a second)." Figure 5.10 shows a representative ALS layout.

## 5.4.1  The Visual Approach Slope Indicator

After the pilot has visually located the ALS system and has aligned the aircraft toward the runway, he/she can then use visual glide slope indicators for the transition to touch down and establish that the aircraft is approaching at the proper glide slope angle. There are two

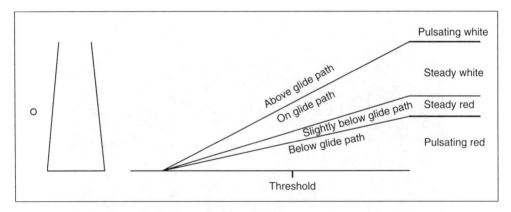

**Figure 5.11**  PVASI approach glide path guidance

typical visual presentations: the VASI and PAPI. The VASI is an optical reference device located on the ground at the sides of the runway. There are a variety of VASI designs dependent upon the desired visual range and the type of aircraft utilizing the runway. Each unit consists of a "bar" of one, two, or three light units, referred to as *boxes*. To the pilot, each box takes on a certain color depending upon the position of the aircraft. If the aircraft is too low on the glide slope, all the bars will appear red; if it is too high, all of them will appear white; and if the aircraft is on the correct path, half the bars will appear red and the other half will appear white. The most common systems found in the United States include VASI-2, VASI-4, VASI-12, and VASI-16. The number following the acronym indicates the number of boxes in the system. The AIM specifies that "Two-bar VASI installations provide one visual GP, which is normally set at 3°."

Three-bar VASI installations provide two visual GPs. The lower GP is indicated by the near and middle bars and is normally set at 3°, while the upper GP, defined by the middle and upper bars, is normally 1/4° higher. This higher GP is intended for use only by high cockpit aircraft to provide a sufficient threshold crossing height. Although normal GP angles are 3°, angles at some locations may be as high as 4.5° to give proper obstacle clearance. VASI lights are typically visible up to 20 miles at night and 3–5 miles during daylight hours. Typically at 4 NM from the airport runway threshold, descent using the VASI will provide safe obstruction clearance within ±10° of the extended runway centerline. There are also variants of the typical VASI including the tri-color VASI and the Pulsating VASI (PVASI). Both of these systems have different visual presentations from the standard VASI system, and Figure 5.11 shows a typical PVASI layout where the light is generated by a single light source indicated by the circle next to the runway representation.

### 5.4.2   Precision Approach Path Indicator

The PAPI is a visual approach slope aid also approved for use in the United States and other countries. This system gives a more precise indication to the pilot of the approach path of the aircraft and utilizes only one bar. The system consists of four lights on either side of the approach runway that change between white and red to provide visual indications of

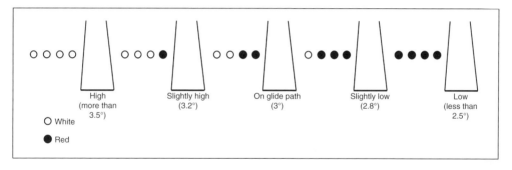

**Figure 5.12**   Precision approach path indicator

the aircraft position. Typically, these are installed on the left side of the runway and provide glide slope indications as shown in Figure 5.12.

## 5.4.3   The Final Approach Runway Occupancy Signal

There have been numerous accidents resulting from aircraft landing while the runway is blocked by another aircraft or vehicle. Indeed, the worst loss of life in an aircraft accident occurred when an aircraft on takeoff ran into an aircraft blocking the runway. With current technology such as the Runway Status Light (RWSL) systems, this accident might have been avoided. These RWSL systems provide visual cues to the crew about when to proceed on taxiways, enter a runway for takeoff, and advise if there is traffic on the runway during approach. An example of this is the Final Approach Runway Occupancy Signal (FAROS), which is part of an RWSL system, is independent of ATC control, and works automatically based on input from airport runway and taxiway surveillance systems. The FAROS system flashes the PAPI lights when an object is occupying protected runway space. The PAPI will stop flashing when the object moves off the runway-protected area. In order for this system to work, the transponder must be in operation while the aircraft is at the airport, as noted in the AIM: "When operating at airports with RWSL, pilots will operate with the transponder 'On' when departing the gate or parking area until it is shutdown upon arrival at the gate or parking area. This ensures interaction with the FAA surveillance systems such as ASDE-X, which provides information to the RWSL." There are also numerous other airport lighting systems to advise pilots as they transition around the airport. Navigation on airport surfaces is covered in AIM, Section 10.

## References

Airbus (October 2001), CAT II/CAT III Operations, Report STL472.3494/95.

Battlett J L *Microwave and RF Avionics Applications*, CRC Press, Boca Raton, FL, 2008.

FAA (May 2002), Order 8260.3B Change 19, United States Standard For Terminal Instrument Procedures (Terps), FAA, Washington, DC.

FAA (August 2013), Aeronautical Information Manual, Change 3, FAA, Washington, DC.

Helfrick A D (2007), *Principles of Avionics*, 4th ed., Avionics Communications, Leesburg, VA.

Honeywell (2001), *DFS Embraces New Precision Approach Technology*, Honeywell.

International Civil Aviation Organization (ICAO) (June 2000) Report 8071, Vols. 1 and 2. *Manual on Testing of Radio Navigation Aids, Volume I: Testing of Ground-Based Radio Navigation Systems*, 4th ed; reprinted June 2000; Amendment 1: 10/31/02. *Manual on Testing of Radio Navigation Aids, Volume II: Testing of Satellite-Based Radio Navigation Systems*, 5th ed, ICAO.

McShea R E (2010), *Test and Evaluation of Aircraft Avionics and Weapon Systems*, AIAA, Reston, VA.

Moir I and Seabridge A (2003), *Civil Avionics Systems*, Professional Engineering Publishing, London.

Moir I and Seabridge A (2006) *Military Avionics Systems*, John Wiley & Sons, Ltd, Chichester.

Thales (2010), ILS 420 System, Technical Manual, Parts 1, 2, and 3. Thales, Neuilly-sur-Seine.

# 6

# Correlated-Extremal Systems and Sensors

Evgeny A. Konovalov[1,†] and Sergey P. Faleev[2]

[1] *Late of the Russian Tsiolkovsky Academy of Cosmonautics, Moscow, Russia*
[2] *State University of Aerospace Instrumentation, Saint Petersburg, Russia*
[†] Deceased

## 6.1 Construction Principles

Aerospace *Correlated-Extremal Systems* (CES) participate in high-precision positioning, speed control, orientation control and in the process of aerospace vehicle arrival (or set point) guidance. They excel in computational speed and in the calculation accuracy of the correlation function (Dickey, 1958; Beloglazov, and Tarasenko, 1974; Baklitskiy and Yuryev, 1982; Iwaki *et al.*, 1989; Collier, 1990; Tanaka, 1990; Degawa, 1992; Scott and Drane, 1994; Watanabe *et al.*, 1994; Bernstein and Kornhauser, 1996; Jo *et al.*, 1996; Kim *et al.*, 1996; Lakakis, 2000; Savvaidis *et al.*, 2000; White *et al.*, 2000).

The CES (Figure 6.1) operates as exemplified by the two images shown in Figures 6.2 and 6.3, obtained using appropriate sensors. Determination of the *Principal Peak-to-Peak Position* (PPP) maxima of the cross-correlation or cross-covariance function or CCF (Figures 6.4 and 6.7) for these images is the main purpose of the *Correlated-Extremal Sensing Device* (CESD). Using that positional data, an aerospace vehicle can be automatically controlled such that the PPP coordinate module output tends to zero as will be explained later.

In Figure 6.1, $t$ is time as usual, and the other symbology is as follows:

$g_{(t)}$ is the primary (or current) image derived from Imaging sensor 1.
$y_{(t)}$ is the secondary (or master) image derived from Imaging sensor 2.
$y_s$ is a second image area selected for orientation.
$g_k$ and $y_k$ are images processed for cross-covariance extension distortion minimization.

---

*Aerospace Navigation Systems*, First Edition. Edited by Alexander V. Nebylov and Joseph Watson.
© 2016 John Wiley & Sons, Ltd. Published 2016 by John Wiley & Sons, Ltd.

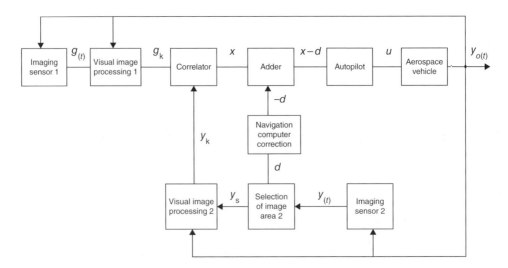

**Figure 6.1**    Cross-correlation flight control system (CFCS)

**Figure 6.2**    Satellite view of a city park

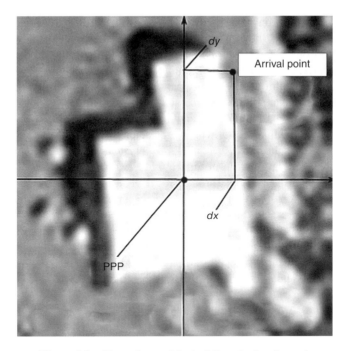

**Figure 6.3**    View of one of the buildings in the city park

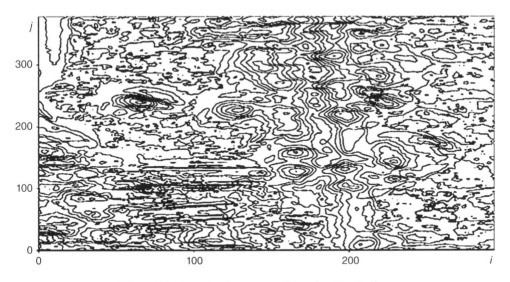

**Figure 6.4**    Family of cross-correlation function isolines

$d$ is the orientation or aiming mark deflection.
$x$ is the maximum PPP data – the orientation, location, or aiming error,
$u$ is the command signal to the aerospace vehicle.
$y_{o(t)}$ are the output parameters of the vehicle motion (location, orientation, speed).

Image sensors transform physical fields into electrical signals and can cause aperture and nonlinear distortions. Hence, the PPP error is increased because visual image processing exacerbates these distortions. The adder following the correlator adds an orientation (or aiming) mark deflection (derived from the output parameters of the vehicle motion) to the PPP data.

Figure 6.2 shows a satellite view of a city park along with one of its buildings (Figure 6.3) showing the vehicle's arrival point. The *Cross-Correlation Function* (CCF) of these images is given as a family of isolines in Figure 6.4, and in 3D form in Figure 6.5. CCF irregularity and the presence of many maxima are shown. The principal (or largest) maximum is called the *global maximum*, whereas the others are called *local maxima*. The region containing the global maximum is shown on a large scale in Figure 6.5b. Two orthogonal sections of the cross-correlation function (CCF) passing through the PPP are shown in Figures 6.6 and 6.7. Figure 6.6 is an orthogonal section shown along the ordinate axis $y$ (see Figure 6.4); and Figure 6.7 along the $x$-axis.

In Figure 6.6, the cross section at $i=67$ (continuous line) passes through a maximum maximorum PPP of the CCF, and the conditionality of the maximum is very high. Cross sections that are separated by only three pixels (left on the dashed line and right on the dotted line) are near the zero argument and are significantly lower than their highs and the PPP in this case is already quite vague.

Figure 6.7 shows that the cross section at $j=102$ (full line) passes through a maximum maximorum PPP of the CCF and the conditionality maximum is not as high. Cross sections that are separated by only three pixels above and below (shown by the dashed line and the dotted line) are significantly lower than the PPP.

It is also evident from Figures 6.6 and 6.7 that the sections along the ordinate axis $y$ have multiple local maxima, whereas a section along the abscissa $x$ has only one maximum.

### 6.1.1   General Information

The fact that the correlation function reaches the PPP at the zero argument underlies the operation of any CES.

Self-correlated orientation region characteristics have a major influence on the functioning of the CES. Sharp peaks in the PPP region greatly exceeding the PPPs of other local maxima in the correlation function (Figure 6.5) conform to the normal mode of CES operation. However, the flat top of the PPP correlation function of Figure 6.8 shows poor localization of the PPP and the commensurate low CES accuracy. The presence of a correlation function's local maxima close to the PPP (the so-called *ravine surface functional*) leads to large anomalous errors in the CES such as the capturing of an arbitrary local maximum.

In addition to the self-correlation orientation properties of the data sensor, the CES functioning is greatly influenced by the methods of operand accessing for the correlation calculations, the attendant noise and the nature of the CESD construction.

Both primary and secondary physical fields are required for CES functioning. The primary orientation field exists independently, but the secondary field needs to be defined by the CES user. For example, starry sky fields are used in star-tracking guidance and geophysical fields are used for guidance in near-Earth space.

Information sensors produce operands for CES correlation calculations by converting fields to two-dimensional images (pictures), or to one-dimensional (line segments), or to single-value (point) form. The cross-correlation properties of the area of orientation in all three variants will still appear.

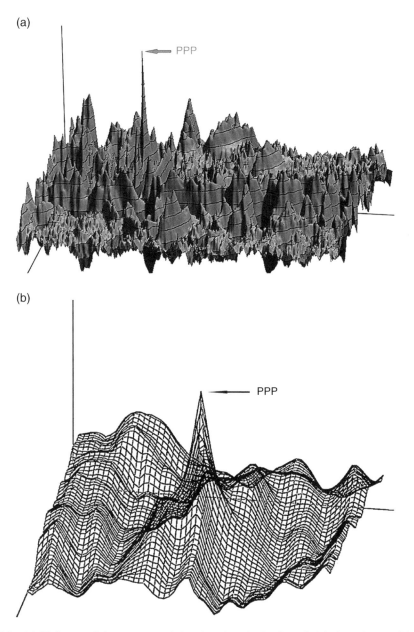

**Figure 6.5**   (a) 3D image of the cross-correlation function. (b) Region of global maximum of the cross-correlation function on a large scale

Control via two coordinates is easily managed using two-dimensional images, but any decrease in correlation calculation complexity afforded by the two lesser options leads to concomitant decreases in the vehicle control capability.

CES information sensors are characterized by operand access time and quality (picture dimensions), self-descriptiveness of the physical field's conversion (see Section 6.2) and

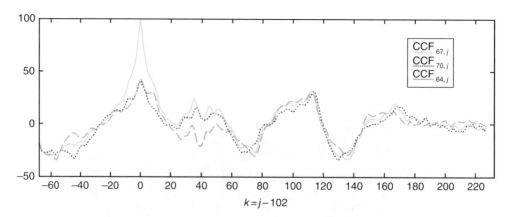

**Figure 6.6**   Orthogonal section of the cross-correlation function

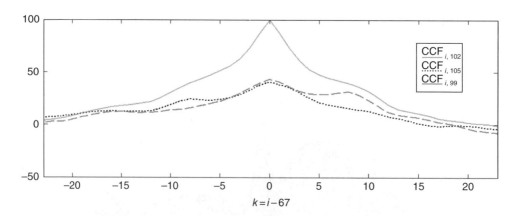

**Figure 6.7**   A second orthogonal section of the cross-correlation function

aperture distortion. Angle distortion, scale difference and mutual operand rotation problems appear if the sensors for the current and master pictures observe different points in space. Changes in the physical fields can manifest themselves if the sensors work at different times. Also, image gain compression effects can appear under various conditions. Finally, sensor output is always accompanied by noise that can adversely influence the cross-correlation characteristics of the images.

Some CES units extract both images simultaneously, whilst others prepare master images in advance. For example, a determination of vehicle speed over clouds can be made according to active images of cloud cover even though in this case master images are relevant only for a split second. During flight over visible territory, however, master images of this territory acquired in advance may be used. Lastly, astro-orientation is based on starry sky images because these do not change over very long periods.

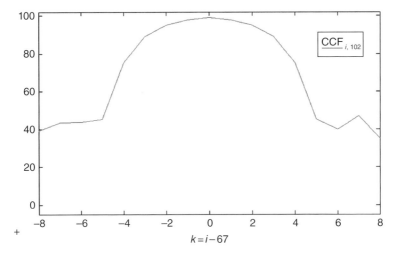

**Figure 6.8**  An ill-conditioned cross-correlation function global maximum

**Table 6.1**  Image correction for cross correlation improvement

| Image procedure | CES type | | |
|---|---|---|---|
| | Astro-orientation | Map matching | Guidance |
| Scaling | No | Seldom | Always |
| Rotation | Possibly | Seldom | Possibly |
| Projective images compensation | No | Very seldom | Possibly |
| Aperture distortion compensation | Very seldom | Very seldom | Sometimes |

CES technology solves several problems in aerospace vehicles control including the following:

- Astro-orientation.
- Position fixing, speed determination and orientation of motion along a surface using various geophysical fields (such systems are termed *map matching* or *terrain reference navigation CES*).
- Vehicle direction towards an arrival point (*guidance CES*).

The last problem is the most difficult one because image processing composed of several operations concerned with each other is required (Table 6.1). Correlation methods for vehicle direction towards an arrival point were originally proposed by Konovalov *et al.* (1960).

Many systems for angle, distance and speed measurement that work with one-dimensional signals also determine the parameters of the CCF maxima, but this is not usually obvious.

The angular locations of the 'fixed' stars are almost permanent and without parallax, so only the initial orientation and the subsequent rotation of the master image during astro-orientation are needed. However, the master image's scaling during a flight route involving different directions and various altitudes must also be known for correct CES operation during guidance over the field. This is clarified in Figure 6.1, where it is understood that before a flight the astronavigation system should be set to a particular part of the sky. The reference site for orientation is typically a hemisphere or even less of the starry arch. Under these circumstances, for reference site orientation, only a turn is required.

When a vehicle approaches its arrival point, the current picture scale changes vastly, and this always requires concomitant changes in the master image scale, too. Even rotation of the master image when flying to the arrival point from different directions may be necessary. Master images made in advance are taken from various angles of arrival, and compensation for etalon projection distortion may also be required during flight up to the arrival point on a real trajectory. Sometimes, during combined guidance (Section 6.4), sensors with poor apertures may be used, in which case compensation for aperture distortion must be employed.

Historically, radio engineering and radio-locating CES with regular signals were designed and constructed. Two-dimensional optical, television and radio-locating CES using star tracking, and guidance using random physical field implementation, subsequently appeared. Currently, quasi-three-dimensional (stereoscopic computer vision) and three-dimensional (holographic and drone acoustical coordinator) CES are being developed.

All CES can be divided into systems (i) that have master images in memory and (ii) that receive both images for correlation processing simultaneously. Some CES employ two or more CESD and integrate them with sensors of different types as required for solving problems or according to the environment involved.

However, all CES have similar mathematical foundations for their functioning, and the efficient production of the various digital elements and computer systems has taken on special significance for CES realization. Furthermore, hybrid CES now combine in their structure digital computers along with elements of different types: optical, acoustical and coherent-optical (holographic) processors, also needing efficient production methods because they can't be replaced – analog elements are less stable and need parametric stabilization, or work in narrow environmental ranges.

### 6.1.2  Mathematical Foundation

The first CES were developed for the speed measurement of vehicles, and their basic functioning can be shown by a simple example, as follows.

Assume that two radars (or two video cameras) are located on board an aircraft, and the antennas (or objectives) of both are directed orthogonally to the surface of the earth. The first radar has an output signal $X_1(t)$ as defined by the underlying surface. The second radar is shifted with respect to the first in the direction of the aircraft velocity vector $V$ on a base $d$. Hence the output signal of the second radar will be $X_2(t) = X_{1(t-\tau)}$ where the delay $\tau = d/V$.

The signals $X_1(t)$ and $X_2(t)$ are applied to the correlator and the correlation function is calculated with variable values of the shift $\lambda$:

$$R(\lambda) = \frac{1}{T}\int_0^T X_1(t-\lambda)X_2(t)dt. \qquad (6.1)$$

When the maximum of this expression is found (at $\lambda = \tau$), it is easy to determine the aircraft velocity $V = d/\lambda$.

For finding the extremum of Function (6.1), the dimensions of the signals $X_1(t)$ and $X_2(t)$ do not matter. Historically, the first CES designed to measure the speed of moving vehicles were radio-based and one-dimensional. Today, both a flat image (two coordinates), and a three-dimensional image area (three coordinates) can be employed. In the most modern development of the particular mathematics, the extremelization of the function $R(\tau)$ can be conducted not only by the selection of $\lambda$ (the unknown $\tau$), but also by a direct computation of the coordinates of the global maximum.

The interval $T$ is limited by tactical considerations, it usually being desirable to know the speed quickly but with acceptable accuracy.

### 6.1.3   Basic CES Elements and Units

CES contain sensors for two images. The sensor for the first, or current, image (see Figure 6.1) works in the real-time mode, and the image formation time is limited, on the one hand, by the sensor's operating speed and, on the other, by the vehicle motion parameter measurement and subsequent motion correction.

The second, or master, image sensor can also work in the real-time mode in some quite ordinary systems. In the majority, the master images are obtained in advance and stored in memory devices. Image processing devices may implement all the necessary image transformations for cross-correlation and PPP increase, but may be omitted in some CES.

The CES image sensor is a non-trivial computing unit. Usually, two methods are used for the determination of the PPP of the CCF of the current and master images. The first method uses correlated discriminator and optimized smoothing chains, and the second method is based on advancing to a PPP position via some sequential decisions.

A Controller implements the vehicle's control law and may also be considered as a computing unit. From the viewpoint of the CES dynamics, the greatest time constant for the entire system is defined by the aerospace vehicle itself, and the relevant model involves integration and various other dynamic procedures depending on the particular type of vehicle. The Controller usually involves an integro-differential unit, and the CES sensor dynamics are nonlinear and can be approximated by an aperiodic unit.

### 6.1.4   Analog and Digital Implementation Methods

There are numerous methods for calculating the CCF such as temporal and spectral. Image distortions (turning, displacement, scale, angle and aperture distortions, and nonlinearity) have considerable influences on the calculated result. Also, some prior information about the PPP location is required for a successful PPP search, otherwise an extensive enumeration of possibilities is needed.

It is possible to simplify a computational procedure by the method of sign correlation to arrive at an image derived from the original where scale and turning can be less harmful.

Finally, in addition to the computer system inside the correlated-extremal device that calculates, or seeks for, the correlation maximum, the CES can include discrete, optical, supersonic and various other additions including hybrid types.

The CES has passed through a semi-centennial evolution. The first analog optical correlators for measuring the correlation function of different television plots presented on photographic films appeared in 1952 (Kretzmer, 1952). Here, the correlator functioning mode was that parallel beams of light passed through combined masks with images $F_1(x,y)$ and $F_2(x,y)$ in the form of different television pictures and focused on the sensitive layer of a photoreceiver. This produced an output current conforming to the expression:

$$R(\xi,\eta) = C \int_s F_1(x,y) F_2(x-\xi,y-\eta) dx dy \qquad (6.2)$$

where $C$ is a constant proportionate to the luminous intensity; $\xi$ and $\eta$ are the magnitudes of the deflection displacements of the masks with respect to each other along the appropriate axes.

As a consequence of the relative displacements of the images on both $X$ and $Y$ coordinates, a spatial correlation function is formed by the photoreceiver's output showing the statistical semantics of the initial television image plots. An extreme value indicates the exact conjunction of these images on the $X$ and $Y$ coordinates.

The maximum magnitude of the correlation function can be determined by the hill-climbing method or by classical correlated discriminator methods.

As mentioned earlier, for the automated seeking of the correlation function extremum and its tracking, a function corresponding to the correlation function's derivative $R'(\tau)$ is required at the correlated discriminator's output.

The value of $R'(\tau)$ corresponds to the following expression:

$$R'(\tau) = \frac{1}{T} \int_0^T X_1(t) X_2'(t \pm \tau) dt \qquad (6.3)$$

So, to acquire this correlation function derivative $R'(\tau)$, it is necessary to differentiate $X_2(t)$ with all the undesirable consequences of increasing high-frequency components and noise levels.

The process of capturing and tracking the correlation function derivative zero is determined by the discriminating characteristic parameters: the linear range dimension and its steepness in the zero region.

The parameters of the correlated measuring index of the relative displacement of the images are formed in compliance with the image's measurement accuracy, noise intensity and motion dynamics. The steepness of the discriminating characteristic is large if there is no noise. However, uncertainty induced by noise results in a typical display of two 'splashes' with different polarities and with a linear middle part in the region of sign reversal. Any increase in the noise level leads to a decrease in the middle part steepness and 'splash' amplitude.

Nowadays, progress in digital computing systems has been so considerable that CES realization difficulties have given way to the difficulties inherent in the understanding of efficient computing algorithm organization. The automated search for the PPP region, with the aim of capturing the PPP for tracking, has been inconvenient until recently, but the best algorithms for correlated discriminator functioning have yet to be found. The question of the relationship between CES accuracy and the necessary correlation distance of the CCF has not yet been completely answered, though this CES accuracy is close to the correlation distance of the CCF as seen in Figures 6.6 and 6.7. The correlation distances, proportionate to the CCF width in the principal maximum region, can vary by an order of magnitude between different axes of one image.

Progress in CES sensors has lead to considerable improvements in image resolution and to a corresponding rise in CES potential accuracy. However, in practice it is difficult to realize this rise because of high CCF irregularity that increases the probability of abnormal errors and a spreading of the noise component in the image spectrum.

## 6.2   Image Sensors for CES

CES include sensors that transform different physical fields to electrical signals that a CESD computing unit is able to process. Hence, CES can be classified according to the various sensor types including radio-location technology, optical (television) techniques, thermal imaging, as well as laser, acoustic, magnetometric, multispectral and gravitational methods.

The analysis of a sensor's influence on CES quality depends on the following major sensor parameters:

- The resolution, or number of independent image elements that are provided by the sensor
- The persistence, or speed with which images can change
- The speed with which the sensor can transduce a physical field to a signal that can be processed by the CES
- The inevitable aperture distortion that leads to the loss of some spectral components of the useful signal
- And finally, one of the most undesirable of all – nonlinear distortion in the sensor.

The number of image elements $N$ determines the comparative level of quantization (sampling) noise. The effect of $N$ on the CES quality is shown in Figure 6.9.

The radar sensor is usually the slowest one, but its persistence restricts the potentialities of only some CES, for example at the terminal phase of guidance at the arrival point. The rest of the sensors have considerably better operating speed reserves and do not influence the CES work dynamics.

The signal-to-noise ratio in the image determines the connection between a sensor's output signal and an initial field, and the current image of the field obtained by the sensor

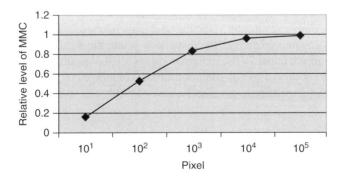

**Figure 6.9**   The effect of pixel numbers on the Maximum Maximorum of Co-variation Function (MMC) level (the average for a large number of images)

is considered here. The main questions concerning a correlated-extremal sensor's informativity quantitative estimation are as follows.

Current and master images are formed from different artificial (secondary) and natural (primary) physical fields. The master image's position is related to an object's motion along a required trajectory in a selected coordinate system. When any deviation from the defined trajectory is not large, and the current and master image displacement does not exceed the radius of their spatial CCF, then preset methods of extremal orientation, navigation and guidance are used. The formation of current images is realized by various sensors that transform different physical fields to electrical (or other) signals that can be handled by a correlated-extremal device's computing unit.

The selection of a CES sensor depends on its signal informativity and this most useful receiving and transmitting information becomes the main property of the sensing element. Useful information from various CES sensors can differ drastically, so a quantitative assessment of their qualitative variability is required.

Consider a model for measuring the value of a pixel and let the image be a two-dimensional Gaussian field. The measured value of a pixel for linear sensors will be simply the amplitude; for quadratic sensors it will be the square of the amplitude; and for essentially nonlinear sensors, the fourth power of the amplitude.

Inaccuracy in measuring the value of a pixel using this model can be reflected in two ways. The first way is traditional and simple – it considers the dispersion of the measured values of the pixel. The second way involves a consideration of the entropy of the distribution of the pixel's measured values. This allows the showing of, first, an informational measure of the measurement noise as determined by the selected evaluation algorithm; and second, the influence of the quantization of the noise level of the pixel values during digital processing. The larger the quantum level (or quantization step size) of the pixel values, the lower are the requirements for hardware implementation of the estimation algorithm – but the worse the informational content of the pixel.

Table 6.2 gives different field transformation algorithms for the sensors $x_i$, $i = 1, 2, ..., 9$; the output signal probability density for the Gaussian field $v_1(x_i)$ (this being the most informative one); their midvalues $m_1$; and dispersions $M_2$ and residual fluctuations $\gamma$. It is obvious that the sensors handling both field components $u$ and $v$ are much the better.

In the table, $D$ is the field dispersion at a sensor input, and $D_{2.3}$ is the appropriate estimated variance (Faleev, 1980). The univariate probability densities of the estimations generated by the corresponding algorithms $x_i$ (for $i = 1, 2, ..., 9$) are placed in the second column, and the next two columns show expectation values $m_1$ and dispersions $M_2$. The normalized variances for the root-mean-square variance $\gamma_1$, the dispersion $\gamma_2$ and the squared dispersion $\gamma_3$ are presented in the columns 5, 6 and 7. The indicated values are the results of $N$-independent values $x_i$ averaged, and are followed by squaring or fourth involution as required.

The value of the normalized variance for any algorithm is expressed by the following formula:

$$\gamma_j = \frac{M^2}{m_1^2}, \quad \gamma_{j+1} = 4\gamma_j, \quad \gamma_{j+2} = 14\gamma_j, \quad j = 1,2,3.$$

One of the most natural performance measures for a sensing element in a complex system is the amount of information acquired. With regard to the estimation of non-random

**Table 6.2** Characteristics of image pixel estimation algorithms

| $x_i$ | $v_1(x_i)$ | $m_1$ | $M_2$ | $\gamma_1$ | $\gamma_2$ | $\gamma_3$ |
|---|---|---|---|---|---|---|
| $x_1 = \|u\|$ | $\dfrac{D}{\sqrt{2\pi D}}*e^{-\left(x_1^2/2D\right)}$ | $\sqrt{2D/\pi}$ | $(\pi-2)D/\pi$ | 0.571 | 2.283 | 7.991 |
| $x_2 = u^2$ | $\dfrac{1}{\sqrt{2\pi Dx_2}}*e^{-\left(x_2/2D\right)}$ | $D$ | $2D$ | — | 2 | 8 |
| $x_3 = u^4$ | $\dfrac{1}{2\sqrt{2\pi Dx_3^{3/4}}}*e^{-\left(\sqrt{x_3}/2D\right)}$ | $3D^2$ | $96D^4$ | — | — | 10.667 |
| $x_4 = \|u\|+\|v\|$ | $\dfrac{4}{\sqrt{\pi D}}*\Phi_0\left(\dfrac{x_4}{\sqrt{2D}}\right)*e^{-\left(x_4^2/4D\right)}$ | $2\sqrt{2D/\pi}$ | $2(\pi-2)D/\pi$ | 0.285 | 1.142 | 3.996 |
| $x_5 = \sqrt{u^2+v^2}$ | $\dfrac{x_5}{D}*e^{-\left(x_5^2/2D\right)}$ | $\sqrt{\pi D/2}$ | $(4-\pi)D/2$ | 0.273 | 1.093 | 3.825 |
| $x_6 = u^2+v^2$ | $\dfrac{1}{2D}*e^{-\left(x_6/2D\right)}$ | $2D$ | $4D^2$ | — | 1 | 4 |
| $x_7 = (u^2+v^2)^2$ | $\dfrac{1}{4D\sqrt{x_7}}*e^{-\left(\sqrt{x_7}/2D\right)}$ | $8D^2$ | $320D^4$ | — | — | 5 |
| $x_8 = D_{2,3}$ at $\dfrac{\varpi^*}{\varpi_*}\to 0$ | $\dfrac{x_8}{4D^2}*e^{-\left(x_8/2D\right)}$ | $4D$ | $D^2/2$ | — | 0.5 | 2 |
| $x_9 = \sqrt{x_8}$ | $\dfrac{x_9^3}{2D^2}*e^{-\left(x_9^2/2D\right)}$ | $\dfrac{3}{4}\sqrt{2\pi D}$ | $(32-9\pi)D/8$ | 0.132 | 0.527 | 1.845 |

parameters of the probability density of a random process, the residual uncertainty of estimation can be used:

$$H_0(y_i) = -\int_B v_1(y_i)\log v_1(y_i)\,dy_i, \quad i=1,2,\ldots,9.$$

Here, $y_i = x/m_1(x_i)$ is normalized to their own average estimation generated by the $i$-th algorithm of Table 6.2, $H_0$ is the entropy of the normalized estimation and $B$ is the domain of the values $x_i$.

The entropy of a probability density for the algorithms shown in Table 6.2 are calculated by the formula

$$H(x_i) = -\int_B v_1(x_i)\log\left[x_{i0}v_1(x_i)\right]dx_i$$

and is shown in Table 6.3.

Here, $x_{i0}$ is the least distinguishable quantum of an $x_i$ variable, proportionate to the root-mean-square value of the noise; $H_0$ is the residual entropy of the normalized values, and the numerical value of $H_0$ is given in bits (for logarithm base 2).

Numerical values for the entropy $H_0$ decrease from 1.37 bit for the $x_1$ algorithm to 0.56 bit for the $x_9$ algorithm that corresponds to a evaluation parameter's residual uncertainty decrease of more than two. Comparing the numerical values of the entropy $H_0$ in the first four

**Table 6.3**  Informativity of pixel estimation algorithms

| $x$ | Entropy $H(x)$ | $H_0$ | $H_0$, bit |
|---|---|---|---|
| $x_1$ | $0.5\log\left(\dfrac{\pi e D}{2x_{10}^2}\right)$ | $0.5\log\left(\dfrac{\pi^2 e}{4}\right)$ | 1.373 |
| $x_4$ | $1.727(7)+0.5\log\left(\dfrac{D}{x_{40}^2}\right)$ | $1.727(7)+0.5\log\left(\dfrac{\pi}{8}\right)$ | 1.053 |
| $x_5$ | $0.5\log\left[\left(\dfrac{D}{2x_{50}^2}\right)*e^{(c+2)}\right]$ | $0.5\log\left[e^{(c+2)}/\pi\right]$ | 1.033 |
| $x_9$ | $0.5\log\left[\left(\dfrac{D}{2x_{90}^2}\right)*e^{(3c+1)}\right]$ | $0.5\log\left[\left(\dfrac{9}{4\pi}\right)e^{(3c+1)}\right]$ | 0.560 |
| $x_6$ | $\log\left(\dfrac{2eD}{x_{60}}\right)$ | $\log e$ | 1.443 |
| $x_2$ | $0.5\log\left[\left(\dfrac{\sqrt{\pi}D}{2x_{20}}\right)*e^{(1-c)/2}\right]$ | $0.5\log\left[\sqrt{\pi}D*e^{(1-c)/2}\right]$ | 1.131 |
| $x_8$ | $\log\left[\left(\dfrac{2D}{x_{80}}\right)*e^{(c+1)}\right]$ | $\log\left[0.5e^{(1+c)}\right]$ | 1.275 |
| $x_3$ | $0.5\log\left[\left(\dfrac{\sqrt{\pi}D^2}{x_{30}}\right)*e^{(1-3c)/2}\right]$ | $0.5\log\left[\left(\dfrac{\sqrt{\pi}}{3}\right)*e^{(1-3c)/2}\right]$ | 1.287 |
| $x_7$ | $\log\left[\left(\dfrac{8D^2}{x_{70}}\right)*e^{(1-c)}\right]$ | $\log\left[e^{(1-c)}\right]$ | 0.610 |

Note: Euler constant $c=0.577$.

algorithms of Table 6.3 with the respective values of normalized variance from Table 6.2 (column 5) it can be seen that the order of the algorithm's preference is the same: $x_9$, $x_5$, $x_4$, $x_1$, which were exhibited in column 1 of Table 6.2. Accordingly, the algorithm of pixel value estimation $x_9$ is better than $x_5$, algorithm $x_5$ is better than $x_4$, etc. But algorithm $x_9$ requires the maximum number of calculation procedures, whereas algorithm $x_1$ requires the minimum number of calculation procedures.

For these algorithms the numerical value of the entropy $H_0$ is smaller if the normalized variance $\gamma_1$ is smaller. In other words, the closer the normalized estimate probability density is to a delta-function, the smaller is the entropy $H_0$, so the smaller normalized variance and the respective algorithm of pixel estimation is better.

## 6.3  Aviation and Space CES

CES have been developed for both the precise measurement of earth satellite track angles, and the orientation of aircraft by heavenly bodies. CES can also participate in measurements via geophysical fields, the horizontal and vertical velocity components of volumetric fields (3D relief maps) and the traversed path and location of an aerospace vehicle.

## 6.3.1  Astro-Orientation CES

The idea of orientation by the Sun, the Moon, the planets and the stars comes from antiquity. Of all the subsequently developed mathematical and astro-orientation devices, CES are the most perfect angular optical-electronic sensors for the precise realization of manoeuvres when allowable errors of astro-orientation are measured in angular minutes or seconds.

It is necessary to have precise information about an aerospace object's orientation in order to decide on the choice of heavenly bodies, and for the preparation of a proper master image of a coelosphere region for correlation and guidance within that region's optical axis with an exactness that will assure star capture for valid CES tracking. Stars in the coelosphere are presented as bright dots, so measurements are taken for the expansion of a cross correlation function of the master image and the determination of a suitable coelosphere region at the initial tracking stage. These can be acquired (for example) by temporarily defocusing the master image.

Aerospace objects are subject to compound motions, and any of several coordinate frames for solving the relevant problems may be used, including earth, orbital, sun, worldwide and static star coordinate frames. The consequence of several coordinate frames being used is that many calculations for interconversion may be needed. That is why any resultant orientation error is influenced by the physical astro-orientation environment of the CES and the calculation accuracy.

CES are installed on high accuracy stabilized angular data platforms; or strapdown computational approaches are used for decreasing the orientation-error dynamic constituent. Under such conditions, high astro-orientation accuracy of the CES is achieved because the radii of cross correlation of the standard and real region of the sky is usually very small.

Aviation astronomical year-books publish special tables that can be used for choosing astro-orientation regions and the initial setting of astrotracker viewing, or *aspect angles*, of flying objects on the basis of information about the exact time and their latitude and longitude.

CES is particularly suited to astro-orientation because its accuracy does not depend on range ability and flight duration. However, it is a complex technique and its failures do have adverse effects on aerospace flight programs.

## 6.3.2  Navigational CES

The prime purpose of navigational CES in aerospace vehicle control systems changes from the correction of accumulated errors arising in self-contained navigation systems, to fundamental navigation. For example, this occurs when a transition to low-altitude high-accuracy flight relevant to surface features is required.

Navigational CES can also solve another important problem concerning *Terrain Reference Navigation* (TRN), or map matching.

Etalon fields and relief profiles prepared prior to a vehicle's departure may lose relevance over time. Modern etalon actualization presents a problem because it usually relates to large-scale maps where changes can occur, including those relevant to human activity, so a broad geophysical field spectrum must be used for near-earth navigation. This means that a vehicle's trajectory scale, start time, the relevant flight region (sea, plain, foothills, mountains, etc.) and accuracy requirements will be defined according to the selection of a suitable field depending on its seasonal and daily stability (e.g. due to aquifer changeability).

Increased informativity, the possibility of fast image actualization, thermal imaging and other geophysical fields may improve CES navigational accuracy. However, the great volume of correlation calculation necessary for map comparison navigation complicates the application of qualitative image processing algorithms. Decreasing the picture size disrupts operation of the CES; and increasing the picture size adversely affects its accuracy. Passive methods of current image extraction predetermine CES operational stealth that is useful in some applications.

Actually, the idea of map compare navigation using radio-location fields was patented in 1944 (US Patent 2,508,562), but it is modern technology that has made the method viable.

### 6.3.3  Aviation Guidance via Television Imaging

One of the most promising high-accuracy guidance methods to a selected arrival point involves CES guidance via television imaging. This tracking procedure commences by capturing a master image of the terminal region and including an arrival point indication.

When navigating solely by television, thermal imaging or radio-location, arrival point information can be missed, and contrast between that arrival point and its surroundings can be low. However, the incorporation of CES technology makes information about any displacement of the arrival point and any contrasting object considerably more accurate (Figure 6.3). Arrival point tracking by a correlation sighting device is then realized using normal master and current image processing, taking into account that accurately known arrival point displacement.

Consider a simple example. Suppose a signal $S(t)$ representing a master image $A$ from a sighting device output is applied to a memory device (with multiple reading capability) at the start of CES operation when the *Sighting Device Axis* (SDA) coincides with the *Direction to the Arrival Point* (DAP). This may therefore mean that any relative image motion is missed (Figure 6.10). The moment at which the master image is recorded is when the sighting device operates at a maximum range limited by the intrinsic noise, which means that $S(t)$ is memorized in the presence of additive noise $n(t)$, which corrupts the signal as it is read into the memory.

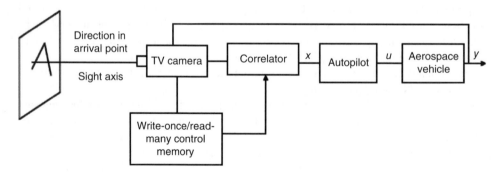

**Figure 6.10**   Simple Cross-Correlation Flight Control System (CFCS) with TV imaging

Under the influence of various random factors such as the vehicle's aerodynamic skewness, any atmospheric heterogeneity, etc., the SDA will deviate from the initial DAP, resulting in the current image $A$ deviating from the memorized master image $A'$ by an angle $\varphi$. If this angle error is small, as measured from one coordinate, it can be phase shifted (delayed) by a time $\tau$ between the memory device signal $S(t)$ and the current sighting device signal $S(t-\tau)$.

Referring to Figure 6.10, the circuit elements consist of a tracking sighting device (a TV camera), a memory device with multiple reading capability and an image motion correlator with a converter amplifier and power driver that provides automated SDA to DAP tracking. Using the correlator output voltage $U_{out}$, the power amplifier controls the autopilot to bring the angle error angle to zero. That is, the entire network is actually a complex feedback system.

The measuring element (correlator) characteristic must be an uneven function of the angle error $\varphi$ for efficient tracking, which means that it must be a differentiated signal with smoothing.

If the output signal of the cross-correlator is equal to the correlator function derivative,

$$R(\tau) = \lim_{T \to \infty} \frac{1}{T} \int_0^T X(t) S'(t-\tau) dt,$$

where $X(t) = S(t) + n(t)$.

When the relative delay becomes zero, the output of the cross-correlator also passes through zero, so a direction-finding navigational CES characteristic is formed, resulting in a high-accuracy angle correction in the automated tracking mode. That is, the output tracking drive voltage initiates an accurate horizontal turn until the angle error $\varphi$ is reduced to zero and the SDA is in the same direction as the DAP.

Extracting information about the image brightness vertically – that is, across the scan lines – can be accomplished by gating those scan lines. The output of the television viewfinder will then simultaneously present two video signals consisting of lines and frames, the temporary positions of which, when compared with the corresponding memory device video signals, will give the angular deviation of the viewfinder axis both horizontally and vertically. These two video signals therefore allow the use of a single correlation measurement method in a two-coordinate viewfinder angular tracking system. Also, to provide viewfinder automatic tracking in another plane the system may be supplemented by an identical control loop with a single actuator.

An important advantage of such an elegant correlation method is the fact that the error signals determining the time shifts of the vertical and horizontal images are calculated over a time equal to the duration of one video frame (0.02 s).

The correlator statistical characteristic for small angle errors, and in the linear part of the characteristic is

$$U_{out} = \kappa_d (\tau + \Delta\tau_n) = \kappa_d \tau + \kappa_d \Delta\tau_n,$$

where $K_d$ is the slope of the curve at a point $\tau=0$, which defines the temporary image's motion efficacy and depends on the autocorrelation characteristics of the orientation region.

Operating in real time, the system signals are accompanied by internal noise and correlator errors $\Delta\tau_n$ that result in a steepness decrease $K_d$. Also, the filter section tracking drive has a carryover factor $K_f$.

Because the angular coordinate $\varphi_p$ is used to define the direction to the aiming point relative to a stabilized platform, an SDA angular coordinate $\varphi_B$ is used.

A television sighting device providing inertialess sending target scanning is used for angle error $\varphi$ to time-delay $\tau$ transformation, and is a linear element with a carryover factor $\mu = \tau/\varphi$. Hence, the general carryover factor $K = \mu \cdot \kappa_d \kappa_f$.

It is possible to simplify the computation of CCFs that appear when quantizing the levels of the input signals, by replacing one of the multiplied signals by an alternating signal of constant amplitude, the sign of which corresponds to the sign of the original signal. This simplifies the calculations, but leads to a systematic error and a deterioration in the equivalent signal-to-noise ratio. The theoretical estimation of such a loss is 1.19 dB. This can be obtained by the comparative modelling of the CCF calculating process for a Gaussian signal in the presence of Gaussian noise by evaluating the error in determining the position of the CCF maximum on the abscissa (Dehtyarenko and Kozubovskiy, 1962).

For short television signal sequences of about 11–20 lines, both theoretical conclusions and experimental data testify to the absence of correlation function maximal shift or zero derivative (when $\tau=0$) due to sign correlator use; and the correlation function ratio error doesn't increase over about 1–2.5%.

During navigational CES work, the absence of equal scales for the viewfinder image and the memory device signals leads to the appearance of *scale errors*. In extreme cases, the PPP, $K_d$ and noise-to-signal ratios decrease under scale divergence and become zero. Factors connected with observation angle modifications and with image turns have similar PPP and noise-to-signal ratio dependences.

Errors in the $\tau$ shift calculations under good conditions (image scale coincidence and input noise disturbance absence) involve only 1–2 television image elements. Under the worst conditions (10% scale divergence and a correlator input noise-to-signal ratio of 1) and optimal integration time, a correlation method error does not exceed 10–15 television image elements.

Experimental results show that the correlation function slope and the linear region of its derivative are essentially constant during a video signal realization of up to four or five lines. Also, experimental checks on the influence of scale divergence show that the correlation function and its derivative maintain acceptable shapes if a scale disparity in the signals does not exceed 10%. Further divergence increase leads to sharp CCF distortion.

An angle aiming modification of several degrees also influences the correlation function. An image turn of 1–2° is acceptable, but if it is 4°, the correlation function maximum becomes diffuse. If the image turn is more than 13°, the system efficiency fails (Figure 6.11).

In the case of a video signal noise level of more than 10 dB, the system remains efficient despite considerable steepness in the operating characteristic decrease. Under these circumstances, the correlator output fluctuations are practically absent.

If an image turn is greater than the angles shown earlier, the CCF begins to disintegrate and the PPP recedes relative to the local maximums; the PPP's conditionality decreases; and the noise component level increases, leading to a PPP search and tracking performance decline.

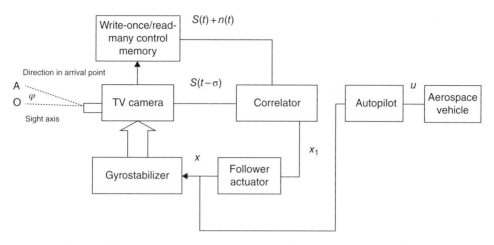

**Figure 6.11**   A developed Cross-Correlation Flight Control System (CFCS)

Similar consequences are observed for large image scale differences and angle distortions exceeding defined limits.

Given proper attention to the aforementioned operating conditions, modern navigational CES with ranges of several kilometres demonstrates root-mean-square deviations from the arrival point of only a few metres (Konovalov *et al.*, 1960; Shahidzhanov, 2002).

## 6.4   Prospects for CES Development

Modern CES are now utilized in many flying vehicles such as satellites and orbital stations; aircraft, including Unmanned Aerial Vehicles (UAVs); and guided rockets of various kinds all of which therefore depends greatly upon the relevant CES properties. Rapid improvements in CES computing systems have allowed the concomitant development of new algorithms for seeking extremums very quickly, so lowering the probability of abnormal mistakes and raising the speed and accuracy of CESDs due to the faster processing of images for determining the PPP of the CCF. Achievement of the limits of technological perfection in some sensors for CES has led to the creation of computing systems for improving the quality of the obtained images. It has been possible to optimize the signal communications from the target sensor by transforming the physical field and thereby compensating for aperture distortions peculiar to the sensor itself. These actions make possible a rise in the informativeness of the images, a reduction in the level of noise, and also promote a quality improvement in the CES.

Miniaturization makes possible many practical applications for combined and integrated CES; and correlation calculations no longer present significant problems.

### 6.4.1   Combined CES

For round-the-clock and all-weather applications of modern CES, they can be equipped with various sensors for providing the reception of current images by day or night and under any weather conditions. For example, combining data from these sensors enable power station operatives to carry out consumer-related tasks under any changing and complex application conditions.

## 6.4.2   Micro-Miniaturization of CES and the Constituent Components

At present, CES micro-miniaturization is based on the non-volatile fixation of functionalized application algorithms in microprocessors and programmed logic matrices (PLMs). These chips incorporate many millions of gates and can contain all the algorithms necessary for CES realization. Separated space programming techniques enhance the working capacity of the modern low consumption PLM – and it weighs only a few grams.

## 6.4.3   Prospects for CES Improvement

In current CES development there are three primary threads:

1. The application of new algorithms specific to PPP for CCF coordinate seeking.
   For such algorithms, the direct calculation of required values arising from new types of data processing implies the repeated raising of the performance of onboard computer equipment.
2. Further micro-miniaturization along with expansion of admissible functioning conditions due to the application of modern constituent components.
   This should enable CES to be duplicated and built into even the smallest flying vehicles.
3. The expansion of CES application conditions.
   This is made possible by qualitative improvements in sensors, their miniaturization and the integration of various types of sensor to permit the overlapping of all available and prospective application conditions in the flying vehicle. Also, by the usage of effective algorithms and devices for preliminary processing to present reference images for minimizing distortions and incrementing the PPP for CCF.

Special interest is being taken in methods of calculating PPP coordinates, and it is sufficient to note that Figure 6.5 defines the PPP of the CCF. The most interesting are methods for the direct calculation of the abscissa and ordinate of the PPP of the CCF, and it appears unnecessary to consider either the correlation discriminator or its optimum smoothing circuits. However, not all technical methods present such insights. An experiment involving the production of a bubble (or plume) of a light gas under a CCF, and the observation of its behaviour, has been carried out. The bubble could rise to a global maximum, so resolving the problem, but could also enter the next local maximum, leading to abnormal mistakes. In fact, for a CCF contained in a neighbourhood with a global maximum and also several local maxima, such abnormal mistakes are practically inevitable. This effect is characteristic of periodic images, the surface of the sea, for example along with ongoing construction or extensive forest planting. A struggle against aperture distortion in sensors is also on the agenda, as are methods of CCF calculation that are not sensitive to image scale changes, to their turnings and to foreshortening distortion. In fact, it is necessary to develop CES which are able to recognize a subject at various distances along with its turnings and inclinations.

## 6.4.4  New Properties and Perspectives in CES

The progress considered earlier on an algorithmic and constituent component basis predicts the prospects of developing new CES properties. Any flying vehicle, even the smallest, can nowadays be equipped with CES, and global communications can deliver reference images of the widest classes of objects to flying vehicles.

Any new evolutions in flying vehicles or objects of navigation or guidance cannot now negate the mutual correlation communications of reference and current images. Already-developed technologies in the reduction of uniform systems of coordinates and the processing of three-dimensional images – and in the long term, volumetric images and the integration of various sensors – are able to provide reliable operation to fulfil tasks in almost all possible CES operating conditions.

Figure 6.12 illustrates aperture and foreshortened distortions, changes of scale and turn, and processable distortion discrimination characteristics, in prospective CES on the basis of (1) normal initial, (2) smoothed and (3) aggravated CCF (Figure 6.13).

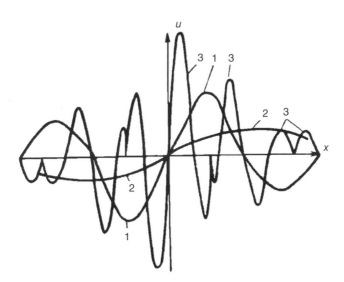

**Figure 6.12**  Discriminator curves for an advanced Cross-Correlation Flight Control System (CFCS): 1, ordinary; 2, smoothed for minimizing abnormal errors; and 3, improved steepness to increase the sharpness of the peak of the global maximum

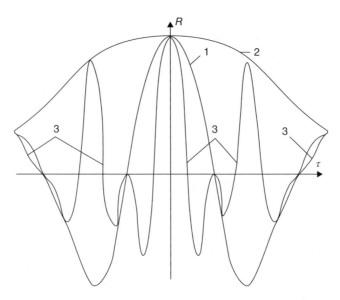

**Figure 6.13**   Cross-correlation functions. 1, ordinary; 2, smoothed; and 3, steeper cross-correlation functions

# References

Baklitskiy V. K. and Yuryev A. N. (1982). *Correlated-Extreme Methods of Navigation – Moscow: 'Radio and Communication'*, 256 pp. (in Russian).

Beloglazov I. N. and Tarasenko V. P. (1974). *Correlated-Extreme Systems – Moscow 'Soviet Radio'*, 392 pp. (in Russian).

Bernstein, D. and Kornhauser, A. (1996). *An Introduction to Map Matching for Personal Navigation Assistants.* Technical report. New Jersey TIDE Center, Princeton University, Princeton, NJ.

Bonner T. W. Position locating method, US Patent 2,508,562, May 23, 1950. Belmont, MA, USA; assigned to USA as represented by Secretary of War Application Serial No. 558,013, October 10, 1944.

Collier, W. C. (1990). 'In-Vehicle Route Guidance Systems Using Map Matched Dead Reckoning,' Proceedings of the IEEE Position Location and Navigation Symposium, pp. 359–363.

Degawa, H. (1992). 'A New Navigation System with Multiple Information Sources,' Proceedings of the Vehicle Navigation and Information Systems Conference, pp. 143–149.

Dehtyarenko P. I. and Kozubovskiy S. F. (1962). The Analysis of Errors of Relay Correlation Functions, Avtomatika, Kiev, No. 3 (in Russian).

Dickey F. R. (1958). 'The Correlation Aircraft Navigator, A Vertically Beamed Doppler Radar,' Proceedings of the National Conference on Aeronautical Electronics, Dayton, OH, May, pp. 463–466.

Faleev S. P. (1980). Calculation and modeling of devices for processing the signals of control systems: Study Guide. Saint Petersburg Electrotechnical University 'LETI' (Saint Petersburg State University of Aerospace Instrumentation), Saint Petersburg, 110 pp. (in Russian).

Iwaki, F., Kakihari, M., and Sasaki, M. (1989). 'Recognition of Vehicle's Location for Navigation,' Proceedings of the Vehicle Navigation and Information Systems Conference, pp. 131–138.

Jo, T., Haseyamai, M., and Kitajima, H. (1996). 'A Map Matching Method with the Innovation of the Kalman Filtering', *IEICE Transactions on Fundamentals of Electronics, Communications and Computer Sciences*, Vol. **E79-A**, pp. 1853–1855.

Kim, J. S., Lee, J. H., Kang, T. H., Lee, W. Y., and Kim, Y. G. (1996). 'Node-Based Map Matching Algorithm for Can Navigation System,' Proceedings of the 29th ISATA Symposium, Florence, Italy, Vol. **10**, pp. 121–126.

Konovalov E. A., Tumanov A. V., Dyatlov Y. M., and Isaev N. S. (1960). Guided Missile Guidance System, Consisting of a Head Image Registration-Combining and Self-Tuning Autopilot. Patent 1840806 SU (in Russian). MPK F41G 7/00.

Kretzmer E. R. (1952). Statistics of television signals, BSII, v. **31**, no. 4.

Lakakis K. (2000). 'Land Vehicle Navigation in an Urban Area by Using GPS and GIS Technologies,' Ph.D. thesis. Aristotle University of Thessaloniki, Department of Civil Engineering, Thessaloniki.

Savvaidis, P., I. Ifadis, and K. Lakakis (2000). Thessaloniki Continuous Reference GPS Station: Initial Estimation of Position. Presented at the EGS XXV General Assembly, September, Nice, France.

Scott, C. A. and C. R. Drane (1994). 'Increased Accuracy of Motor Vehicle Position Estimation by Utilizing Map Data, Vehicle Dynamics and Other Information Sources', Proceedings of the Vehicle Navigation and Information Systems Conference, pp. 585–590.

Shahidzhanov, E. S., editor (2002). Problems of creation of guided and corrected aerial bombs. Engineer, Moscow, p. 528.

Tanaka, J. (1990). 'Navigation System with Map-Matching Method,' Proceedings of the SAE International Congress and Exposition, pp. 45–50.

Watanabe K., Kobayashi K., and Munekata F. (1994). 'Multiple Sensor Fusion for Navigation Systems,' Proceedings of the Vehicle Navigation and Information Systems Conference, pp. 575–578.

White, C. E., Bernstein, D., and Kornhauser A. L. (2000). Some map matching algorithms for personal navigation assistants. *Transportation Research Part C* vol. **8**, pp. 91–108.

# 7

# Homing Devices

Georgy V. Antsev[1,2] and Valentine A. Sarychev[2]
[1] *Concern Morinformsystem-Agat JSC, Moscow, Russia*
[2] *Radar-mms Research Enterprise JSC, Saint Petersburg, Russia*

## 7.1 Introduction

The term "homing" refers to the autonomous guidance of an aerospace vehicle to a target via comprehensive onboard terrain sensors capable of responding to topology (including height), temperature, and other physical parameters of the target and the surrounding terrain. This essentially involves the augmentation and dynamic mismatch correction of both preprogramed data from a flight director algorithm and radio control, which is more accurate. It also includes the possible movement of that target relative to both the terrain and the homing vehicle.

This chapter primarily considers the autonomous homing systems (Bolkcom, 2004; Isby, 2004; JP 1-02, 2004) as employed by, mostly, Unmanned Airborne Vehicles (UAVs) (Shephard's Unmanned Vehicles, 2005). Such autonomous control has been applied to a wide range of military vehicles such as guided weapons of various levels of sophistication (JP 1-02, 2004).

The various position and dynamic conditions exhibited by an aerospace vehicle occupy points in a *phase space*, which is a space in which all possible states of the system exist. That is, each possible state of the system corresponds to one unique point in that phase space. Their coordinates include both attitude coordinates (angle of attack, roll, yaw, skid, etc.), linear and angular velocities, accelerations, and the performance characteristics of the homing devices.

An autonomous vehicle can evaluate the actual coordinates of such points for *secondary processing*, $PC_{AEV}$. Similarly, the target phase coordinates, both absolute and relative to the vehicle, $PC_{TG}$, can be used to establish any discrepancies between them and their required values $PC_{AEVO}$ (the "O" standing for "ordered"). These may then be used to establish the necessary control signals.

The rule for the shaping of these control signals (also referred to as *mismatch parameters*) is called the *flight director control algorithm* (Shneydor, 1998). The mismatch parameters are given by

$$MIS_i = \pm\left(PC_{AEVIi} - PC_{AEVOi}\right), \tag{7.1}$$

*Aerospace Navigation Systems*, First Edition. Edited by Alexander V. Nebylov and Joseph Watson.
© 2016 John Wiley & Sons, Ltd. Published 2016 by John Wiley & Sons, Ltd.

where $i$ is the relevant control signal number. The plus sign is used in guidance mode analysis and the minus sign is used in homing mode analysis. Most commonly, only two control signals are taken into account, corresponding to the shaping of two mutually orthogonal control planes comprising steering controls.

Historically, the main purpose of combat-oriented homing systems was for aerospace vehicles and their weaponry for targeting air, ground, underground, space, subsurface, and surface sea objects (Gunston, 1979; Ozu, 2000). However, civil applications of homing methods have included the implementation of guided vehicle navigation, and landing upon prepared or nonprepared facilities, when it is required to deliver various objects into a specified area. Also included are the transport of medicines, sensors, reactors, and reagents inside a biological object to its organs; when creating internal diagnostic systems for man-made entities (e.g., inside aircraft engine pipelines, nuclear power plant equipment, and chemical production facilities such as gas and oil pipelines); machine vision systems; motor vehicle motion monitoring sensors, and sensors for various security systems.

Homing systems are of the utmost importance because their parameters and characteristics determine the efficiency of the whole aerospace vehicle to a considerable extent. At the time of writing, the most urgent problems involve the provision of efficient aerospace vehicle's homing system operation under conditions of intermittent data exchange between external control systems and guided vehicles due to intense moisture and dust-and-smoke mixtures, for example, and in the military field, electronic countermeasures.

A guided aerospace vehicle travels in two planes, the horizontal and the vertical, which means that the guidance must cover these two planes. Hence, at least two guidance loops (which may be identical) will be required. This and other aspects define the *dimensionality* of the overall system. Among the other aspects are the vehicle movements resulting from the conversion of those phase coordinates that take into account both relative and absolute movement of both the vehicle and the target. That is, the system dimensionality is really determined by the number of phase coordinates describing the phase trajectory along which homing is performed.

An increase in homing dimensionality results when the number and value of the phase coordinates to be followed up by the homing device increases, for example, when homing to super-maneuverable targets where it is also necessary to evaluate lateral accelerations. In such cases it is necessary to follow the unified control principles of the aerospace vehicle and the homing device taking into account the overall dynamic response of the radio guidance loop component parts. In radar homing devices, for example, vehicle acceleration significantly and often unpredictably changes the efficiency of such systems. The uncertainty function of the signal received by these systems is due to the occurrence of additional frequency modulation with corresponding spectrum broadening and a decrease in spectral density. In turn, for super-maneuverable homing the appearance of derivative-tracked phase coordinates (range, velocity, observation angles), the order of which exceeds the astatism of the servo network order, forces a change to multiple-loop control in which the accuracy and stability requirements can only be provided by several loops.

For military applications, homing devices may operate in either single or multiple target tracking modes. In the first case, often called the *continuous tracking mode*, the homing device data channels are directed to only one target, which allows for the continuous measurement and evaluation of the relative movement coordinates and subsequent modification of the vehicle control commands. In the second, or *space scanning* case, all targets are individually highlighted within the coverage area. In this mode, the coordinates of multiple targets are continuously evaluated based on extrapolations of their trajectories. Correction of these extrapolation results is performed using measurements at the moments of arrival of signals from the targets. Consequently, it becomes possible not only to control the aerospace vehicle flight but also to simultaneously modify target designation commands with respect to the multiple targets. This case is typical for group operations (see Section 7.2.1).

Homing devices make possible the autonomous operation of various robotic aviation and space objects far from their control centers, and in the military field, Network-Centric Warfare (NCW) has been described by Cebrowski (2004), Stein, Garska, and McIndoo (2000), Kruse, Adkins, and Holloman (2005), Ling, Moon, and Kruzins (2005), Phister and Cherry (2006), and "Net-Centric Environment—The Implementation" (2005).

Most military aerospace homing devices (Ben-Asher and Yaesh, 1998) operate under conditions of intense mechanical impacts and vibrations, are limited in terms of weight and size, and often operate over short periods. Nevertheless, they must implement fully automatic modes of operation and accommodate large maximum to minimum range ratios. Despite these constraints, it has recently become possible to package them into single units because of progress in microelectronics, MEMS, optoelectronics (Jiang, Wolfe, and Nguyen, 2000), and nanotechnology (Maluf and Williams, 2004). There has also been considerable progress in the provision of sufficiently localized external phase coordinates (for single targets).

In recent years, homing devices have evolved considerably and are now commonly divided into the following categories:

*Category G1* encompasses semi-active homing devices using analog receivers and with unconditional operation logic determined by the hardware design—that is, having no airborne computer or integrated circuit components.

*Category G2* also refers to semi-active homing devices, but using an airborne computer and having integrated circuit components.

*Category G3* comprises both active and semi-active homing devices incorporating analog receivers and digital, reprogrammable, control machines used for the secondary processing of data from that receiver. This involves the buildup of logic for the capture, tracking, and antijamming of self-protected countermeasure signals. Microminiature devices, integrated microcircuits, and microstrip techniques are widely used and are protected against basic jamming, but are not fully protected against jamming outside loitering zones and have limited range in the rear hemisphere.

*Category G4* takes account of active, semi-active and combined homing devices incorporating digital receivers and high-speed digital signal processors. The main differences from G3 systems are longer effective ranges and better jamming immunity (including from lateral lobes). However, these homing devices have limited Doppler frequency analysis zones as determined by input analog filters, and they do not have RN tracking, but only angular tracking and middle Pulse Repetition Rates (PRRs).

*Category G5* systems provide for digital processing beginning with the high frequency or the first intermediate frequency in the broadband. They have minimal, and sometimes completely missing, analog path lengths, and include high-performance signal processors. Consequently, these homing devices operate at high and medium PPR and provide for the parallel analysis of target detection, velocity, and range. They also implement antijamming algorithms against both jamming generated directly by targets and external jamming generated by other sources.

The target capture range is increased, and due to the middle repetition pulse rates the range of target capture in the rear hemisphere is also increased.

*Category G6* encompasses all intelligent homing devices (see Section 7.8).

## 7.2    Definition of Homing Devices

### 7.2.1    Homing Systems for Autonomous and Group Operations

An aerospace vehicle in the context of homing devices may be assessed dialectically: it can be either a single vehicle or a group of vehicles. The simplest case of homing by a group of vehicles is when separate aerospace vehicles in a group are capable of performing independent homing in the absence of interaction with other aerospace vehicles. For example, in the military context, the use of several identical guided weapons increases the chance of a successful hit.

However, if the interaction is used to increase the efficiency of the homing device functions, this is referred to as *group operation*. Group operation includes the use of target environment data exchange (i.e., interaction with external data sources for each aerospace vehicle) to enable re-aiming and flight path correction, and also for shaping commands for changing operation modes such as the actuation of a jamming source on a single or group of aerospace vehicles, as well as applying reconnaissance and target designation functions to separate aerospace vehicles. In this case, a group is controlled only by its own airborne homing devices and interchanges information both relative to the external situation and with its own current status and intentions. For example, when attacking maneuvering targets, the issue of up to 30 control commands per second would be required. In order to issue each control command, it would usually be necessary to process at least four target image frames so that the exchanged data stream rate should be no less than 120 frames per second. The relevant data communication equipment included in homing devices should perform the functions of shaping, processing, controlling, storing, archiving, and distributing the motion imagery for collecting data, performing reconnaissance, and observing the battlefield.

In group interaction, it is possible to establish two- and multiposition observation of the target environment (Derham *et al.*, 2007; Donnet and Longstaff, 2006; Gray and Fry, 2007; Haimovich, Blum, and Cimini, 2008; Lehmann *et al.*, 2006; Sammartino, Baker, and Griffiths, 2006; Liu *et al.*, 2014), and by any one of the following circuits: SISO, SIMO, MISO, MIMO, where S = single, M = multi, I = input, and O = output. In essence, when carrying out group operations, a set of aerospace vehicles using information interaction performs the functions of a kind of virtual aerospace system that implements distributed control principles (Hume and Baker, 2001; Teng *et al.*, 2007). Such a virtual system can perform automatic tracking of targets in a scanning mode. Here, the necessity of receiving precise information on the coordinates of relative movement of aerospace vehicles and targets becomes ever more important, especially when new targets appear in the coverage area of the group.

Group operations are inherently peculiar to the organization of network-centric operations and their performances. The information domains of network-centric operations have network structures where each information element is a node. Only under such conditions can another aerospace vehicle be a "participant" in NCF. The information domain is expected to efficiently consolidate all the aerospace vehicles in a combat space. This combat space is necessary for the establishment of aerospace vehicle homing in NCF to ensure the success of organized and properly performed combat operations. In this case, homing must be informatively supported sufficiently far away from the location of combat control centers within a wide range of dynamically changing parameters and conditions.

Summarizing, a homing device may receive data from similar devices to support group operations and to assist in the execution of the particular homing function currently being implemented. This homing system, functioning under group operation conditions, will be referred to hereunder as a *homing device* or *homing system*. However, the term "device" is preferable for a homing system that exists on a separate aerospace vehicle.

## 7.2.2 Guidance and Homing Systems

To understand the specific features of homing devices, it is relevant to consider their differences from radio control systems that only guide aerospace vehicles via a preset phase trajectory using appropriate control commands. The first such systems to serve aerial vehicles appeared during World War I, and were called "aerial torpedos," but they proved inefficient (Augustine, 2000). In principle, radio control systems meet the definitions of homing systems completely only if "attachment" is done independently by airborne means. However, using this method, it is almost impossible to hit fast-moving targets. Today, over the major parts of their flight trajectories, most homing devices behave exactly in this way if the phase trajectory is fixed or is provided by other navigational aids such as inertial, Doppler-inertial, air data–monitored heading hold, Doppler air data–monitored heading hold, or via satellite, etc. In such cases, aerospace vehicle's flight directions must be assessed in addition to accelerations and velocities relative to the Earth. The route covered by the vehicle is calculated by double integration of acceleration or single integration of velocity. Flight correction is performed by homing devices using data channels that record the current target environment at intermediate points of the trajectory and especially on the final segment, where the main function of the vehicle is implemented after correction of the accumulated navigational errors. That is, homing devices exist only to serve the final part of flight trajectory that culminates at the target. Hence, only radio control systems properly called *homing devices* are those which at any flight segment (most often, the final one) perform independent guidance into a space–time domain determined by some description, presentation, or image specified earlier or shaped during flight.

Homing devices providing aerospace vehicle flight along preset trajectories only are sometimes called *radio control guidance systems or devices*, or simply guidance systems. It is guidance according to shaped images of the vehicle environment that enables it to efficiently arrive at moving targets via flight corrections, consider changes in jamming and target environments, change airborne radio facility operation modes, and perform efficient data exchange during group operations. These are the so-called correlation-and-extreme homing systems (Lennox, 2002) used as component parts of homing devices for long-range aerospace vehicles homing to stationary ground targets.

Homing using correlation-and-extreme systems is performed according to information retrieved from geophysical fields (gravitational, radiation, magnetic, radar distribution, heat, and optical contrast, plus terrain relief along the flight route), the parameters of which are closely connected with particular areas on the ground surface (see Chapter 6). The location of the aerospace vehicle is determined by comparing the current field distribution captured during flight to the reference distribution of the same field captured beforehand. This enables terrain binding to be performed with high accuracy.

During preparation for combat, and in the course of their missions, state-of-the-art (and prospective) military aerospace vehicles must undergo automated selection of the type of attack and maneuvering procedures to be implemented for the vehicle to arrive at a target within the shortest time possible.

Homing is rigidly limited by the availability of guidance procedures related to a specific image of a target environment, not necessarily to a target itself but to a specified space–time domain. The homing function must also be able to distinguish the homing guidance procedures of other vehicles such as fighters, ground attack aircrafts, helicopters or satellites, and still take it to its designated space–time domain.

Summarizing, a homing device has the following characteristic features: installation on board an aerospace vehicle; independent guidance according to a specific image of the environment (most often a target environment); control of the vehicle and the operation of its systems; robotization of functions; structural, functional, and informational integrity; and the possibility of providing for group operations. In short, a homing device must take the leading role in the homing procedure having regard to the intended purpose of the entire aerospace vehicle.

## 7.2.3   Principles and Classification of Homing Devices

Homing devices are distinguished by the types of target vehicle involved—fighters, bombers, air-to-air missiles, air-to-surface missiles, guided aerial bombs, space objects, Unmanned Combat Aerial Vehicles (UCAVs), and target drones. Missiles installed on airborne carriers are called Air-Launched Cruise Missiles (ALCMs), which allows further classification based on subdivisions of aerospace vehicles including medium-range, short-range, and long-range, tactical, strategic, antiaircraft, ballistic, cruise, subsonic, nonstrategic ballistic, supersonic cruise missiles, super-maneuverable and hypersonic aircraft, UCAV lifting capacity, etc. This list may also be expanded to include mobile objects autonomously injected into a specified space–time domain. The most complex are homing devices for fighters and air-to-air missiles (irrespective of which is the target and which is the weapon), and it is these that are characterized by the most intricate target and aerospace vehicle trajectories. Hence, their weights and dimensional parameters impose the strictest requirements. In recent times, homing devices for multipurpose aircraft capable of functioning as fighters, ground support aircraft, and bombers have become ever more widespread.

Some homing devices are Anti-Ship Capable Missiles (ASCMs) (Pace and Burton, 1998), and others are missiles targeting ground, subsurface, sheltered, and space domains. Missile delineation indicates the locations of vehicle deployment and that of the target, for example, "space-to-ground," "air-to-ship," and "Surface-to-Air Missile (SAM)."

Homing devices are also subdivided by the methods and sources of receiving data channel information, including nonautonomous (*In-sensors*), autonomous (*Out-sensors*), and combined (*In + Out-sensors*). Nonautonomous homing devices are those for which at least part of the information required for shaping the control law is retrieved from signals coming from the target. The availability of a data channel that functions only as an Out-sensor means that the guidance procedure is implemented entirely within the aerospace vehicle. Mostly, autonomous guidance implies the use of inertial systems and sensors, or Inertial Measurement Units (IMUs), and Inertial Navigation Systems (INS). The latter can be platform mounted or strapdown. Also used are accelerometers that measure gravitational acceleration, and angular velocity sensors that measure the angular deflections of a gyro axis during rotation of the aerospace vehicle on which it is mounted. It is virtually impossible to employ deliberate jamming of these sensors because that would require introducing distortions into the Earth's gravitational field or changing the rotation rates of the gyros.

It is not always necessary to restrict homing devices to purely autonomous or nonautonomous systems. A class of autonomous homing devices (Out-sensors) has been designated as that which may be "permitted" to interact with the target environment (using, e.g., satellite navigation technologies and long-range and short-range navigation system signals) until the control law can be modified by direct interaction with information about the vehicle's motion as received from its own sensors without contact with either the target or via commands from a control post. It is here that autonomous homing commences.

Correlation-and-extreme guidance methods are provided by autonomous homing devices because, even in the target area, such a homing device uses a pictorial image of the environment containing the target. However, any aerospace vehicle's flight correction procedure immediately renders the homing device nonautonomous. From the combat command and control center (or from other aerospace vehicles in the case of group operations), an aerospace vehicle can receive updated images of both the target environment and the geophysical field, according to which the homing device performs autonomous guidance. In the case of information received from other aerospace vehicles, the use of nonautonomous homing devices corresponds to the autonomous operation mode of homing of a particular aerospace vehicle receiving this information. A similar situation occurs when an aerospace vehicle is required to hit a specified point on a target when approaching it, using the more precise data channels (follow-up channels) of the homing device.

The advantages of autonomous homing devices include very long effective ranges and high jamming immunity due to the provision of concealment. Their disadvantages are their limitation to stationary targets only, poor guidance accuracy due to errors in the vehicle's own location definition, and very high operational reliability requirements. Combined homing devices (Biezard, 1999) represent an aggregate of autonomous and nonautonomous systems.

The earlier examples indicate a trend in homing device development consisting of improvements in the combined functioning of data channels that make possible longer effective ranges, closing-in velocities, and angular velocities with respect to the target's line of sight, leading to better accuracy. Better jamming immunity is implemented by the provision of concealment technology, and the reliability of homing devices in general continues to improve. However, all these improvements are implemented at the expense of information redundancy and mutual correction. The most common integration option is based on the predominant use of an autonomous system during the initial stages of control and a nonautonomous homing system during the final segments of the guidance trajectory.

There are also alternative definitions of autonomous and nonautonomous homing systems. In autonomous systems, jamming immunity is ensured due to concealed operation only, while in nonautonomous operation it may also be due to jamming resistance. Under these definitions, the class of nonautonomous systems includes the so-called semi-active homing systems with external illumination. Here, the aerospace vehicle is guided to its target by signals received from that target, these being generated through the illumination of the target from another aerospace vehicle, from the ground, or in space. Passive systems of homing to radiating sources also fall within this class. Such systems stand out as canonical for the implementation of Low Probability Intercept (LPI) technologies (Pace, 2009). For nonautonomous and combined homing systems, this technology includes the organization of measures for decreasing homing system observability within the current data channel used by the system.

Homing devices may include sensors employing different physical phenomena such as inertial systems using accelerometers, air velocity sensors, and gyros (Titterton and Weston, 2004), altimeters (Cruise missile radar altimeter, 1994), satellite navigation systems, Doppler velocity and drift radar, radar terrain awareness warning systems (terrain and obstacle avoidance), correlation-and-extreme systems (Skolnik, 1990), and means for establishing interaction during group operations. It is important to realize that all these systems are usually engaged in serving the aerospace vehicle control channel during homing while also participating in other automatic control loops. For this reason, competent jamming of the information serving the homing devices can successfully disrupt the homing procedure.

In these circumstances, mention is often made of *corresponding signal fields*, which in the case of autonomous systems are understood as arising from own radar, thermal radiation, the Earth's magnetism, target emissions, navigation system signals (including from satellites), or generated by various active nonautonomous homing devices correlated with targets and particular areas on the ground surface. Such fields can also be signals from the altimeter along the flight route. In principle, corresponding signal fields can be detected by any of their characteristics at any stage of conversion in the course of processing within the homing device, given the possibility for such transformation via the presence in that homing device of an appropriate sensor at the beginning of the signal conversion chain. Homing device signal fields have dimensionalities not lower than the number of sensors included.

A signal field can be shaped from the characteristics of an electromagnetic wave detected at the homing device input antenna, and a signal field for a radar homing device is often provided with polarization patterns, which in principle allows the analysis of homing polarization errors. In the ultimate case, it may be possible to use the characteristics of the target environment as signal field properties: the position and brightness of glitter points on the target surface, the distribution of altitudes along the flight path, geophysical fields, etc. When an autonomous homing device is on a sustained flight, the role of this *physical field* is played by its own physical fields representing the Earth's surface and various natural reference points. Conditions for the recognition of the spatial-temporal distribution of characteristics by signal fields suitable for a homing device are as follows: the possibility of a "follow-up" signal in the homing device channel, the use of corresponding signal characteristics in the course of homing device functioning, the genetic association of signal fields with the characteristics of a target or target environment within the target area, and the structure of characteristics according to spatial-temporal coordinates sufficient for the development of contrast and gradient in the signal properties against the observed target environment.

These notions allow the postulation of the following chain of conversions for homing devices: physical field $\rightarrow$ signal field $\rightarrow$ phase coordinates $\rightarrow$ phase trajectory. In essence, the phase coordinates are the estimates of the physical field. However, sometimes this set of concepts is simplified by either improving the chain or introducing information entities in a somewhat different way. For example, a different concept for the navigation field is often used: it is basically just another name for the signal field, though it is more correct to consider it as a field shaped by phase coordinates.

Recent work in this area has highlighted the conversion of one field into another, by which means information on the relation between the physical field and the characteristics of the target (and target environment acting as the target) with the signal field can make possible the shaping of the target radar characteristics (Kozlov, Logvin, and Sarychev, 2005–2008) for use in the operation of intelligent radar homing devices. Hence, for distributed targets, a corresponding physical field might be structured in the form of range, velocity, and goniometric portraits; or arrangement diagrams of glitter points (scattering centers), marked by their Radar Cross Sections (RCS).

Combined homing devices of serial, parallel, and mixed types are distinguished depending on the methods of interaction between autonomous and nonautonomous sensors. In serial systems, autonomous and nonautonomous sensors operate serially in time: at the initial stage inertial guidance is performed, at the medium stage radio control commands come into play, and at the final stage autonomous homing takes over. In parallel systems, autonomous and nonautonomous sensors operate simultaneously throughout the entire guidance period. In mixed systems, autonomous and nonautonomous sensors and systems are integrated according to the flight stages and within these stages. Under simultaneous operation, sensors having different physical natures can operate independently or be combined (integrated) into a single complex system such as group operation.

The primary objectives for homing device development are improvements in guidance accuracy, expansion of combat application conditions, and improvements in jamming immunity. Indeed, most efforts are directed toward the creation of a unified, target-universal, combined terminal guidance radar/optoelectronic system. The types of physical fields involved determine whether radar, thermal, laser, TV, magnetometric, or other homing devices are more appropriate. For example, precision homing devices intended for ground targets are usually built as combined types that allow for good jamming immunity improvement. In this context, the injection of an aerospace vehicle into a precomputed point in space is usually performed with the help of an inertial system or a satellite radio navigation system. However, guidance errors using these systems can be great and to decrease them the method of Digital Terrain Mapping (DTM) correction can be used, usually for the area adjoining the target, which is entered into the homing device processor memory in advance. As a result, after shaping the required terrain area image by a more precise—but no longer autonomous—system, correlation-and-extreme processing tie-in of the received image to the DTM is performed, and the coordinates of the image frame center are found. This information makes it possible to determine the aerospace vehicle's own coordinates and subsequently plot a trajectory to the target. While approaching the target along such a calculated trajectory, it is necessary to perform surveillance of the terrain in the target area and ensure continuous measurement of the aerospace vehicle's own coordinates.

Integrated homing devices may include heterogeneous sensors, both structurally and informationally. Typical examples include inertial/satellite homing devices—integrated sensor systems built using an aggregate of radar and nonradar (Infrared or IR, magnetometric, TV, satellite, inertial, etc.) sensors. For establishing the mutual logistics of heterogeneous sensor operation, the architecture of homing device informational resources must allow for stagewise increases in capabilities and perform modifications without significant limitations. In the first instance, and quite often, homing device signal and data antennas and processors serve several heterogeneous subsystems (sensors), which is indicative of a wider degree of homing device integration. For example, the provision of group operations can be realized by commands transmitted via a radar channel. In this respect a homing device acts as an integrated sensor system.

Combining autonomous and nonautonomous information equipment into a single integrated system on board an aerospace vehicle makes possible significant improvements in the weights and dimensional characteristics of such homing devices. The appearance of optoelectronics and developments in computing technology have also contributed to such improvements. Hence, a global reconstruction of homing devices based on the application of high-speed processors capable of performing simultaneous multichannel data processing from all data sensors located on the aerospace vehicle is required (Lewis, Kretschmer, and Shelton, 1986; Pace, 2000). Developments in Super High-Frequency (SHF) solid-state electronics have led to the design and production of transmitter/receiver modules for various wave-bands from millimetre to decimetre, as well as those using Ultra-Wideband (UWB) signals. Other developments have included new antenna arrays that allow overlapping octaves in the operating frequency band. These latter two developments have made possible the creation of integrated, active, phased antenna systems (APAA) types, or more precisely, digital APAA (Neri, 2001; Richards, 2005; Rodrigue, Bash, and Haenni, 2002; Stutzman and Thiele, 1997; Tsui and Stephens, 2002; Wirth, 2001). This enables the creation of homing systems with spatial arrangements of digital APAA modules.

Recently, when creating combined homing systems, developers have tended to apply Software-Defined Radio (SDR) technology that allows adjustment to random frequency bands, receiving different types of modulated signal, and using equipment with programmable guidance (SDR Forum, 2007). For homing systems employing such technology, future prospects are associated with the application of UWB-signals (Astanin and Kostylev, 1997; Astanin et al., 1994; Kozlov, Logvin, and Sarychev, 2005–2008; Li and Narayanan, 2006; Sun, 2004; Zetik, Sachs, and Thoma, 2007; Kulpa, 2013). Special hopes in connection with these signals are dependent on the possibility of so equipping aerospace vehicles for sheltered targets (Narayanan and Dawood, 2000).

It should be noted that descriptions of specific homing devices can be encountered in the literature that, while remaining within the frames of the usual classifications, also have features that distinguish them from other similar homing devices. Examples include homing devices for aerospace vehicles exhibiting low radar visibility, homing devices for highly maneuverable aerospace vehicles, homing devices with target tracking in scanning mode adaptive homing devices, intelligent homing devices, multiposition homing devices, and precision weapons homing devices.

Currently, the main data sensors in homing devices that ensure aerospace vehicle's operation in any weather or time-of-day are autonomous radio-technical sensors. These can be subdivided into two groups—autodyne and radar. In some cases, the allowable volume of autonomous homing devices is limited to several cubic centimetres, much of which is taken up by the actuating circuit devices, the power supply, and the safety-and-arming mechanism. This is the reason why many radio-technical homing devices have been based on the autodyne principle (Piper, 1995), under which the active element of the transmitter in terms of control circuits is also a receiving device. The use of autodyne technology leads to maximal simplification and, as a consequence, makes possible the microminiaturization of the reception/transmission module. However, the price paid for simplicity is low jamming immunity and hence low efficiency, because autodyne construction does not allow for the provision of selection by range. So, an obvious alternative to autodynes is the miniaturization of radar homing devices, initially the transmitter–receiver part.

Usually, primary and secondary signal processing are distinguished in radar homing devices. Primary processing is understood as an aggregate of procedures for searching, detecting, and identifying a target environment and, in general, shaping ambiguous range measurements, closing-in velocities, and angular coordinates. Secondary processing is understood as an aggregate of algorithms for receiving assessments of all the phase coordinates required for controlling the aerospace vehicle, the categorization of targets by their degree of importance, issuing commands to the aerospace vehicle and the interfaced systems including the communications surveillance means, the radio countermeasures, and various other onboard sensors when performing network-centric operations.

In state-of-the-art guided radar homing devices, it is the radar sensor that determines such major system parameters as the resolution of all measured parameters and all-weather, round-the-clock capability, along with immunity from all types of jamming, an effective range of operation (detection, selection, tracking), and weight and dimensional characteristics. It is data exchange between the autonomous homing device and the navigation and reconnaissance devices that provides the aerospace vehicle with good performance quality in terms of firing range, accuracy, jamming immunity, and overall efficiency. The homing device also stops controlling the aerospace vehicle after a direct hit, at the actuation of a remote proximity fuse, at self-destruction after passing the target, and at the moment of crossing the dead zone boundary.

## 7.3   Homing Device Functioning in Signal Fields

### 7.3.1   Characteristics of Homing Device Signal Fields

As mentioned earlier, the signal fields of autonomous homing devices are not directly associated with the target. Signal fields and principles of autonomous homing device operation are in many ways universal for any means of guidance and homing, and take little account of the specifics of homing-to-target, so as such they will not be considered hereinafter.

Signal fields for nonautonomous homing devices can be shaped by signal properties such as an amplitude (including normalizing for the range, too), Doppler and temporal shifts (for UWB-signals), assessments of the observed polarization status of received , and surveillance object scattering properties (Kozlov, Logvin, and Sarychev, 2005–2008; Xu and Narayanan, 2001), the brightness, energy, spectrum widening, surveillance object contour, range and velocity portraits, secondary modulation effects, magnetic field gradient, direction of arrival, etc. Operation by signal fields implies that jamming accompanying the homing process does not greatly distort that signal field, and in some cases the jamming source can act as a target if the homing mission is intended to destroy that source of interference.

The performance of nonautonomous homing devices depends largely on the frequency band selected for operation. Sheltered targets (having no obvious navigational contrast reference points) are best detected in the long-wave range, and follow-up channels in the millimetre wave range (Klein, 1997; Lukin, 2001; Zhouyue and Khan, 2011), though now in the terahertz range as well (Naftaly, 2014). Homing to radiation sources inherent in targets may then be possible (Fan, RuiLong and Xiang, 2001a, b; Neng-Jing, 1995; Wang and Zhang, 2001). Vehicles using such technology include the Anti-Radiation Missile (ARM), High-Speed ARM (HARM), Destruction of Attack Module (HDAM HARM), Targeting System (HTS HARM), Advanced Anti-Radiation Guided Missile (AARGM), and ARM with

Intelligent Guidance & Extended Range (ARMIGER). Vehicles that are home to radio signals depend on the LPI principles introduced by Schrick and Wiley (1990), Wiley (1985), and Lee (1991). In both the jamming case and under guidance to a radiating target, information about times of transmission is missing, making the direct measurement of distance to the source of radiation or jamming impossible. Therefore, to find the range to a radiating target, and establish the closing-in speed, when signals are received from several spaced-apart aerospace vehicle's flight path points, methods of indirect assessment must be used.

In principle, any specific signal field can be destroyed by external means, primarily by jamming. Therefore, it is reasonable to consider its stability, reliability, and strength in this context. The main method for improving signal field stability consists of increasing its dimensions by using more homing device sensors (Jie, Xiao-ming, and You, 2006; Klass, 1998). In accordance with basic information theory, each component of such a multidimensional signal field should deliver separate data about the target environment. Mostly, this separation is provided by the integration of partial fields of different physical entities initially separated (e.g., by frequency bands) into a single multidimensional signal field (Jankiraman, Wessels, and van Genderen, 2000). Such methods of multidimensional signal field generation correspond to the appearance of integrated homing devices as previously noted.

The signal field for a homing device is not only structured by the current target environment, but its anisotropy also depends on priorities generated during guidance. This signal field, along with a structure facilitating the generation of homing algorithms (e.g., via the deletion of excessive information), is also the homing device function, which is usually combined with identification. The first priority of a homing device is the shaping of tracking channels by direction, that is, by angular coordinates. Given the correct operation of that homing device, such guidance is capable in principle of taking the aerospace vehicle to a specified point without help from other channels such as range and speed. "Picture-based" homing also comes down to directional homing, but as compared with the range and velocity guidance channels, the angular channel does suffer from significant guidance errors. Nevertheless, the sole use of nonangular guidance channels does not allow the efficient execution of such a homing task, at least under the real limitations of all types of resources available in aerospace vehicles and homing devices. Nowadays, to provide informational support to aerospace vehicle's homing devices, it is necessary to have sufficiently accurate assessments of the range-to-target, the bearing angles, the closing-in velocity, and the line-of-sight angular velocities in the vertical and horizontal planes.

Because most of the guidance errors arise in the homing device goniometric channel, special measures should be taken to increase the accuracy of its operation during homing, especially to maneuvering targets, which implies an increase in the corresponding signal field sensitivity to changes in target environment observation angles. This particular feature highlights the necessity for improving the antenna decoupling from the aerospace vehicle's angular oscillations, these being due to position and speed corrections. Guidance according to these parameters makes it possible to significantly improve the guidance accuracy. When using a PAA especially an APAA), combined correction signals relevant to changes in the angular and linear aerospace vehicle motions make it necessary to take into account changes in the APAA beam shaping laws. The use of an APAA also provides for short intervals between addressing the most important targets due to the application of programmable scanning of the area of responsibility. Moreover, APAA provides for target homing within an area having no obvious navigational contrast reference points (a sheltered target) when establishing the multiple beam tracking of reference points located outside that area.

## 7.3.2   Optoelectronic Sensors for Homing Devices

For many years, most homing devices have used laser homing (including semi-active types, with external laser target illumination) or TV and IR image homing, and because of their unique features, optoelectronic aiming systems merit special attention. The optoelectronic system within homing devices provides facilities for searching, detecting, tracking, coordinate finding (including range, using a laser range finder, for example) and target motion parameters, their identification, and the solution of aiming and homing tasks along with the homing device. A TV-commanded missile homing system comprises a gyro-stabilized TV homing device with variable-length focal lens, and a guidance and autonomous control system intended for the reception and storage of flight mission data and the generation of missile stabilization and control signals. As the missile approaches a target, the stages of target detection and identification begin.

Optoelectronic sensors for homing devices using the IR band have become widespread. Their task is to discern thermal emissions from a specified object (target) against a background of environmental emissions, for example, from other heated bodies. Hence, an IR homing device must be equipped with a fairing for protection against ram airflow and thermal impact, but which must also be transparent to the IR-band wavelengths involved. IR homing devices also involve radiation-reflecting materials and surfaces (for mirror and mirror-lens antennas) in addition to materials and surfaces for absorbing radiation and converting it into electric signals.

The laws of thermodynamics determining the thermal emissions of heated bodies make it possible to implement the frequency selection of observed objects having different temperatures and to design IR homing devices for "cold" targets (400–500 K), usually at the surface; "medium" targets (500–1000 K), typically helicopters; and "hot" targets (over 1000°K), such as jet aerospace vehicles of various types and ballistic missile heads. With the help of such frequency selection, Doppler correction can be considered using the radiation received when a homing device closes in on a target. All such considerations can be taken into account for both passive and active IR homing devices.

In the "predigital" period of radio development, use was made of IR homing devices that converted object sighting angles into proportional displacements at the focal plane of the corresponding image. This involved a rotating disc modulator located at the focal plane and having a scan pattern of transparent and opaque areas, so modulating the received radiation. Over 20 variants of such modulation disc scan patterns are known via which, depending on the image pattern, information on the target angular position is converted into signal parameters having different types of modulation.

At the present time, rotating elements are almost never used, and the sensing element receives the full radiation energy coming from the observed object. Figure 7.1 shows a block diagram of an IR homing device with a square sensing element consisting of a single semiconductor chip for implementing monopulse tracking (Skolnik, 1990). This is divided into four quadrants, each with a sensing electrode, so that the signal produced at each will depend on the area of the relevant quadrant irradiated by the image. The signal amplitude ratios thus determine the angular displacement of the target image from the optical axis.

If the number of sensing elements is significantly increased, the sensor can approach close to a model of the human eye. By this means, the location of the illuminated part of the sensing element (or group of sensing elements) can indicate both the sighting angle value and the

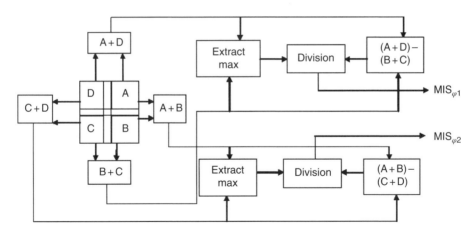

**Figure 7.1**   IR homing system schematic

sighting plane position, leading to a measurement of the observed object's position in both elevation and azimuth. In the case of a large number of sensing elements (dissection elements or pixels), an outline image of the object can be obtained and the IR homing device processor can then perform identification and authentication of the observed image against stored samples.

Combined IR and radar homing devices may be encountered in aerospace vehicles intended for long flight phases. At high altitudes, the relevant homing device operates using a mid-IR spectrum sector of 3.3–3.8 µm, while at low altitudes and under cloudy conditions it will operate within the radio frequency band. A far IR band sector at 7–10 µm may also be utilized, in which case an IR-transparent uncooled sapphire window must be employed. For use at this waveband, an appropriate sensing element—typically indium antimonide or platinum silicide—must be employed, and the focal array may be located on in a biaxial gimbal suspension.

For military applications, a missile warhead is usually conical, and provision is made in the imager for line-of-sight angular displacement relative to the longitudinal axis, to which end a three-mirror optical system in a Dewar vessel may be used.

### 7.3.3   Radar Homing Devices

Radar homing devices provide for target environment radar surveillance and depend on natural phenomena and technologies that have been understood and used successfully for over a hundred years (Kanashchenkov and Merkulov, 2003; Levanon, 1988; Nathanson, 1991; Schleher, 1991; Skolnik, 1990, 2001; Stimson, 1998).

On the whole, radar homing devices are to a large extent similar to modern radar systems (Skolnik, 1990) except that they are intended to support steep and fast maneuvers of friendly aerospace vehicles or the tracking of highly maneuverable targets. However, unlike aviation radars, the semi-active mode is used much more often. This is similar to two- or multiposition observance as well as the passive mode that provides tracking and hence guidance to operating radio facilities. It is in the latter case that corresponding homing devices are called passive, but they do not employ the principles of radiometry (radio/heat location). Homing devices using optical detection principles are executed according to a "radar" diagram, where a laser and collimator are used as the signal source. Compared to radar, such a homing device is characterized by higher accuracy in determining coordinates and velocity but is inferior in terms of all-weather capability and object search speed.

The current generation of radar homing devices implements practically all the achievements of modern radio electronics including broadband high-frequency operation, and may have minimal or completely absent analog paths—that is, they may be entirely digital. Their high-performance digital signal processors are capable of operating in any repetition modes and can track multiple targets during group operation, performing parallel analysis and determining both target velocity and range. They can also implement antijamming algorithms for the selection and identification of modes using assessments of the scattering and radiating properties of the targets (Kozlov, Logvin, and Sarychev, 2005–2008) and various other emitted signals including those for concealment control, by employing the antenna aperture synthesizing mode and controlling radar visibility by using adaptive multibeam APAAs. These processors make the entire radar data path digital, and APAA enables the conversion of flat antennas into conformal ones so that the transmitting and receiving array elements may be situated on the surface of the aerospace vehicle. This provides for a wide instantaneous field of view and an aperture with increased gain factor and also helps to minimize the RCS of the whole aerospace vehicle.

Radar homing devices, being effectively miniature radars (particularly those for military purposes), have some important features that distinguish them from the corresponding airborne and ground-based radar systems. Short-term and one-time operation regimes enable these homing devices to use transmitters having no forced cooling, but capable of operating throughout the vehicle flight time based on the thermal inertia of the structure. The relevant energy-intensive equipment, not being protected by radiators or coolers, together with miniaturization requirements for fitting into allocated volumes within the aerospace vehicle and with limited weight parameters, lead to the cautious use of current CAD/CAM/CAE methods and also the taking of manual design techniques into full account. These techniques are based on the accumulated experience of development schools active in design methods common in avionics. It is in homing device aerospace vehicle systems design, plus a high degree of software implementation, that represents the main know–how of the homing device manufacturer.

Depending on the location of the primary source (transmitter), active, semi-active, passive, and combined homing devices may be distinguished. In the active homing case, the target is irradiated by an onboard transmitter, and the reflected signal from the target environment is then used to control the aerospace vehicle. Under semi-active homing, the target is irradiated by a transmitter remote from the homing device and a scattered signal is received by an onboard receiver. There may also be situations where the target itself is a source of electromagnetic emission, for example, from active radio facilities or via thermal radiation from a running engine. In these cases it is possible to use passive homing by reference to a signal radiated by the target itself, when the homing vehicle requires only a receiver. Mixed-type nonautonomous systems include various combinations of homing devices. Depending on the situation, use is made of the signal from the target itself, and this ensures the best guidance characteristics in terms of accuracy, jamming immunity, and effective range. Generally, the use of combined homing devices increases the effective range and improves the jamming immunity, guidance accuracy, and operational reliability. Depending on the frequency ranges of the electromagnetic waves involved, the relevant homing signals may be radio-electronic (radar), thermal (IR), and optical.

The integrated nature of radar homing devices has made itself felt in the transition to architecture based on combined active/passive radar systems. Guidance to a specified target area may be performed using a high-resolution channel integrated into a radar homing device such as a millimetre-waveband radar channel with high spatial resolution. Also involved are the antenna aperture synthesizing mode (Pasmurov and Zinoviev, 2005) and UWB signals (Astanin and Kostylev, 1997), the latter being capable of influencing the jamming environment, too. In some cases an alternative to a high-resolution radar channel can be optical or IR imaging, the laser channels being included in the homing device.

Good target environment image resolution can be achieved through the use of the Doppler Beam Sharp (DBS) mode or by the use of the synthesized antenna aperture method as mentioned earlier, both modes providing for high azimuthal linear resolution (directional range). However, their use in homing devices is associated with certain difficulties, the main ones being discrepancies between the requirements for providing good linear resolution in azimuth and target guidance accuracies in addition to the necessity of compensating for trajectory instabilities during aperture synthesizing.

To achieve high linear resolution in the DBS or the synthesized antenna aperture modes, an aerospace vehicle should move at an angle of at least 10–15° to the target. In this connection, new guidance methods are required, which by means of corresponding trajectory bending would simultaneously meet both the high resolution and high guidance accuracy requirements.

The tasks undertaken by a homing device during the final homing segment are as follows (Kanashchenkov and Merkulov, 2003):

1. Controlled surface scanning over a selected angle range within the target area
2. Inner-scan and interscan processing of received signals
3. Detection, classification, identification, and selection of a target
4. Target capture and autotracking
5. Measurement of target coordinates and their derivatives, and the shaping and issuing of phase coordinate assessment vectors for target homing
6. Target fragment identification for provision of selective and high-accuracy guidance into the specified target area
7. Analysis of the jamming situation and, if required, the activation of antijamming facilities or aerospace vehicle re-aiming toward the active jammer
8. Generation of preparation and control signals for a radio proximity fuze.

An analysis of this list of tasks reveals that for the majority, the methods, equipments, and algorithmic procedures providing solutions close to optimal have now been developed. Using high-resolution data channels such as those with synthesized antenna apertures, or via the application of UWB-signals, it appears possible in principle to guide an aerospace vehicle into an area of higher target vulnerability, so allowing a significant improvement in the operational efficiency.

The fineness of the image resolution received by a radar homing device, as determined by the dimensions of the resolution elements in terms of radial and directional range, should be coordinated with, and protected against, possible errors in the homing procedures. It is therefore necessary to itemize the interrelationships between the trajectory type and the achieved resolution when using different modes of radar surveillance. Specifically, when the aerospace vehicle is caused to move along a flight path by flight director surveillance control, it should be in such a way that not only are the tactical missions fulfilled but also the best conditions for target environment observation are created.

The cross-linear resolution $\delta$ ($\delta$RC, where RC is the range cross) at a target range RN in the DBS mode is as follows:

$$\delta RC = \sqrt{\frac{\lambda \times RN}{2}} \tag{7.2}$$

Here, $\lambda$ is the wavelength of homing device radar channel.

This is much less (by more than an order of magnitude) than the same parameter of a traditional antenna homing device beam with an aperture size AP:

$$\delta RC = \frac{\lambda}{AP} \times RN \tag{7.3}$$

Consequently, the use of the DBS mode will ensure high-precision guidance of an aerospace vehicle to a target.

Still higher values of $\delta$RC can be achieved when implementing the antenna aperture synthesizing mode. In principle, this mode makes it possible to obtain almost any resolution, determined only by the length of the synthesis interval. Concurrently, the synthesizing mode allows the accuracy of the aerospace vehicle true flight path presentation to be more stringent, too. In turn, the requirements of the micronavigation sensor (used when implementing the synthesizing mode) specifications also become more stringent, resulting in the necessity of reaching a compromise between the obtainable values of $\delta$RC and the requirements of the micronavigation sensors.

It follows that the most expedient mode of trajectory signal processing for high-precision homing is the DBS mode. However, application of this mode in homing devices requires plotting a curved flight path that would satisfy the requirements of both guidance accuracy and of obtaining the required resolution. Usually, a curved flight path is only executed during the initial segment of an aerospace vehicle flight, and the final segment is usually a precision homing to an immobile surface target or to a moving target. With such flight paths, the power consumption of aerospace vehicle control devices increases, and in this context the issue of homing procedure efficiency improvement becomes urgent. Usually, transition to homing in the real beam mode occurs only if the tracked target or its components are resolved in azimuth.

To handle the homing task using the DBS mode, several different homing trajectories can be suggested, differing in the position of the active segment initial point and, consequently, having different initial relative bearings, target heading parameters, and trajectory lengths, but having the same ranges to the target.

Given the accumulation of signals reflected from the observed terrain segment, the required Time of Processing (TPR) may be estimated:

$$TPR = \sqrt{\frac{\lambda \times RN}{2 \times V_{AEV}^2 \times \sin^2 (PEL)}} \tag{7.4}$$

Here, the PEL is the current relative bearing of the target, and $V_{AEV}$ is the aerospace vehicle flight speed.

Consequently, the ratio for Resolution (7.2) is made more precise:

$$\delta RC = \frac{\lambda \times RN}{2 \times V_{AEV} \times TPR} \tag{7.5}$$

During flight along a specified path, the target's relative bearing decreases, which leads to a gradual turn of the aerospace vehicle toward that target. Hence, the vehicle performs no abrupt maneuvers accompanied by fast target bearing changes, which simplifies the homing task and makes the requirements of the aerospace vehicle aerodynamic parameters less stringent.

It is generally presumed that to detect targets against actual backgrounds, it is necessary to limit the values of linear azimuth resolution $\delta RC$ and range $\delta RN$ to no more than 5–10 m.

Linear resolution in a range $\delta RN$ will require a Width of Band Signal (WBS) given by the following equation:

$$WBS = \frac{c}{2 \times \delta RN} \tag{7.6}$$

Here, $c = 3 \times 10^8$ m/s.

To obtain a fine-detail image of a ground surface area, the control of a homing device antenna pattern comes down to setting a pattern maximum in the direction of the center of this area and holding it in that direction during the aerospace vehicle's motion along its flight path. Given sufficiently accurate settings and stabilization of the antenna, the use of telescopic scanning becomes possible.

To provide an unambiguous measurement of the maximum Doppler shift of the target-reflected signal frequency, a coherent pulse sequence should be used as a sounding signal and the Pulse Repetition Frequency (PRF) should be no less than

$$FRP = \frac{2 \times V_{AEV}}{\lambda}, \tag{7.7}$$

giving the maximum unambiguous range $RN_{MUA}$ as follows:

$$RN_{MUA} = \frac{C}{2 \times FRP} \tag{7.8}$$

Usually, homing complies with the condition $RN_{MUA} < RN$, and so-called blind zones appear that impose limitations on the selection of particular values of the frequency of pulse repetition, PRF. A signal reflected from the same range zone is made to fall within the same time interval (range gate) throughout the time of TPR processing. To achieve a still better resolution (by approximately an order of magnitude), the $\delta RC$ during trajectory signal processing—the squared phase taper—should be considered. The compensation of trajectory instabilities of an object in motion can be performed based on auto focusing methods.

The radar image obtained by means of synthesizing is characterized by peculiarities connected with speckle-noise. Image spottiness can significantly degrade the probabilities of detection and identification of small-sized ground targets. The impact of this speckle-noise is attenuated by means of the incoherent accumulation of several independent images of the same terrain area. The impact of this speckle-noise is attenuated by means of the accumulation of several independent incoherent images of the same terrain area.

A major error in homing device antennae (e.g., monopulse types) adjustment does not necessarily affect the accuracy of target (aiming point) coordinate measurement. It is only on the final segment near the target that the required accuracy is attained with the help of the homing device monopulse antenna. To attain the required accuracies in the measurement of target coordinate corrections in azimuth and elevation, and for further target tracking, it is necessary to use the method of target angle measurement relative to the ground speed vector. This may be accomplished by comparing the target signal Doppler frequency against the maximum frequency of the background signal spectrum in the same resolved range element as the target. The most important advantage of this method is the independence of angle estimations from errors in the navigation system of ground speed vector measurement, errors in the antenna pattern tie-in to the aerospace vehicle's construction line, and errors in the coordinate recalculations from one system to another. Selection of high repetition frequency sounding signals ensures unambiguous frequency measurement, and the concomitant angle measurement accuracy is achieved by the accuracy of the target and background signal Doppler frequency shift measurements (Kanashchenkov and Merkulov, 2003).

Corrections to aiming point coordinates may then be determined, after which the homing device autotracking system is changed to the aiming point capture mode, which results in the initial conditions being entered into the tracking system in the form of aiming point range coordinates, angular coordinates, and closing-in velocity. Then, the initial capture error follow-up is performed, wherein information on the constituents of the closing-in velocity, line-of-sight angular velocity, and angular position of the target in both planes is used in the homing algorithms.

Software-based implementation of the autotracking algorithm can be presented in the form of two main procedures: first, concerning the output signals from multidimensional frequency-sensitive detectors, which are angular deviation functions of the tracked target from the aerospace vehicle velocity vector in azimuth, elevation angle, and range; and second, for the filtering of the target coordinate parameters. To provide homing, a flight director control algorithm with linear resolution stabilization in azimuth and range (the area of the resolution elements) can be used, in which case the mismatched parameters required for shaping control signals (CNT) in azimuth $\text{CNT}_{AZ}$ and elevation $\text{CNT}_{EL}$ planes are determined by the following formulas, respectively (Kanashchenkov and Merkulov, 2003):

$$\text{CNT}_{AZ} = K_{AZ}\left(\hat{AZ}_{TG}(t) - \hat{AZ}_{TGO}(t)\right) = K_{AZ}\left(\hat{AZ}_{TG}(t) - \arcsin\frac{\hat{RN}(t) \times \lambda}{2 \times \hat{V}_{AEV}(t) \times \text{TPR} \times \delta\text{RC}_{AZ}}\right)$$

$$(7.9)$$

$$\text{CNT}_{EL} = K_{EL}\left(\hat{EL}_{TG}(t) - \hat{EL}_{TGO}(t)\right) = K_{EL}\left(\hat{EL}_{TG}(t) - \arcsin\frac{\hat{RN}(t) \times \lambda}{2 \times \hat{V}_{AEV}(t) \times \text{TPR} \times \delta\text{RN}}\right)$$

$$(7.10)$$

In (7.9) and (7.10), $K_{AZ}$ and $K_{EL}$ are factors of proportionality taking into account the weight of control signals over azimuth and elevation channels, respectively; the superscript points out corresponding smoothed estimates of the current values of target coordinates, azimuth $AZ_{TG}(t)$, elevation $EL_{TG}(t)$, range $RN(t)$, and the ground speed of the aerospace vehicle $V_{AEV}$; $AZ_{TGO}(t)$, elevation $EL_{TGO}(t)$. Required values (O meaning "order") of the target azimuth and elevation provide consistency in the linear resolutions of azimuth $\delta RC_{AZ}$ and range $\delta RN$.

When the aerospace vehicle is guided in accordance with the flight director control algorithms (7.9) and (7.10), its flight is performed along a curved path and a constant resolution area is provided.

According to the scheme of control signal reception (7.9) and (7.10), the ratios may be rewritten as follows:

$$CNT_{AZ1} = K_1 \times MIS(AZ), \quad CNT_{EL1} = K_2 \times MIS(EL) \tag{7.11}$$

Here, MIS (mistake) is a tracking mistake by some target characteristic. The "multidimensional" procedures for control are then shaped (Kanashchenkov and Merkulov, 2003) as follows:

$$\begin{aligned} CNT_{AZ2} &= K_3 \times MIS(AZ) + K_4 \times MIS(\omega_{AZ}), \\ CNT_{EL2} &= K_5 \times MIS(EL) + K_6 \times MIS(\omega_{EL}) \end{aligned} \tag{7.12}$$

$$\begin{aligned} CNT_{AZ3} &= K_7 \times MIS(AZ) + K_8 \times MIS(\omega_{AZ}) + K_9 \times \omega_{AZ,TG} \\ CNT_{EL3} &= K_{10} \times MIS(EL) + K_{11} \times MIS(\omega_{EL}) + K_{12} \times \omega_{EL,TG} \end{aligned} \tag{7.13}$$

$$\begin{aligned} CNT_{AZ4} &= K_{13} \times MIS(AZ) + K_{15} \times MIS(\omega_{AZ}) + K_{16} \times \omega_{AZ,TG} + K_{17} \times \frac{d\omega_{AZ,TG}}{dt} \\ CNT_{EL4} &= K_{18} \times MIS(EL) + K_{19} \times MIS(\omega_{EL}) + K_{20} \times \omega_{EL,TG} + K_{21} \times \frac{d\omega_{EL,TG}}{dt} \end{aligned} \tag{7.14}$$

$$\begin{aligned} CNT_{AZ5} &= K_{22} \times MIS(AZ) + K_{23} \times MIS(\omega_{AZ}) + K_{24} \times \omega_{AZ,TG} + K_{25} \times \frac{d\omega_{AZ,TG}}{dt} + K_{26} \times \frac{d^2\omega_{AZ,TG}}{dt^2} \\[2mm] CNT_{EL5} &= K_{27} \times MIS(EL) + K_{28} \times MIS(\omega_{EL}) + K_{29} \times \omega_{EL,TG} + K_{30} \times \frac{d\omega_{EL,TG}}{dt} + K_{31} \times \frac{d^2\omega_{EL,TG}}{dt^2} \end{aligned}$$
$$\tag{7.15}$$

In (7.12)–(7.15), $\omega$ designates corresponding angular velocities of the target's line of sight.

Procedures (7.14)–(7.15) determine the issuance of control signals when homing to a maneuvering target.

## 7.4   Characteristics of Homing Methods

### 7.4.1   Aerospace Vehicle Homing Methods

Autonomous homing is essentially a method for reckoning a path relative to the known initial position of an aerospace vehicle. Since such reckoning can be performed for long-distance flights, the selection of a coordinate system is very important. In homing procedures engineering, use is made of geocentric, geodetic, orthodromic, and other coordinates involving a zero point aligned with the Earth's center. The software for such long-range homing usually includes algorithms for converting coordinates from one system into another. Geocentric and geodetic coordinates for polar areas lead to instability of the differential equations describing the aerospace vehicle motion, so the number of such equations has to be significantly increased. For this reason, homing devices in the polar areas use the so-called all-latitude path reckoning algorithms.

Homing executed at small distances from a target is based on the assumption that the Earth is flat, which makes possible the selection of rectangular coordinate systems such as normal Earth-referenced, normal, and trajectory (Kanashchenkov and Merkulov, 2003). The zero point of the Earth-referenced coordinate system is aligned with some conditional reference point on the Earth's surface. When describing aerospace vehicle flight dynamics in the atmosphere, Earth-referenced coordinate systems are normally considered inertial. Under such assumptions, the absolute velocity vector of the vehicle's center-of-mass motion is replaced by the ground velocity vector. Similarly, the absolute angular velocity vector is replaced by the angular speed vector relative to the normal Earth-referenced coordinate system—that is, the ground angular velocity. The directions of the two axes in this coordinate system are selected as unchangeable relative to the Earth—that is, they are tangential to the Earth's meridian and parallel. Hence, the third coordinate axis will be directed upward along the local vertical.

A normal coordinate system is a moving coordinate system, the zero point of which is usually aligned with the center of mass of the aerospace vehicle, one of the axes being directed along the local vertical.

The zero point of a trajectory coordinate system is aligned with the aerospace vehicle's center-of-mass, with one of its axes coinciding with the vehicle's ground velocity vector and another again directed along the local vertical. The vehicle's motion trajectory curvature radii in the horizontal and vertical planes can be considered in this coordinate system.

The homing procedure proper, wherein primary attention is paid to the measurement of the target motion coordinates and parameters, additionally uses a number of rectangular coordinate systems, including the adjustment, antenna, and aerospace vehicle–referenced coordinate systems, the zero points for which are aligned with the aerospace vehicle's center of mass. The adjustment coordinate system is set to an angle corresponding to the selected zero direction of the homing device data channel. The antenna coordinate system sets the direction of one of the axes along the antenna system target's line of sight. The aerospace vehicle–referenced coordinate system, as its name implies, is aligned with the main axis of the vehicle, usually with the construction line.

If it is required to provide a group sweep of aerospace vehicles for military purposes, the vehicle groups are divided into closed, open, and dispersed formations. Sometimes, the terms are used to specify the form of a group of aerospace vehicles such as front formation, v-formation, convoy, and bearing, among others. Intervehicle navigation terms may also be used as applicable: plane-to-plane, between UCAV, between space objects, and between cruise missiles. Of course, the ultimate case here is the docking of aerospace vehicles; while for military applications, the availability of decoy targets having dynamic characteristics that may differ greatly from those of the combat units is of particular importance for combat formation arrangements.

In Equations 7.9–7.15, an important role in the generation of control commands is played by the procedure for obtaining the required parameter values, which obviously depends on the selected aerospace vehicle guidance law to the target. For this reason, the required ratios must be substantiated, so linking them to the implementation of the homing procedure.

Only two-point guidance methods are implemented in homing. A phase trajectory can be plotted for any homing method: the motion of a nonmaterial point (the aerospace vehicle) relative to another nonmaterial point (the target). Consideration of the response times of both the aerospace vehicle and the target makes it possible to plot a dynamic trajectory in phase space, but the vehicle's actual trajectory is also influenced by jamming and noise affecting the closed control loop.

For all two-point guidance methods, the target bearing must be estimated on board the aerospace vehicle, for which purpose the angular phase coordinates AZ and EL as applied in Equations 7.9–7.15 are used to determine the target angular position relative to the longitudinal axis of the vehicle.

In the case of direct guidance (Kanashchenkov and Merkulov, 2003), the aerospace vehicle's longitudinal axis will always be directed toward the target. Consequently, the required values of the relative bearings will be equal to zero, $AZ_{TGO} = EL_{TGO} = 0$, whence the airspeed vector of the vehicle will be aligned with the direction-to-target. Sometimes this is called the "pursuit method," and it may be applied for guidance to a stationary or slow-moving target in the absence of crosswind.

Advantages of the direct guidance method are its invariance with respect to the guidance range and altitude of the target, and also the simplicity of the homing device, which is essentially a clinometer that records the target bearings $AZ_{TG}$ and $EL_{TG}$ in its own coordinate system. This method implies an essentially straight-line homing trajectory for any aspect angle to an observed stationary target, and the actual curvature of the guided vehicle's flight path is larger for a lower vehicle speed and a higher wind velocity.

In the case of moving targets, the direct method provides for guidance along the "curve of pursuit." Such a method requires finding in two planes the sliding angle and the angle of attack, respectively, for estimating the vector of the aerospace vehicle's speed relative to its longitudinal axis.

Fast-moving targets generate a curve of pursuit with the curvature so large that an aerospace vehicle may be rendered incapable of making a turn within its allowable range of transverse g-loads. Flight along such a curve of pursuit also leads to unacceptably great guidance errors, a decrease in the effective range of the homing device, and an increase in guidance time. All these effects contribute to major homing disruption under electronic countermeasures.

As can be seen from Equations 7.9–7.15, the laws of control in azimuth and elevation bearings have similar structures. Hence, the use of notation subscripts 1 and 2 makes it possible to tie in the control law with the plane (not necessarily those for azimuth or elevation) in which the vehicle performs control. In this case bearings $PEL_1$ and $PEL_2$ may be used in expressions like Equation 7.4.

The method of pursuit with complementary lead angle is sometimes called continuous lead guidance (Kanashchenkov and Merkulov, 2003). Here, the control signals, together with summands proportional to a ground object relative bearing projections, also contain projections of the complementary lead angle, which is selected to be proportional to the line-of-sight angular velocity. When using this continuous lead guidance method, the flight director control algorithm can be presented in the following form:

$$CNT_{1,2} = K_{PEL1,2} \times PEL_{1,2} + K_{\omega 1,2} \times \omega_{1,2}. \tag{7.16}$$

With the help of acceleration $\omega_{1,2}$ the condition for the minimization of the aerospace vehicle ultimate miss under direct guidance may be written so that the required value of the line-of-sight angular velocity would satisfy the condition $\omega_{1,2\text{TGO}} = 0$.

From homing to fast-moving targets, variants of guidance to a set-forward point are used, most often the proportional guidance method (Kanashchenkov and Merkulov, 2003), wherein control signals contain summands proportional to the line-of-sight angular velocity:

$$\text{CNT}_{1,2} = N_0 \times \left( -\frac{d\text{RN}}{dt} \right) \times \omega_{1,2} - a_{1,2\text{AEV}}. \tag{7.17}$$

When using the proportional guidance method, a required lateral acceleration $\mathbf{a}_{1,2}$ and excess load $\text{EXS}_{1,2}$ in the control plane must be proportional to the line-of-sight angular velocity $\omega_{1,2}$ and the velocity of the vehicle closure to the target $\mathbf{V}_{\text{CL}}$:

$$\mathbf{a}_{1,2} = N_0 \times \mathbf{V}_{\text{CL}} \times \omega_{1,2}$$
$$\text{EXS}_{1,2} = \frac{N_0}{g} \times \mathbf{V}_{\text{CL}} \times \omega_{1,2}, \tag{7.18}$$

where $N_0$ is the navigation parameter (dimensionless), $g$ is the acceleration of free fall, and $\text{EXS}_{1,2}$ is dimensionless. Most often, $\mathbf{V}_{\text{CL}}$ and $\omega_{1,2}$ are estimated by radar sensors (velocity autoselector and clinometer), and acceleration by control plane accelerometers. The main attributes of this method are all-aspect and all-altitude capability, and almost straight-line guidance trajectories.

In the course of homing by the proportional guidance method, the homing device target's line-of-sight $\mathbf{r}_{\text{HDT}}$ tends to move in parallel with itself. Here the lead angle is a variable depending on the range to target. Under this guidance method, it is not necessary to measure the vector of the target motion velocity or to determine its lead position. The target velocity vector changes the angular position of the line-of-sight $\mathbf{r}_{\text{HDT}}$ in space, or more precisely the tangential constituents of the target velocity vector and the angular position of the line of sight. To implement parallel closing-in, it is necessary to determine the derivative of angle $\text{PEL}_{\text{HDT}}$ and change the position of the homing device velocity vector proportionally to this in such a way that the target's line of sight moves in parallel with itself. The homing device is consequently rendered more complex through the addition of a system of gyros that record the angular position of the line of sight as early as at the start. This method is well-conjugated with all guidance methods used by homing devices and does not respond to the impact of wind loads on the aerospace vehicle.

The greater the navigation parameter $N_0$, the straighter the flight path of the aerospace vehicle and the longer the effective range of the onboard avionics with the same fuel reserve. At the same time, the course of the homing parameter $N_0$ will be converted into guidance errors (misses). With $N_0 = 3$, the guidance trajectory becomes practically a straight line. Given such values, steady guidance is also ensured for any interception aspect angles.

A serious drawback of the proportional guidance method is poor controllability of the homing device at long ranges because the initial aiming (target designation) errors will not be eliminated. When homing to maneuvering targets, variants of the proportional guidance method with a shift are generally used (Kanashchenkov and Merkulov, 2003), where an additional summand is introduced that is proportional to the increments of line-of-sight angular velocity in corresponding planes caused by a ground object maneuver:

$$\text{CNT}_{1,2} = N_0 \times \mathbf{V}_{CL} \times \left( \omega_{1,2} + \text{MIS}\omega_{1,2} \right) - \mathbf{a}_{1,2} \qquad (7.19)$$

where $\text{MIS}\omega_{1,2}$ are the increments of line-of-sight angular velocity measured by the system clinometer and conditioned by the target maneuvers.

One of the variants of the flight director control algorithm provides all-aspect homing to highly maneuverable targets. This is the continuous lead guidance algorithm:

$$\text{CNT}_{1,2} = a_{1,2\text{TGO}} - a_{1,2\text{TG}} = \frac{N_{\text{PEL}1,2}}{\hat{V}_{CL}} \times \left( \text{PEL}_{\text{TG}1,2} - \text{PEL}_{\text{AEV}1,2} \right) + \frac{N_{1,2\omega}}{\text{RN}} \times \hat{\omega}_{1,2} - \hat{a}_{1,2\text{AEV}} \qquad (7.20)$$

At the present time, homing devices often employ the technique of augmented proportional navigation based on extrapolating noise-adaptive and range-dependent schemes. In this case, the line-of-sight angular velocity is converted into homing device acceleration transverse to the line of sight.

In homing devices with loops of the follow-up type, the main role is played by the homing angular channels, as was mentioned earlier. Selectivity improvements in the goniometric position indicators are connected with the necessity for narrowing the antenna system pattern, which causes increases in the weight and dimensional characteristics of the antenna devices. The arrangement of such devices on small aerospace vehicles can be a challenging task, for which reason additional measures are used in homing devices for the spatial selection of targets falling within the receiving antenna directional pattern opening. In systems using pulsed radiation, follow-up range finders are used as additional selectors, and in systems operating in continuous radiation modes, follow-up selection by velocity is employed.

The main purpose of range-finding selection is in the unblanking of the homing device servo system receiver only for the time of arrival of signals reflected from the selected target. For the rest of the time, the receiver is blanked so that signals from other targets do not pass through the receiving channel and so do not affect the angular position indicator. In this way, aerospace vehicle's directional guidance is performed only toward those targets that are captured and tracked by the range tracking channel. The pulse signal generated by this channel that opens the homing device receiving channel for a short time is often referred to as a "range gate."

The autotracking of targets by velocity provides the possibility of receiving continuous information on the motion velocity of that tracked target and enables the selection of moving targets against a background of stationary or slow-moving objects in the target environment. An important function of this automatic velocity tracking channel (as with the tracking-by-range channel) is the additional spatial selection of group targets falling within the limits of the homing device receiving antenna directional pattern. Differences in the radial constituents of the group individual target velocities will produce different frequency increments in the received signals, which can then be separated by narrow-bandwidth filters.

The range- and range-derivative finding channels featured in the homing devices provide for the shaping of the range estimates for a specified point or object, the closing-in speed, the target's absolute velocity, and the automatic selection of objects by range and velocity. These estimates

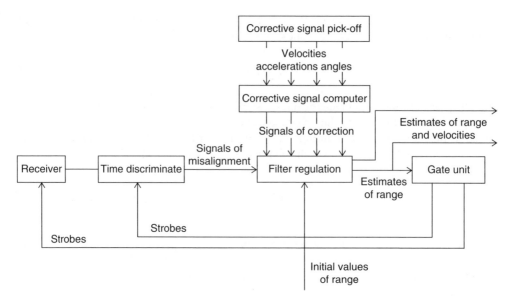

**Figure 7.2**   The range and velocity auto selector scheme

are necessary for the generation of efficient aerospace vehicle status control commands, and the relevant sensors are often called velocity and range autoselectors. These autoselectors increase the selectivity of the homing devices as well as their jamming immunity because of the reduced opening state times of the receiving devices. Usually, an autoselector is made according to a schematic that takes into account filter delay times and derivatives and is comprised of a timing discriminator (the sensitive element) and a control circuit (the filter and governor). For shaping the estimates of range and its derivatives, a gate arranging device plays the role of an actuator, a sensor for correction signals, and a correction signal computer (Figure 7.2).

The main problems that need solving in the creation of homing devices are concerned with the implementation of maximum and minimum ranges in target detection, provision for high degrees of jamming immunity, provision for high guidance accuracy with emphasis on prospective low-observability and high-maneuverability targets, the fulfillment of requirements to minimize weight and dimensional characteristics, and the generation of estimates of all the phase coordinates required for aerospace vehicle control.

## 7.4.2   Homing Device Dynamic Errors

Homing devices are located on aerospace vehicles that are moving bodies performing in-flight angular oscillations in orthogonal control planes and in roll. Hence, control signals transmitted to such vehicles are shaped by the measured line-of-sight angular velocities, and it is obvious that the homing device antenna must be stabilized in space, otherwise body oscillations would add together with the line-of-sight angular velocity and the vehicle would be controlled by false signals. Another false signal source is antenna beam distortion due to the influence of the homing device radome, called synchronous errors (Kanashchenkov and Merkulov, 2003). These may have very strong influences on homing accuracy.

The homing channel of a homing device is a complex, nonlinear, nonstationary stochastic system whose dynamic (inertial) characteristics are directly projected onto the homing errors (Siouris, 2004). Control signal filtering based on the calculation of acceleration control is usually designed for noise smoothing when computing line-of-sight angular velocity, for damping any residual movements in the aerospace vehicle body (conditioned by synchronous errors), and for minimizing the effects of data issue variable frequencies.

The navigation parameter depends on the time of flight to the target and the control input weight ratio. An estimate of target maneuver acceleration is based on measurements of the homing device line-of-sight angular data. Additionally, homing devices have to adapt to current conditions in accordance with *a priori* information on the target type, the dynamic characteristics of the aerospace vehicle on the final segment of the trajectory, and the quality of information on the range and homing device measurements. Homing to air targets requires knowledge of the closing-in speed, the acceleration vectors of both the target and the aerospace vehicle, and the line-of-sight angular velocity vector. Moreover, ranges of homing start and finish as determined by the aerospace vehicle performance and are preset into the control law as parameters. Homing devices implementing control procedures for ground targets comprise shapers for the estimates of range, velocity, drift angle, relative bearing, the line-of-sight angular velocity vector, and the vehicle's own acceleration vector.

Currently, homing devices are frequently equipped with antenna aperture–synthesizing devices. Here, linear resolution improvement by angular coordinates is only achieved when an aerospace vehicle moves at a rather large angle to the line-of-sight, whereas the aerospace vehicle tracking line must coincide with that line-of-sight in order to arrive at the target. Hence, the flight director control algorithms inherent in homing tasks must simultaneously satisfy contradictory requirements to preset angular resolution and miss value. Here, homing is accompanied by the vehicle's flight along a curved path, whereas the antenna aperture synthesizing is performed on the initial segment and the corrective turn toward the target on the final one.

Many of the listed problems are solved by semi-active homing with target illumination from an additional source located remotely to the homing device.

Deviations of the actual homing device flight path from the preset flight path occur due to inaccuracies in the performance of procedures for assessing the control parameters required for homing, and are considered as manifestations of the corresponding dynamic errors.

## 7.5    Homing Device Efficiency

Homing device efficiency is understood to mean the degree of compliance with the requirements of the purpose, that is, the efficiency of the control system over the aerospace vehicle status in accordance with its allotted tasks. As with any human-triggered systems, homing system efficiency may be assessed by means of indices (quantitative assessments) and criteria (rules according to which the degree of compliance with the purpose is determined). Distinguished among homing device quality factors are *final* (global), *individual*, and *composite* indices. Final indices assess the results of the homing device performance having regard to the effects of all the stages and modes of the vehicle's operation. Individual indices characterize the operational efficiencies of all the homing device subsystems and units (the execution of all the separate operations) in order to use these assessments for passing judgment on the overall homing device efficiency. The most common individual

quality factors for the whole aerospace vehicle include the following performance characteristics: the range of altitudes, velocities and effective ranges, the number and types of
controls, the ranges of allowable overloads, the response rate characteristics, the flight
path, and the dynamics equations. In the military context, if homing devices are for weapon
delivery, the most important performance indices are the dimensions of application and hit
zones. Of course, many composite or combined indices also exist.

Homing device performance indices include the following: controllability, reliability,
response time, the set of control signals, bandwidth, and observability. For the information
and computation system, the probabilities of correct target detection or false alarm, the filing
of view parameters, the accuracy and resolution, the jamming immunity, and the reliability,
etc. are usually used. In turn, these indices depend directly on more specific indices: the target
illuminating signal parameters, the antenna directional characteristics, the receiver sensitivity,
the transmitter power, the signal path passband, the gain factor, and the signal processor capabilities. A hierarchy of the individual indices influencing the final indices is usually compiled.
Particular indices that declare themselves at all stages of homing device operation are especially singled out, as opposed to those characterizing only certain operations. Examples of
all-stage indices are operation accuracy, jamming immunity, reliability, cost and convenience
of operation.

Since the particular indices depend on time and affect the efficiency of the whole homing
device operation in different ways, composite indices (also called *resultants*, *performance
functions* or *quality functions*) are often used. These include *extreme*, *limiting* (threshold),
*fixing*, and *mixed* criteria. Extreme efficiency criteria determine the rule in which the best
homing device is considered as that which provides extremes of particular indices. Limiting
criteria define a preference for those homing devices (or their operational modes) in which
indices no worse than specified ones are implemented. Fixing criteria require the implementation of indices equal to specified values. Mixed criteria represent certain combinations of
the first three.

For a maximum performance system criterion, a minimum homing error (miss) is used as
the extreme criterion. An example of a limiting criterion for a missile is the rule according to
which the homing error should not exceed the efficient hitting range. An example of a fixing
criterion is simply a zero miss (direct hit) criterion. The Neyman–Pearson criterion serves as
an example of a mixed criterion, requiring provision for the maximum probability of task execution given fixed resources. The final quality factors of homing devices are usually based on
threshold criteria. The optimization of homing devices via individual indices uses extreme and
limiting criteria.

## 7.5.1  Homing Device Accuracy

To assess homing accuracy, which is one of the most important individual indices for homing
devices, use is made of both instantaneous (current and isolated) and integral indices (Kee,
1992). Misses and control errors are used as instantaneous indices for characterizing homing
device accuracy at particular moments in time. Here, a miss is the distance between a target
and the homing device in the *scattering plane*. The scattering (or picture-wise) plane is
understood as a plane passing through the target's center of mass (or through a preset point
in an area where the aerospace vehicle is to be delivered), which is perpendicular to the

relative velocity vector of the homing device, the velocity vector being the difference between the absolute velocity vector of the homing device and that of the target.

A miss is determined not only by the parameters of the homing device itself but also by the conditions of its application. Assuming that the control channels of the homing device itself and the systems supported by it are ideal and not affected by each other, the dependency of the current miss on the guidance conditions is determined by the following ratio:

$$\text{MISS}_{1,2} = \frac{\text{RN}^2 \times \omega_{1,2}}{V_{\text{CL}}} \tag{7.21}$$

Both the velocity and the range are valid for the current moment in time, after which it is conventionally assumed that the homing device moves to the target entirely inertially. The final miss is determined in a similar way, but after the homing device has stopped issuing any control commands.

## 7.5.2   Homing Device Dead Zones

A dead zone is a phenomenon inherent in any aerospace vehicle's radio control system and is determined by the distance to a target or destination point at which the homing device stops functioning normally. Such dead zones occur for various reasons determined by the aerospace vehicle itself, the automatic adjustment closed loop system, and the target environment observance characteristics. In the dead zone the aerospace vehicle rudders are usually set into their neutral positions and the further direction of motion is determined by the velocity vector up to the moment when control stops.

Given that various specific reasons account for dead zone occurrence, in general each reason is matched by its own dead zone—but the dead zone radius is always the maximum value amongst all the dead zone radii. Thus, the miss value of the whole homing device is determined by the maximal value of the dead zone radius.

The main reasons for dead zone occurrence are as follows:

1. Instantaneous errors in homing devices exceeding some allowable value of a miss, which can be interpreted as being caused by random factors such as noise or jamming existing in a closed control loop;
2. Kinematic trajectory lift-off caused by limited aerospace vehicle maneuverability wherein the process of guidance method implementation is disturbed;
3. Kinematic trajectory lift-off caused by the response time of a closed loop;
4. Instability of an automatic adjustment closed loop system (often due to an increase in a kinematic link transmission factor being inversely proportional to the distance to the final point, which requires the implementation of special distance measurements);
5. The response time of the tracking system to the target or destination point;
6. Various forms of target noise (angular, range, velocity, polarization) caused by the granularity of scattering centers (Ostrovityanov and Basalov, 1985) the number and geometry of which continuously change during the approach;
7. Limitation of the dynamic range of the homing device at short distances to a target, where the received signal level increases significantly.

Signal limitation leading to increasing target designation errors at the minimum range of radio control arises due to the necessity of temporarily blocking the receiver after sending the probe radio signal. This is because, if the range is too small, the reflected signal will arrive too quickly and the receiver will be still blocked at that moment. One possible way of solving the receiving and transmitting antenna isolation problem would be to fit the transmitting part in the nose and the receiver processing circuit and safety-and-arming devices in the tail, which would maximize the distance between the homing device and the target.

At short ranges of approach to the guidance area, target noises are converted into target portraits (Kozlov, Logvin, and Sarychev, 2005–2008) so that the object outlines can be assessed and identified. Such channels are provided by false signals, including the UWB (Astanin and Kostylev, 1997; Kozlov, Logvin, and Sarychev, 2005–2008) signals, use of the antenna aperture synthesizing mode (Pasmurov and Zinoviev, 2005), and transition to the millimetre waveband (Clark, 1992).

## 7.6   Radio Proximity Fuze

The use of a radio proximity fuze represents one type of homing device, and usually involves a channel intended for the generation of a one-time command for military purposes, the operation of which is based on a target environment analysis for detonation of the warhead of a guided missile (Kanashchenkov and Merkulov, 2003; Zarchan, 2002). However, it should be noted that a precision homing device is not a radio proximity fuze and does not contain one in its structure, since it ensures a miss which is smaller than the allowable one. Currently, a radio proximity fuze fulfills the task of controllable destruction, where a missile warhead detonation ensures the required degree of hit. Here, it is important to consider the characteristics of the warhead itself so that in essence the radio proximity fuze selects an approach phase trajectory to the target in order to implement a specified mode of detonation in the target area. Typical examples include an anti-ship warhead detonation below the waterline, or detonation inside a bunker for the destruction of a heavily armored target.

A "canonical" radio proximity fuze is comprised of the following: a radio-frequency module, a program module with setting via an inductive channel, a receiver of the external signals used for trajectory correction, a brake unit for trajectory correction, a microelectromechanical safety-and-arming device, and a power source.

A radio proximity fuze may be included structurally in the homing device or sometimes separately, and it serves to operate in the dead zone of the "parent" homing device. It should have a hitting area that exceeds the value of the miss caused by the homing device dead zone radius, taking into account both isotropic and anisotropic types of warhead bursts. This *dynamic hitting area* is shaped after taking into account the deformation of the hitting area due to the velocities of the missile and the target. For an isotropic burst, the dynamic hitting area sphere center shifts forward along the aerospace vehicle's longitudinal axis, and for an anisotropic burst the nose angle is decreased relative to the initial (static) hitting area cone. Hence, the dynamic hitting area defines the radio proximity fuze actuation area because a one-time command for warhead detonation will be shaped only when the target is inside that area.

The dynamic hitting area is normally "translated" into the parameters of a filter procedure adjusted to Doppler frequency domains for the signal received from the target, which corresponds to the dynamic hitting area. A radio proximity fuze can be implemented as an amplitude,

Doppler or pulse type if the target spatial-temporal position information is extracted from amplitude, Doppler frequency or reflected signal delays, respectively (Kanashchenkov and Merkulov, 2003). Radio phase proximity fuzes are missing due to the presence of random phase constituents originating from the reflections of the sounding signals from the target surface. Currently, intelligent radio proximity fuzes are designed, the operation of which depends on the received target image and environment. Such radio proximity fuzes take destruction decisions autonomously by analyzing the situation in the target area and selecting a detonation point.

Amplitude radio proximity fuzes are usually actuated using the second derivative of the time of radar signal envelope change and are normally used along with isotropic warhead projectile bursts. Pulse radio proximity fuzes have sounding pulse durations of about one nanosecond and dead zones of about 15 cm. However, the possibility is being considered of designing a radio proximity fuze by applying short-pulse versions of UWB-signals, since in this case the dead zone is essentially included and a very high PRF with a duration of less than 200 ps can be used.

The Doppler radio proximity fuze is controlled by the Doppler frequency change, that is, the derivative of the low-frequency signal phase. It generates the destruction command when the Doppler frequency is comparable with a threshold value determined by the angle of the dynamic hitting area burst debris. Clearly, the radio proximity fuze match with the dynamic hitting area should be accompanied by setting the required direction of its antenna pattern maximum. When using the APPA, such a setting is carried out almost instantaneously and along with the generation of a corresponding component beam with parameters different from those of the target tracking beam (if a sheltered target is to be destroyed (see Section 7.3.1.1).

A Doppler radio proximity fuze block diagram is given in Figure 7.3. All its elements are typical for Doppler narrow-band radar with an element set for a dynamic hitting area. It contains two identical channels operating on various carrier frequencies for an actuation area overlapping in two mutually perpendicular planes passing through the aerospace vehicle's longitudinal axis. Two transmitting and two receiving antennas are located on opposite sides of the aerospace vehicle body for the purpose of mutual screening. The safety-and-arming

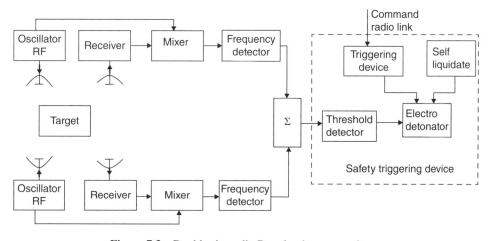

**Figure 7.3**  Double site radio Doppler detonator scheme

device provides for the safe storage and transportation of the radio proximity fuze, its arming upon command, and warhead detonation as it reaches the dynamic hitting area and self destruction.

## 7.7   Homing Device Functioning Under Jamming Conditions

The susceptibility of homing devices to the impact of jamming is determined by the use of signal fields, which can be perceived by homing device sensors. Under the influence of jamming signals a significant decrease in the detecting, measuring, and selecting characteristics of homing devices takes place, and this has been confirmed by a long history of radar and optoelectronic countermeasures. The influence of jamming has proved to be very efficient if the homing device operates according to an instantaneous field of view. Today, the organized jamming of homing device operation is mandatory when conducting military operations. The ongoing development of radar and optoelectronic warfare for the creation of an organized jamming and the associated application methods is a dynamic factor that will exert a significant and increasing influence on the result of combat operations. Combinations of jamming, counter-fire, and maneuvering measures may now be considered efficient. Homing device structures, especially for combat, comprise hardware and software that provide homing operation under real conditions of electronic warfare, and at the present time development is skewed toward software growth, usually performed within the framework of homing device updating.

During homing the main data channel is angular, functional failure in which is caused by the following types of jamming: coherent, polarization, blinking, discontinuous, redirecting, jamming from a dedicated space point, and amplitude modulation on the monopulse meter switching frequency (Chrzanowski, 1990; Kanashchenkov and Merkulov, 2003). Autonomous, semi-active, and passive homing devices are more jam-resistant. For vehicles using homing by the proportional navigation method, angular jamming is required, which causes major changes in the tracking angular velocity. For disrupting range and velocity autoselectors, the main types of jamming required are as follows: deflecting, aiming, and frequency-based barrage jamming, noise, passive, repeater jamming, and combinations thereof covering blinking and masking.

If homing devices are intelligent (see Section 7.8) and perform target environment identification, an account should be taken of the possibility of setting the corresponding jamming to cause a decrease in the RCS of real targets, the distortion of radar characteristics (Astanin et al., 1994), the simulation of large RCS with small-sized false targets, the simulation of the radar characteristics spectral structures, the simulation of false targets of natural origin, a combined (optical and passive radar) false target, the simulation of radar characteristics temporal patterns (Kozlov, Logvin, and Sarychev, 2005–2008), and the application of complex signals.

Appropriate individual declutter technologies are available for each type of homing device jamming. Also, to some extent there is a universal approach to providing homing device jamming immunity based on the execution of power, structural, and informational concealment, especially for cruising (hopping) path areas and for the creation of multidimensional excess informational signal fields stable in terms of jamming damage. This stability is implemented by combining active, semi-active, and passive sensors and also IR, optical, and radar methods using both heterogeneous and homogeneous physical fields different, for example, in frequency bands, types of signal, and polarization.

A single type of jamming cannot suppress all homing devices, for which reason special types of jamming signals are used that affect certain types of homing devices and channels. Moreover, types of jamming signals different from each other are used to suppress homing devices of the same class, but using different types of signals and processing methods. The classification of jamming following various criteria is available in Kanashchenkov and Merkulov (2003).

Depending on their origin, types of jamming are subdivided into nonorganized (natural, non-man-made, and electromagnetic) and organized (man-made artificial). Obviously, it is an organized jamming that plays the most important role during homing countermeasures.

Jamming devices are divided into active and passive types. Active jamming is created by jamming transmitters and radiated into the spatial area where homing devices are located. Passive jamming is performed by the reflection of sounding signals from suppressed homing devices using, for example, artificially created reflectors, dipole clouds, the changing of radio wave distribution properties, or a decrease in the RCS of protected objects.

Suppressing and simulating jamming methods are differentiated according to the nature of the jamming. Masking jamming suppresses homing devices by degrading their receiver properties and creating a background of clutter that significantly impedes, or completely disables, target detection and identification. It does this by singling out useful signals reflected from targets and making it impossible to measure the parameters of signals carrying target information such as spatial position, motion parameters, etc. with the required accuracy. Masking capacity increases with increases in jamming power.

Simulating (or deception) jamming imposes signals on homing device inputs similar to useful ones, but containing false values for some data parameters. This leads to the loss of part of the useful information and decreases the system bandwidth. The action of this form of suppressing jamming is based on the fact that homing device receiver amplification paths have limited input signal dynamic ranges. It is therefore possible to create a jamming signal power value at a homing device input where the receiving channels lose the ability to perform their functions of separating out useful information. Suppressing jamming can cause homing devices to lose a target or switch a tracking system from tracking mode to search mode.

For tactical applications, jamming is divided into *self-cover jamming* and *external cover jamming* (created for group protection). Self-cover jamming is where a target under attack carries the jamming source itself, whereupon so-called individual protection of the object is implemented. In the external cover case the jamming source is installed in a separate jammer included within a combat group to provide group protection. Jammers often patrol beyond the range of air defense systems and create so-called *out-cordon jamming*, which significantly increases protection and survivability.

Self-cover jamming is the most difficult because the jamming must be created and maximally aligned with the homing device signal field characteristics. In this case all (or the majority) of the source coordinates align with, or are converted into, virtual separation recordings via measurement channels as actually observed. The efficiency of self-cover jamming (and barrage jamming—see later) is significantly high due to the excess use of the homing vehicle's power resource in the presence of such jamming. Successfully matching assessments of both signal and jamming source coordinates significantly facilitate homing (usually, by direction) during relatively powerful jamming.

External cover jamming is most often utilized for IR homing devices and, in principle, may be chosen because of differences in the coordinates of useful signal sources and of jamming sources. Currently, jamming is usually set in combination according to types and sources (including self-cover and external cover jamming) that significantly expand jamming under the conditions in which the homing devices operate.

*Barrage* and *aimed* jamming are divided according to frequency band overlapping. Barrage jamming occupies a wide frequency band that many times exceeds the suppressed receiver bandwidth. Jamming of this type can suppress several homing devices operating at close frequencies and located in the same area. To create such jamming, it is sufficient to have knowledge about the approximate operating frequency bands of the suppressed homing devices, which is why the surveillance equipment controlling the jammers can be rather simple. The main disadvantage of barrage jamming is low power efficiency in the jamming transmitter that arises because only an insignificant part of that transmitter power is delivered to the suppressed receiver inputs of the homing devices.

Aimed jamming occupies a relatively narrow frequency band comparable with that of the suppressed device bandwidth. The jamming signal spectrum frequency median should approximately match the carrier frequency of the suppressed homing device. It follows that the transmitter power for aimed jamming is used more efficiently, but it is necessary to accurately know the carrier frequency of the suppressed homing devices. This complicates the jammer control system and requires the application of high-frequency generators with fast carrier frequency readjustment within a wider bandwidth.

In the case of the same jamming signal spectral power densities for aimed jamming creation, significantly lower weights and dimensional characteristics of the relevant equipment are possible compared with those for barrage jamming implementation. Hence, such jamming is most often used for the individual protection of homing devices.

Jamming can be continuous or pulsed. Continuous jamming consists of high-frequency continuous oscillations modulated (usually by a random law) in amplitude, frequency, and phase, and sometimes these modulations are used simultaneously. Depending on the type of modulation, Amplitude-Modulated (AM), Frequency-Modulated (FM), and Amplitude–Frequency-Modulated (AFM) jamming can differentiated. Noise can also be used as a modulating source, in which case continuous noise is implemented. Pulse jamming is created as a series of modulated or nonmodulated high-frequency pulses.

Homing devices include systems together with data and controlling channels that can provide targets for tracking in direction, range, velocity, and also radio-technical data transmission systems via a command radio line (e.g., to establish group interactions). Hence, homing device radio countermeasures must also have an integrated nature and impact various systems and channels, completely suppressing them or decreasing their efficiencies down to specified levels.

Radio countermeasure tactics should vary for different stages of control system operation so that the warhead detonates in an area safely away from the target. To aid this, continuous noise and pulse jamming can be applied for the cruising portion of a homing device flight.

Continuous noise jamming involves chaotic, random law oscillations in amplitude, frequency, and phase and is often called *fluctuation noise*. At a homing device input, this can be regarded as a random law process having a uniform frequency spectrum within the receiver bandwidth. Such jamming makes possible the masking of useful signals having any structure and shape and within a solid angle and a corresponding range. Because the structure of the noise is similar to internal fluctuation noise in receivers, it is difficult to detect and hence to take measures to diminish its impact on the homing device. Because of this, the homing device resolution degrades significantly and the target coordinate detection accuracy is decreased.

Direct noise jamming is usually the result of the amplification of noises generated within the amplifying devices themselves. It has high spectral uniformity and a sufficiently wide frequency bandwidth that enables it to overlap the wide signal spectrum band.

Modulation-type jamming is created using transmitters in which a carrier oscillation is modulated by a noise voltage in amplitude, frequency, or phase. In practice, combined amplitude–frequency or amplitude–phase modulation is most often used, in which changes in amplitude and frequency or amplitude and phase occur in the radiated signal at the same time.

The development of digital technologies makes it possible to use so-called noise-type signals for jamming. These involve shaped and reradiated signals having the special function of spreading the signal spectrum by carrier modulation. This spreading function is based on pseudorandom code sequences, usually binary, and these are often used either for the phase modulation of radiated signals or for the pseudorandom readjustment of the operating frequency. M-sequences, Gold, and Kasami sequences and others are used as code sequences.

Pulse jamming refers to the simulating class of jammers and is created for homing devices that operate in the pulse radiation mode. *Synchronous multiple pulse jamming* and *asynchronous chaotic pulse jamming* are different, as follows. The principle of synchronous pulse jamming is that the jammer receives signals radiated from homing devices and reradiates them on their carrier frequencies to the homing devices, but with delays. Several response pulses can be reradiated for each received signal and should correspond to the received signal pulses in shape, range, and power.

Using such jamming, it is possible to simulate nonexistent targets, and a group of such pseudotargets with similar angular coordinates, but located at different ranges, can be simulated. Given sufficient response jamming power, when reception is performed by the suppressed radar sidelobes, the angular coordinates of targets that do not match those of the jammer or other protected targets can be simulated. Not only false target motion with any headings but also different types of maneuver can be simulated by programed jamming radiation control. All this leads to the situation presented to homing devices becoming complex, making it necessary to process a large array of information, which dilutes the combat-oriented power intended for true targets.

Asynchronous chaotic pulse jamming refers to a sequence of radio pulses, the carrier frequency of which is close to that radiated by a homing device, and in which the amplitude, duration, and intervals between pulses are changed according to a random law. Such signals create false targets randomly scattered across the homing device field of view. This significantly complicates the selection of true targets in the presence of a large number of jamming marks appearing and disappearing at other ranges and other azimuths. Under certain circumstances the jamming impact on the suppressed homing device receiving antenna sidelobes is to make the field of view more complicated and with many false targets.

The influence of chaotic pulse jamming on command radio channels controlling various objects is actually *barrage jamming by code*. This can cause the complete or partial suppression of transmitted commands, change values of subcarrier modulation parameters, and generate false commands.

The main difficulty in multiple jamming creation is that the radiation of a response pulse series must be performed on the carrier frequencies of radiated signals at those moments when those signals are missing at the jammer receiving antenna input. For this reason, the radiated signal carrier frequency must be stored for a relatively long time and be approximately equal to the radiation pulse repetition period.

A significant decrease in channel numbers can be obtained by multichanneling using the so-called matrix method of radiated signal frequency storing, and the implementation of such a method is performed using multiple (stepped) conversion of input signal frequencies. The general disadvantage of multichannel systems is the possibility of mutual channel impacting that may result in ambiguities in the identification and storing of the radiation signal frequencies. However, this disadvantage may be eliminated by means of special logic designs that remove ambiguity by the application of special isolation circuits and by improvements in the characteristics of frequency-selective filters. Recently, digital signal storing methods ensuring the precise playback of target radiation signal structures have been developed.

The jamming of automatic tracking channels for angular coordinates, range, and velocity of an object's motion in space—that is, disrupting their normal operation—is one of the most important tasks of radio-electronic countermeasures. In general, jamming should change the characteristics of tracking channel converters including degradation of the perceived dynamic properties of targeted objects, the encumbering of protected target tracking and capture systems with intolerably large errors, and switching tracking systems from tracking mode into search mode. However, as already mentioned, universal jamming with identically efficient impacts on different automatic tracking channels, and its acceptable implementation from a technical and economical viewpoint, does not exist at present. Hence, optimum jamming sets have been developed for each channel.

Automatic target direction tracking systems and channels are the main components in any homing device loop. In most cases, any loss of information on the target's angular coordinates can lead to a mission failure, or at least a homing interruption. Hence, it is essential that this information continuously reaches the tracking system, and it does so using radio bearing methods.

In homing devices, directional automatic tracking channels involving monopulse direction-finding methods are used because these offer higher accuracies in target direction determination as compared with other methods. This tracking is performed by means of a control unit and actuator performing shifts in space in accordance with the direction finder output signals. Sometimes, the procedure for angular tracking is modified by the aerospace vehicle rolling around its axis, especially if such a rotation is provided aerodynamically or by means of a low-power reactive engine.

The task of radio countermeasures to directional automatic tracking by an aerospace vehicle lies in the shaping of jamming signals simulating false targets, the directions to which do not match the true target direction.

In the case of group protection jamming radiated from two or more points in space, a group target "wander" energy center can be created and this center may be located outside the target volume. Such wandering has a special name—"target noise"—and this may be displayed in distance, angular coordinates, and velocity, thus increasing an aerospace vehicle's homing errors that can lead to the interruption of tracking by angular coordinates.

Methods for jamming homing devices depend on the types of directional automatic tracking utilized. At the present time there are two main types of equisignal systems for directional automatic tracking: systems with sequential comparison signals (amplitude–phase systems) and systems with simultaneous comparison signals (monopulse systems).

In the sequential comparison case, jamming aimed at the homing device scanning frequency is effective provided that the scanning frequency of the antenna pattern is known or determined during radio countermeasures. Sometimes, in order to attenuate the action of such jamming, so-called hidden conical scanning is used by the homing device where the transmitter and receiver are in separate locations (semi-active homing with target highlighting). The transmitting antenna directional pattern does not perform scanning and the antenna pattern scanning takes place in the homing device receiver. For such a homing device operating system, it is difficult to determine the scanning frequency during radio countermeasures and create jamming aimed at this scanning frequency. In this case barrage jamming over the band of possible scanning frequencies is used. One method for creating such barrage jamming aimed at the homing device's angular tracking channel is to use that channel's carrier frequency. The radiated signals are amplitude modulated by low-frequency noise having a uniform spectrum overlapping the band of possible scanning frequencies. This type of jamming is sometimes called "low-frequency noise" jamming.

Another type of barrage jamming for countering directional automatic tracking can be created by modulating a carrier using a harmonic signal the frequency of which is smoothly and linearly rearranged within the band of possible scanning frequencies of the automatic tracking system's receiving antenna pattern. Furthermore, to obtain maximal amplitude modulation depth, harmonic signals may be replaced by rectangular signals of the "meander" type. Such modulation is significantly simpler to realize, and it is known that the first harmonic signal amplitude of such a shape is approximately 1.3 times larger than the amplitude using purely harmonic modulation.

Monopulse automatic directional tracking in homing devices provides for the acquisition of target angular information after the reception of only one reflected pulse. Currently, monopulse radar detection and ranging methods are successfully used in systems with continuous radiation. Both phase and amplitude monopulse channels are currently used in homing devices, and are more resistant to jamming transmitted from one point in space than are automatic directional tracking systems with sequential comparison signals. An efficient impact on homing device monopulse channels can be achieved by "blinking" jamming created by means of several scattered jammers. A special case of blinking jamming is discontinuous interference, which is a periodic sequence of low duty cycle pulses radiated by one jammer. Such jamming is intended for the disruption of automatic amplification adjustment operation according to the suppressing jamming principle.

Measurement of the distance to a target is of great importance in homing devices because such measurements make possible the correct assessment of the target environment and the detection of the most dangerous targets. The jamming of homing device tracking channels can cause errors not only in range measurement but also by disrupting the normal operation of the goniometric channel. Disruption of the range tracking channel can be performed by the continuous noise method and by jamming that deflects the range gate.

Pulse deflecting jamming is essentially simulating jamming for homing device tracking channels by range. Here, sequences of response pulses are radiated on the homing system's carrier frequency, and the delays relative to the radiated signals change smoothly from zero to some specified value. The delay change law (or "drift law") can be linear or else parabolic with low acceleration at the initial stage of jamming creation. Such jamming deflects the range gate in homing devices and distorts information about the current RN and velocity as well as disrupting the continuity of the target angular coordinate information streams.

Broadband barrage jamming of homing device tracking channels by velocity is inefficient because it is difficult to create a significant spectral jamming signal power density in the relatively narrow tracking filter passband. That is why narrowband noise, the spectrum of which overlaps the band of possible input signal Doppler frequencies, is used for jamming homing device velocity tracking channels.

One of the most common methods for the simulating jamming of homing device tracking channels by velocity consists of deflecting the velocity gate. This form of jamming results in the electromagnetic field at the homing device receiving antenna aperture having the same nature as would signals from two targets moving with equal velocities. At the start of the deflecting cycle, one of these appears to begin maneuvering. That is, the jamming radiation simulates a nonexistent target moving with a velocity different from that of the protected target so that the homing device velocity tracking channel therefore switches to tracking the nonexistent target provided that the relevant signal has the greater intensity. The homing device thus receives distorted information about the target velocity and acceleration for each cycle of deflecting jamming, and so loses the real target during intervals between cycles, leading to the disruption of continuity of the target coordinate information stream.

## 7.8   Intelligent Homing Devices

Currently, all homing device development prospects are concerned with the incorporation of artificial intelligence system attributes to a greater or lesser extent (G6 homing devices—see Section 7.1), (Bonabeau, Dorigo, and Theraulaz, 1999). In both the literature and advertizing media, developed homing devices are most often referred to as being intelligent or cognitive devices. However, the appearance of intelligence is based on a long history of preceding homing device development (Andrea, 2012; Gini and Rangaswamy, 2008; Gini, Stinco, and Greco, 2012; Haykin, 2006, 2012; Rangaswamy, 2012). As already mentioned, homing devices already incorporate integrated systems in their structures, including various sensors, for target environment analysis. Moreover, homing to super-maneuverable targets lead to an increase in tracking channel dimensionalities (see Section 7.4.1). Actually, homing devices became multipurpose systems a long time ago (Hughes and Choe, 2000; Stove, 2004; Tavik et al., 2005) and provide for such functions as guidance to specified objects, analysis of dynamic target environments, selection of attack objects and their most vulnerable parts, group operations, electronic warfare, efficient warhead detonation, re-aiming, self-destruction, flight path selection, and the provision of control and operation reliability. Sometimes, an epigrammatic name is used such as Observe, Orient, Decide, and Act (OODA) homing devices.

Integration provides both a hardware and a software basis for the implementation of multifunctional performance requirements leading to machine intelligence, whence homing devices become capable of acting in a "noncanonical" manner and not under predetermined conditions, albeit with some loss of efficiency. Such a situation becomes the norm for aerospace vehicles participating in network-centric operations. It is evident that operational conditions will require intelligence in homing devices sufficient to enable them to vary their paths, particularly on the final segments. They must also be capable of generating corresponding priorities for data channel operation, especially with regard to the peculiarities of target maneuvers (see Section 7.1). Moreover, such target maneuvering can provide one of the most efficient algorithms leading to the disruption of homing device operation, and here the homing device must use its whole "intelligence" to track any such unexpected target maneuvers. Serious hopes are currently pinned on the implementation of SDR technologies in homing devices, as mentioned in Section 7.2.3.

The aforementioned "noncanonical" conditions of homing device operation arise, first of all, from the high responsiveness needed in situations wherein the operation has to be performed. The continuous growth and ever more comprehensive effects of jamming (and indeed its nomenclature) were identified in Section 7.7. A typical situation for combat homing devices is where targets become more maneuverable along with low observability and use of flight altitude and terrain relief optimally. The necessity for implementing homing device functions for detecting surprise targets is also leading to a new methodology.

Today, an intelligent homing device structure will include a scene analysis unit (the "scene analyzer") and an Environmental Dynamic Database (EDDB) that can be updated from both onboard and external data sources. An EDDB is the main element in knowledge-based algorithms.

In intelligent or cognitive homing devices, waveform selection from a previously shaped library is implemented, taking account of feedback from the receiver to the transmitter. The receiver controls the transmitter after analyzing data about the environment to better extract and identify assigned targets. This marks the difference from the typical radar homing device, where adaptability is limited by the receiver. In the intelligent homing device, heterogeneous data is necessary for receiving target environment data for forming an adequate performance relating to the intentions and threats from the surveyed target. An adaptive transmitter in an intelligent homing device that ensures the efficiency of the feedback channel from the receiver to the transmitter.

In accordance with general dynamic programming schemes, the Cognitive Waveform Selection (CWS) algorithm enables target path forecasting to be minimized for any specific time interval. Antenna aperture direct and inverted synthesizing modes (Yan, Li, and Zhou, 2011) are provided to permit the application of complex signals in the form of discrete frequency sequences. Analyses of flight test results and of modern onboard equipment capabilities (micronavigation, APAA, etc.) are also employed.

Recently, an acute problem has been identified—the capability of homing device operation during so-called information wars (Karlof and Wagner, 2003; Kopp, 2002; Levin, 2001; Pietrucha, 2006). This question is critical due to the development of the "information weapon" concept, which is obviously completely different in principle from traditional weapons, especially since an information war may be conducted unannounced and surreptitiously, in which case homing device data channels may be said to be part of the information battlefield. Information weapons include means of influencing information, such as computer viruses, logic bombs, and mines, which may be actuated upon command or at dedicated times during electronic warfare. Consequently, the information warfare situation during homing guidance demands high levels of information safety, which is the exclusive province of homing device artificial intelligence and is provided at the information level.

The operation of homing devices assumes that they are capable of modifying their flight paths and observation aspect angles to provide the best informative contact with the environment, especially in the terminal parts of flights. This is particularly important during the organization of electronic warfare countermeasures, the implementation of antenna aperture synthesizing modes, and an increase in vehicle survivability. For example, the identification and tracking of group targets and the jamming situation assessment can be significantly simplified if not only the electronic characteristics of the system are changed but also the target observance angle and the path and approach dynamics are intelligently selected. Such tactics are usually implemented when each function is working at the limit of its capabilities (e.g., due to massive jamming) and "help" is required from other functions that have already exhausted their resources. Such a problem is prominently revealed by highlighting the semi-autonomous observance and an analysis of two-position radar characteristics. The accomplishment of tasks for shaping real-time efficient flight path selection is based on the identification and assessment of a rather quickly changing tactical situation and is therefore dependent on the implementation of the corresponding intelligence technologies.

Intelligent homing devices are intended to ensure flexibility in the tasks to be performed such as software optimization for flight tactics and of flight control including the identification of the target environment, the assessment of impact efficiency, and in-flight aerospace vehicle retargeting. The necessary intelligence properties of such homing devices imply verification of these properties and assessment of their quantitative characteristics. Here, all the traditional test methods for radio-electronic systems having multiple nonintelligent characteristics are relevant and provide a basis of comparison with those systems. However, in order to describe homing device intelligent properties, appropriate test conditions must be reproduced. Here, combinations of all test technologies are required—full-scale, semirealistic, control loops based on simulation modeling, etc.—for the creation of all possible situations in which homing devices have to demonstrate their intelligence capabilities. For organizing such tests, special attention must be paid to the playback of potential situations that may be met in future real actions involving homing devices. Such situations require different combinations of homing device functions and these are provided in an integrated manner by simulation modeling, usually with the implementation of intelligent technologies. This situation is made more complex when the assessment of homing device intelligence capability for group operations is required.

At present, the intelligence properties of a homing device are capable of forcing the aviation facility, and itself, to significantly change operating modes. Hence, the sampling of scripts for homing guidance and for the actions of homing devices upon them becomes statistically dissimilar and in general the corresponding statistical methods—that is, the assessment of characteristics received during tests—become ineffective and unreliable and so cannot be used for the assessment of the situation. Here, the descriptions of homing devices assume the use of the appropriate semantics and pragmatics.

Homing devices reveal another fact: for similarly complex but fixed-location multipurpose systems with smart logic and complex operational dynamics, there is also a necessity for current updating during their whole life cycles, mainly by software modification. This is quite similar to homing devices parrying challenges from continuously improving countermeasures, and equally continuously improving their own countermeasures to negate them.

# References

Andrea D. M. (2012) *Introduction to Modern EW Systems*, Artech House, London.
Astanin, L. Y. and Kostylev, A. A. (1997) *Ultrawideband Radar Measurements: Analysis and Processing*, The Institution of Electrical Engineers, London.
Astanin, L. Y., Kostylev, A. A., Zinoviev, Y. S., and Pasmurov, A. Y. (1994) *Radar Target Characteristics: Measurements and Applications*, CRC Press, Boca Raton/Ann Arbor/London/Tokyo.

Augustine, N. R. (2000) The engineering of systems engineers. *IEEE Aerospace & Electronic Systems Magazine*. Jubilee Issue, October.

Ben-Asher, J. Z. and Yaesh, I. (1998) *Advances in Missile Guidance Theory*, American Institute of Aeronautics and Astronautics, Reston, VA.

Biezard, D. J. (1999) *Integrated Navigation and Guidance Systems*, American Institute of Aeronautics and Astronautics, Reston, VA.

Bolkcom, B., (2004) *Military suppression of enemy air defenses (SEAD): Assessing future needs*, CRS Report for Library of Congress, Washington, DC, USA, September 23, RS21141.

Bonabeau, E., Dorigo, M., and Theraulaz, G. (1999), *Swarm Intelligence from Natural to Artificial Systems*, Oxford University Press, New York.

Cebrowski, A. K. (2004) *Implementation of Network-Centric Warfare*, Office of Force Transformation, Arlington, VA.

Chrzanowski, E. J. (1990) *Active Radar Electronic Counter Measures*, Artech House, Inc., Norwood, MA.

Clark, W. H. (1992) Millimeter Wave Seeker Technology. AIAA92-0992 Naval Air Warfare Center, China Lake, CA.

Cruise missile radar altimeter, Jane's Radar, July 17, 1994.

Derham, T. E., Doughty, S., Woodbridge, K., and Baker, C. J. (2007) Design and evaluation of a low-cost multistatic netted radar system, *IET Radar, Sonar & Navigation*, 1(5), 67–73.

Donnet, B. J. and Longstaff, I. D. (2006) *MIMO radar, techniques and opportunities*, Proceedings of the Third European Radar Conference, Manchester, UK, pp. 112–115.

Fan, W., RuiLong, H., and Xiang, S. (2001a) *Anti-ARM technique: Distributed general purpose decoy series (DGPD)*, Proceedings of the International Conference on Radar, Beijing, China.

Fan, W., RuiLong, H., and Xiang, S. (2001b) *Anti-ARM technique: Feature analysis of ARM warning radar*, Proceedings of the International Conference on Radar, Beijing, China.

Gini, F. and Rangaswamy, M. (eds.) (2008) *Intelligent Adaptive Signal Processing and Knowledge Based Systems (KBS) for Radars: Knowledge Based Radar Detection, Tracking and Classification*, John Wiley & Sons, Inc., Hoboken, NJ.

Gini, F., Stinco, P., and Greco, M. S. (2012) *Dynamic Sensor Selection for Cognitive Radar Tracking*, Universita di Pisa, Pisa.

Gray, D. A. and Fry, R. (2007) *MIMO noise radar: Element and beam space comparisons*, Proceedings of the International Waveform Diversity and Design Conference, Pisa, Italy.

Gunston, B. (1979) *The Illustrated Encyclopedia of Rockets and Missiles*, Salamander Books Ltd., London.

Haimovich, A. M., Blum, R. S., and Cimini, L. J. (2008) MIMO radar with widely separated antennas, *IEEE Signal Processing Magazine*, 25 (1) 116–129.

Haykin, S. (2006). Cognitive radar: A way of the future, *IEEE Signal Processing Magazine*, 23(1), 30–40.

Haykin, S. (2012) *Cognitive Dynamic Systems: Perception-Action Cycle, Radar and Radio*, Cambridge University Press, Cambridge, UK.

Hughes, P. K. and Choe, J. Y. (2000) *Overview of advanced multifunction RF system (AMRFS)*, Proceedings of the IEEE International Conference on Phased Array Systems and Technology.

Hume, A. L. and Baker, C. J. (2001) *Netted radar sensing*, Proceedings of the CIE International Conference on Radar.

Isby, D. C. (2004) *US Air Force are evaluating characteristics of new modernization of Raytheons*. AIM—120 AMRAAM. Jane's Missiles & Rockets, January 15.

Jankiraman, M., Wessels, B. J., and van Genderen, P. (2000) *Pandora multifrequency FMCW/SFCW radar*, Record of the IEEE International Radar Conference.

Jiang, R., Wolfe, K. W., and Nguyen, L. (2000) *Low coherence fiber optics for random noise radar*, Proceedings of the IEEE Military Communications Conference (MILCOM), Los Angeles, CA, USA, October.

Jie, S., Xiao-ming, T., and You, H. (2006) *Multi-channel digital LPI signal detector*, Proceedings of the International Conference on Radar, October.

JP 1-02. Department of Defense Dictionary of Military and Associated Terms, April 12, 2001, as amended through, 2004.

Kanashchenkov, A. I. and Merkulov, V. I. (ed.) (2003) *Radio Electronic Homing Devices*, Radiotechnika, Moscow (in Russian).

Karlof, C. and Wagner, D. (2003) *Secure routing in wireless sensor networks: Attacks and countermeasures*, Proceedings of the IEEE International Workshop on Sensor Network Protocols and Applications.

Kee, R. J. (1992) *Estimation for Homing Guidance*, Queen's University of Belfast, Belfast.

Klass, P. J. (1998) New anti-radar missile uses dual-mode seeker, *Aviation Week and Space Technology*, October 26.

Klein, L. A. (1997) *Millimeter-Wave and Infrared Multisensor Design and Signal Processing*, Artech House, Inc., Norwood, MA.

Kopp, C. (2002) Support jamming and force structure, *The Journal of Electronic Defense*, 4, 24–32.

Kozlov, A. I., Logvin, A. V., and Sarychev, V. A. (2005–2008) *The Polarization of Radio Waves*, vols. 1–3, Radiotechnika, Moscow (in Russian).

Kruse, J., Adkins, M., and Holloman, K. A. (2005) *Network centric warfare in the U.S. Navy's fifth fleet*, Proceedings of the IEEE 38th Hawaii International Conference on Systems Sciences.

Kulpa, K. (2013) *Signal Processing in Noise Waveform Radar*, Artech House, Norwood, MA.

Lee, J. P. Y. (1991) *Interception of LPI radar signals*, Defence Research Establishment, Ottawa, Technical Note 91–23.

Lehmann, N. H., Haimovich, A. M., Blum, R. S., and Cimini, L. (2006) Highresolution capabilities of MIMO radar, Record of the 40th Asilomar Conference on Signals, Systems and Computers.

Lennox, D. (2002) Cruise missile technologies and performance analysis, *Jane's Strategic Weapons Systems* 38, 43–50.

Levanon, N., (1988) *Radar Principles*, John Wiley & Sons, New York.

Levin, R. E. (2001) *Electronic warfare—Comprehensive strategy needed for suppressing enemy air defenses*, United States General Accounting Office Report to Congressional Requesters, Washington, DC, GAO-01-28.

Lewis, B. L., Kretschmer, F. F., and Shelton, W. W. (1986) *Aspects of Radar Signal Processing*, Artech House, Norwood, MA.

Li, Z. and Narayanan, R. M. (2006) Doppler visibility of coherent ultrawideband random noise radar systems, *IEEE Transactions on Aerospace and Electronic Systems*, 42(3), 904–916.

Ling, F. M., Moon, T., and Kruzins, E. (2005) Proposed network centric warfare metrics: From connectivity to the OODA cycle, *Military Operations Research*, 10(1), 5–14.

Liu, H., Zhou, S., Su, H., Yu, Y. (2014) Detection performance of spatial frequency diversity MIMO radar. *IEEE Transactions on Aerospace and Electronic Systems*, 50 (4), 3137–3155.

Lukin, K. A. (2001) *Millimeter wave noise radar applications: Theory and experiment*, Proceedings of the Fourth International Symposium on Physics and Engineering of Microwave, Millimeter Wave and Submillimeter Waves, Kharkov, Ukraine.

Maluf, N. and Williams, K. (2004) *An Introduction to Microelectromechanical Systems Engineering*, Artech House, London.

Naftaly, M. (2014) *Teraherz Metrology*, Artech House, Norwood, MA.

Narayanan, R. M. and Dawood, M. (2000) Doppler estimation using a coherent ultrawide-band random noise radar 48 (6), 868–878.

Nathanson, F. E. (1991) *Radar Design Principles*, McGraw-Hill, New York.

Neng-Jing, L. (1995) Radar ECCM new area: Anti-stealth and anti-ARM, *IEEE Transactions on Aerospace and Electronic Systems*, 31(3), 1120–1127.

Neri, F. (2001) *Introduction to Electronic Defense Systems*, Artech House, Norwood, MA.

Net-Centric Environment—The Implementation, Joint Functional Concept, Version 1.0. Arlington, VA, April 7, 2005.

Ostrovityanov, R. V. and Basalov, F. A. (1985) *Statistical Theory of Distributed Targets—A Polarimetric Approach*, Artech House, Dedham, MA.

Ozu, H. (2000) *Missile 2000—Reference Guide to World Missile Systems*, Shinkigensha, Tokyo.

Pace, P. E. (2000) *Advanced Techniques for Digital Receivers*, Artech House, Norwood, MA.

Pace, P. E. (2009) *Detecting and Classifying Low Probability of Intercept Radar*, Artech House, Boston, MA/London.

Pace, P. E. and Burton, G. D. (1998) Antiship cruise missiles: Technology, simulation and ship self-defense, *Journal of Electronic Defense*, 21(11), 51–56.

Pasmurov, A. and Zinoviev, J. (2005) *Radar Imaging and Holography*, The Institution of Electrical Engineers, London.

Phister, P. W. Jr. and Cherry, J. D., (2006) *Command and control concepts within the network-centric operations construct*, Proceedings of the IEEE Aerospace Conference, March 4–11.

Pietrucha, M. (2006) *"Starbaby": A quick primer on SEAD*, Defense IQ Airborne Electronic Warfare Conference, London UK.

Piper, S. O. (1995) *Homodyne FMCW radar range resolution effects with sinusoidal nonlinearities in the frequency sweep*, Record of the IEEE International Radar Conference.

Rangaswamy, M. (2012). *SAM challenges for fully adaptive radar*. Seventh IEEE Workshop on Sensor Array and Multichannel Processing (SAM-2012), Hoboken, NJ, USA, June.

Richards, M. (2005) *Fundamentals of Radar Signal Processing*, McGraw-Hill, New York.

Rodrigue, S. M., Bash, J. L., and Haenni, M. G. (2002) *Next generation broadband digital receiver technology*, Fifteenth Annual AESS/IEEE Symposium.

Sammartino, P. F., Baker, C. J., and Griffiths, H. D. (2006) *Target model effects on MIMO radar performance*, Proceedings of the IEEE International Conference on Acoustics, Speech and Signal Processing.

Schleher, D. C. (1991) *MTI and Pulsed Doppler Radar*, Artech, Boston, MA.

Schrick, G. and Wiley, R. G. (1990) *Interception of LPI radar signals*, IEEE International Radar Conference.

SDR Forum, SDRF cognitive radio definitions, working document. SDRF-06-R-0011-V1.0.0, 2007.

*Shephard's Unmanned Vehicles Handbook* (2005) The Shephard Press, Burnham.

Shneydor, N. (1998) *Missile Guidance and Pursuit*, West Sussex.

Siouris, G. M. (2004) *Missile Guidance and Control Systems*, Springer, New York.

Skolnik, M. (1990) *Radar Handbook*, McGraw-Hill, New York.

Skolnik, M. I. (2001) *Introduction to Radar Systems*, McGraw-Hill, Boston, MA.

Stein, F., Garska, J., and McIndoo, P. L. (2000) *Network-centric warfare: Impact on army operations*, EUROCOMM 2000 Information Systems for Enhanced Public Safety and Security, IEEE/AFCEA.

Stimson, G. W. (1998) *Introduction to Airborne Radar*, Scitech Publishing Inc., Mendham, NJ.

Stove, A. G. (2004) *Modern FMCW radar—Techniques and applications*, European Radar Conference, Amsterdam, the Netherlands.

Stutzman, W. L. and Thiele, G. A. (1997) *Antenna Theory and Design*, John Wiley & Sons, Inc., New York.

Sun, H. (2004) Possible ultra-wideband radar terminology, *IEEE Aerospace and Electronic Systems Magazine*, 19(8), 38.

Tavik, G. C., Hilterbrick, C. L., Evins, J. B., *et al.* (2005) The advanced multifunction RF concept, *IEEE Transactions on Microwave Theory and Techniques*, 53(3), 1009–1020.

Teng, Y., Griffiths, H. D., Baker, C. J., and Woodbridge, K. (2007) Netted radar sensitivity and ambiguity, *IET Radar Sonar and Navigation*, 1(6), 479–486.

Titterton, D. and Weston, J. (2004) *Strapdown Inertial Navigation*, The Institution of Electrical Engineers, London.

Tsui, J. B. Y. and Stephens, J. P. (2002) Digital microwave receiver technology, *IEEE Transactions on Microwave Theory and Techniques*, 50(3), 699–705.

Wang, S. and Zhang, Y. (2001) *Detecting of anti-radiation missile by applying changeable-sample ratios technology in the AEW*, Proceedings of the International Conference on Radar.

Wiley, R. G. (1985) *Electronic Intelligence: The Interception of Radar Signals*, Artech House Publishers, Dedham, MA.

Wirth, W. D. (2001) *Radar Techniques Using Array Antennas*, IEE, London.

Xu, Y. and Narayanan, R. M. (2001) Polarimetric processing of coherent random noise radar data for buried object detection, *IEEE Transactions on Geoscience and Remote Sensing*, 39(3), 467–478.

Yan, S., Li, Y., Zhou, Z. (2011) Real-time motion compensation of an airborne UWB SAR. 2011 European Radar Conference (EuRAD), pp. 305–308.

Zarchan, P. (2002) *Tactical and Strategic Missile Guidance*, American Institute of Aeronautics and Astronautics, Reston, VA.

Zetik, R., Sachs, J., and Thoma, R. S. (2007) UWB short range radar sensing, *IEEE Instrumentation & Measurement Magazine*, 10(2), 39–48.

Zhouyue, P. and Khan, F. (2011) An introduction to millimeter-wave mobile broadband systems *IEEE Communications Magazine*, 49(6), 101–107.

# 8

# Optimal and Suboptimal Filtering in Integrated Navigation Systems

Oleg A. Stepanov
*Concern CSRI Elektropribor, JSC, ITMO University, Saint Petersburg, Russia*

## 8.1   Introduction

This chapter focuses on the design of filtering algorithms for integrated navigation systems and includes four sections. Section 8.2 gives a brief overview of the main approaches and methods used to solve estimation and filtering problems. Section 8.3 describes the statements of the filtering problems solved in integrated navigation systems. Section 8.4 details these statements as applied to the most popular Inertial Navigation Systems (INS) integrated with global Satellite Navigation Systems (SNS), the so-called INS/SNS. Section 8.5 presents an example of designing filtering and smoothing algorithms as applied to the problems of gravimeter and satellite data processing.

## 8.2   Filtering Problems: Main Approaches and Algorithms

The methods of estimation and filtering theory are extensively used in applied problems of measurement data processing. The foundations of this theory are presented and various aspects of their evolution are discussed in abundant literature (Stratonovich, 1960, 1966; Kushner, 1964; Bucy and Joseph, 1968; Van Trees, 1968; Medich, 1969; Jazwinski, 1970; Sorenson, 1970, 1985; Bucy and Senne, 1971; Sage and Melsa, 1971; Mehra, 1972; Gelb *et al.*, 1974; Kailath, 1974, 1980; Lainiotis 1974a, b; Liptser and Shiryaev, 1974; Kurzhanski, 1977; Anderson, 1979; Maybeck, 1979; Tikhonov, 1983, 1999; Fomin, 1984; Pugachev and Sinitsyn, 1985; Anderson and Moore, 1991; Mathematical System Theory, 1991; Grewal and Andrews, 1993; Yarlykov and Mironov, 1993; Brown and Hwang, 1997; Kailath *et al.*, 2000;

---

*Aerospace Navigation Systems*, First Edition. Edited by Alexander V. Nebylov and Joseph Watson.
© 2016 John Wiley & Sons, Ltd. Published 2016 by John Wiley & Sons, Ltd.

Bar-Shalom *et al.*, 2001, 2011; Basar, 2001; Rozov, 2002; Ristic *et al.*, 2004; Daum, 2005; Sinitsyn, 2006; Stepanov, 2011, 2012, and many others).

The fundamentals of filtering theory are presented later, and consideration is given to the main variants of filtering problem statements and algorithms used for their solutions in an approximately chronological order.

## 8.2.1 The Least Squares Method

Modern filtering theory was preceded by the work of Carl Friedrich Gauss (1777–1855), a German mathematician, astronomer, geodesist, and physicist and Adrien-Marie Legendre (1752–1833), a French mathematician who created the Least Squares Method (LSM) (Sorenson, 1970). Both considered the problem of estimating the time-invariant vector of unknown parameters, which is similar to the problem of finding an $n$-dimensional vector $x = (x_1, \ldots, x_n)^T$, using $m$-dimensional measurements $y = (y_1, \ldots, y_m)^T$

$$y = Hx + v, \tag{8.1}$$

where $H$ is an $m \times n$-dimensional matrix and $v = (v_1, \ldots, v_m)^T$ is an $m$-dimensional vector describing the measurement errors. Without assuming the probabilistic nature and statistical characteristics of the vectors of estimated parameters and measurement errors, application of the LSM can be justified using the deterministic approach. In this case, the estimate is found by minimizing the following criterion (Sorenson, 1970; Gibbs, 2011):

$$J^{\text{LSM}}(x) = (y - Hx)^T Q (y - Hx), \tag{8.2}$$

characterizing the difference between the calculated and measured values. It can be seen that the algorithm for calculating the estimate minimizing (8.2) is given by the following equation:

$$\hat{x}^{\text{LSM}}(y) = (H^T Q H)^{-1} H^T Q y. \tag{8.3}$$

In the simplest case, for example, when the scalar parameter $x$ is estimated by the measurements

$$y_i = x + v_i, \quad i = \overline{1.m},$$

assuming that $Q = I_m$ is an identity matrix, an equation for this estimate can easily be obtained in the form of the arithmetic average of the obtained measurements:

$$\hat{x}^{\text{LSM}}(y) = \frac{1}{m} \sum_{j=1}^{m} y_j,$$

which follows from (8.3) with $H^T = (1, \ldots, 1)^T$. This formula can also be presented in recurrence form as follows:

$$\hat{x}_i^{\text{LMS}}(y_i) = \hat{x}_{i-1}^{\text{LSM}}(y_{i-1}) + \frac{1}{i}(y_i - \hat{x}_{i-1}^{\text{LSM}}(y_{i-1})) = \hat{x}_{i-1}^{\text{LSM}} + K_i(y_i - \hat{x}_{i-1}^{\text{LSM}}). \tag{8.4}$$

In a more complicated nonlinear vector case, when $y_i = s_i(x) + v_i, i = \overline{1.m}$, the LSM criterion can be presented as

$$J^{LSM}(x) = (y - s(x))^{\mathrm{T}} Q(y - s(x)), \tag{8.5}$$

where $s(x) = (s_1(x), \ldots, s_m(x))^{\mathrm{T}}$.

To minimize this criterion, nonlinear programming methods should be applied, as used in the design of global extremum searching algorithms (Gibbs, 2011).

While developing the LSM, Gauss also considered the estimation problem from the probability viewpoint. In this case, the errors in various measurements were assumed to be independent and Gaussian. This LSM interpretation contributed considerably to the creation of the Maximum Likelihood Estimation (MLE) method as proposed by a famous English mathematician Ronald Fisher (1890–1962), who is considered to be the forefather of modern statistics (Fisher, 1925).

Assuming that the measurement errors are Gaussian with covariance matrix $R > 0$, the likelihood function can be written as follows:

$$f(y/x) = \frac{1}{(2\pi)^{\frac{m}{2}} \sqrt{\det R^{-1}}} \exp\left(-\frac{1}{2}(y - s(x))^{\mathrm{T}} R^{-1}(y - s(x))\right).$$

It can easily be seen that maximizing this function, or minimizing its logarithm (which is the same), is equivalent to the minimization of (8.5) with $Q = R^{-1}$. Specifically, in the linear case with $s(x) = Hx$

$$\hat{x}^{MLE}(y) = (H^{\mathrm{T}} R^{-1} H)^{-1} H^{\mathrm{T}} R^{-1} y.$$

## 8.2.2 The Wiener Approach

The next significant step in the development of estimation and filtering theory was made by two prominent scientists: A.N. Kolmogorov (1903–1986), an outstanding Soviet mathematician, the founder of modern probability theory, and the great American mathematician Norbert Wiener (1894–1964), whose name is usually associated with the origin of cybernetics (Kolmogoroff, 1939, 1941; Wiener, 1949; Basar, 2001). Whereas Gauss and Legendre dealt with the estimation problem of time-invariant vectors, Kolmogorov and Wiener solved the problem of estimating time-varying sequences and processes. These problems and solution algorithms are considered briefly further.

Assume that discrete scalar measurements are given in the form

$$y_i = x_i + v_i, \quad i = 1, 2\ldots, \tag{8.6}$$

where $x_i$ and $v_i$ are the estimated (valid) signal and noise, being the values of zero-mean Gaussian random sequences with known correlation $k_x(l,\mu) = E\{x_l x_\mu\}$, $k_v(l,\mu) = E\{v_l v_\mu\}$ and cross-correlation $k_{xv}(l,\mu) = E\{(x_l v_\mu\}$, $l, \mu = \overline{1.i}$ functions. Hereinafter $E$ denotes the mathematical expectation.

The problem was to estimate $x_j$ using measurements $y_l$, $l = \overline{1.i}$ accumulated over the time $i$, that is, the $i$-dimensional measurement $Y_i = (y_1, \ldots, y_i)^{\mathrm{T}}$, based on the minimization of criterion as follows:

$$\sigma^2_{apost} = E\left\{\left(x_j - \hat{x}_j\right)^2\right\}. \tag{8.7}$$

It was shown that the estimate minimizing this criterion can be determined as follows:

$$\hat{x}_{j/i}\left(Y_i\right) = K_i^j Y_i, \tag{8.8}$$

Here, $K_i^j$ is $1 \times i$ row matrix satisfying the following equation:

$$P^{x_i Y_i} = K_i^j P^{Y_i}. \tag{8.9}$$

Here, $P^{x_j Y_i} = \{E(x_j y_\mu)\} = \{k_x(j,\mu) + k_{xv}(j,\mu)\}$, $\mu = \overline{1.i}$ is the row matrix determining correlation of $x_j$ with measurements $Y_i$ and $P^{Y_i} = \{E(y_l y_\mu)\} = \{k_x(l,\mu) + 2k_{xv}(l,\mu) + k_v(l,\mu)\}$, $l, \mu = \overline{1.i}$ is the covariance matrix for vector $Y_i$. If this matrix is nonsingular, $K_i^j$ can be found as follows:

$$K_i^j = P^{x_j Y_i}\left(P^{Y_i}\right)^{-1}. \tag{8.10}$$

The minimal variance is determined as follows:

$$\sigma^2_{apost} = \sigma_0^2 - P^{x_j Y_i}\left(P^{Y_i}\right)^{-1} P^{Y_i x_j}.$$

From (8.8)–(8.10), it follows that to obtain an estimate minimizing criterion (8.7), a row matrix defining the correlation between the estimated and measured values and covariance matrix for the measurement vector should be known.

Various relationships between times $i$ and $j$, for which the estimate is sought, provided different names for the estimation problem: with $j = i$, filtering problem; $j < i$, smoothing or interpolation problem; and $j > i$, prediction problem (Medich, 1969).

Actually, the problem of estimating $x_i$ by measurements (8.6) is the problem of extracting the valid signal against the background of the measurement errors.

For continuous time, measurements (8.6) can be written as

$$y(t) = x(t) + v(t),$$

where $x(t)$ and $v(t)$ are the valid signal and noise, which are zero-mean random processes with known correlation $k_x(t_1, t_2) = E\{x(t_1)x(t_2)\}$, $k_v(t_1, t_2) = E\{v(t_1)v(t_2)\}$, and cross-correlation $k_{xv}(t_1, t_2) = E\{x(t_1)x(t_2)\}$ functions. In this case, the estimate minimizing the criterion

$$\sigma^2_{apost} = E\left\{\left(x(t_1) - \hat{x}(t_1)\right)^2\right\} \tag{8.11}$$

can be presented as follows:

$$\hat{x}(t_1) = \int_0^t h(t_1, \tau) y(\tau) d\tau, \tag{8.12}$$

where $h(t_1, \tau)$ is the weighting function specified for all $t_1, \tau \in [0,t]$. The problem was initially considered for time-invariant processes in steady-state modes over infinite time. In this case, the estimate is given by

$$\hat{x}(t) = \int_0^\infty h(t-\tau) y(\tau) d\tau.$$

It was shown that the weighting function $h(u)$ for the steady-state mode satisfies the integral Wiener–Hopf equation (Van Trees, 1968; Gibbs, 2011)

$$k_{xy}(\tau) = \int_0^\infty h(u) k_y(\tau - u) du, \quad 0 < \tau < \infty, \tag{8.13}$$

where $k_y(\tau)$ and $k_{xy}(\tau)$ are the correlation functions for the measurements and cross-correlation functions for the measurements and the estimated processes.

The solution of this equation can be obtained in the frequency domain as a frequency transfer function using factorization and separation of the Power Spectral Densities (PSD), which are determined as (Van Trees, 1968; Gibbs, 2011) follows:

$$S(\omega) = S^+(\omega) S^-(\omega), \quad S(\omega) = \left[ S(\omega) \right]_+ + \left[ S(\omega) \right]_-.$$

Here, $\left[ S(\omega) \right]_+$ and $S^+(\omega)$ do not have zeros and poles in the lower half-plane (the function of time is nonzero only for $t > 0$). Using the accepted denotations, the equation for the frequency transfer function can be written as

$$W_0(j\omega) = \frac{1}{S_y^+(\omega)} \left[ \frac{S_{xy}(\omega)}{S_y^-(\omega)} \right]_+, \tag{8.14}$$

where $S_y(\omega)$ and $S_{xy}(\omega)$ are the PSD of the measurements and cross-spectral PSD of the measurements and the estimated processes. Obviously, estimation algorithms given by the transfer function (8.14) determine some time-invariant dynamic system. This system is often called the Wiener filter, and the approach described earlier to designing filtering algorithms is referred to as the Wiener approach. Evidently, a filter of such a type can be effectively used only in the steady-state mode for the estimation of time-invariant processes. If the assumptions of a time-invariant process and an infinite measurement time are omitted, then to find the weighting function, instead of (8.13) it will be necessary to solve the Fredholm integral equation of the first kind (Van Trees, 1968; Gibbs, 2011):

$$k_{xy}(t_1, u) = \int_0^t h(t_1, \tau) k_y(\tau, u) d\tau, \quad 0 < u < t, \tag{8.15}$$

which can be treated as an extension of (8.9) for continuous time. The need to solve this equation significantly complicates the filter design.

## 8.2.3   The Kalman Approach

The aforementioned limitations were relaxed to a great extent in the methods and algorithms proposed by R. Kalman and R.L. Stratonovich (1930–1997), who are rightfully considered to be the founders of modern filtering theory. Kalman's first paper devoted to filtering described the algorithm of the estimation problem solution using the state-space approach (Kalman, 1960; Basar, 2001). In this case the estimated sequence is described by a shaping filter:

$$x_i = \Phi_i x_{i-1} + \Gamma_i w_i, \tag{8.16}$$

and $m$-dimensional measurements related to this sequence are written as follows:

$$y_i = H_i x_i + v_i. \tag{8.17}$$

Here, $x_i$ is the $n$-dimensional Gaussian space-vector; $\Phi_i$, $H_i$, and $\Gamma_i$ are the known matrices of dimensions $n \times n$, $m \times n$, and $n \times p$, respectively; $x_0$ is a zero-mean Gaussian vector with covariance matrix $P_0$; $w_i$ and $v_i$ are the $p$ and $m$-dimensional Gaussian zero-mean discrete white noise with known covariance matrices $Q_i$ and $R_i$, respectively, that is, $E\{w_i w_j^T\} = Q_i \delta_{ij}$, $Q_i \geq 0$ and $E\{w_i w_j^T\} = R_i \delta_{ij}$, $R_i > 0$. For simplification, it was assumed that the vectors $x_0$, $w_i$, and $v_i$ are not correlated with each other: $E\{x_0 w_i^T\} = 0$, $E\{w_i v_i^T\} = 0$, and $E\{x_0 v_i^T\} = 0$.

It was shown that the filtering algorithm, further called the Kalman filter, minimizing the criterion

$$P_i = E\left\{(x_i - \hat{x}_i)(x_i - \hat{x}_i)^T\right\}, \tag{8.18}$$

is given by

$$\hat{x}_{i/i-1} = \Phi_i \hat{x}_{i-1}, \tag{8.19}$$

$$P_{i/i-1} = \Phi_i P_{i-1} \Phi_i^m + \Gamma_i Q_i \Gamma_i^m, \tag{8.20}$$

$$\hat{x}_i = \hat{x}_{i/i-1} + K_i \left(y_i - H_i \hat{x}_{i/i-1}\right), \tag{8.21}$$

$$K_i = P_{i/i-1} H_i^m \left(H_i P_{i/i-1} H_i^m + R_i\right)^{-1}, \tag{8.22}$$

$$P_i = P_{i/i-1} - P_{i/i-1} H_i^T \left(H_i P_{i/i-1} H_i^T + R_i\right)^{-1} H_i P_{i/i-1} = (I - K_i H_i) P_{i/i-1}, \tag{8.23}$$

where $I$ is a unit matrix.

It should be noted that minimizing covariance matrix (8.18) is understood here as minimization of the corresponding quadratic form. In (8.21) and (8.18), $\hat{x}_i$ and $P_i = E\{(x_i - \hat{x}_i)(x_i - \hat{x}_i)^T\}$ are the optimal estimate and their error covariance matrix, and $\hat{x}_{i/i-1}$, $P_{i/i-1} = E\{(x_i - \hat{x}_{i/i-1})(x_i - \hat{x}_{i/i-1})^T\}$ are the prediction estimate and corresponding error covariance matrix. Here, the connection can be seen between estimate equation (8.21) and recurrence formula (8.4), corresponding to the simplest problem of estimating a constant scalar value.

For the continuous case, the filtering problem is formulated as follows. Suppose that an $n$-dimensional Markov process is described as

$$\dot{x}(t) = F(t)x(t) + G(t)w(t), \quad x(t_0) = x_0 \tag{8.24}$$

and $m$-dimensional measurements are written as

$$y(t) = H(t)x(t) + v(t), \tag{8.25}$$

where $F(t)$, $G(t)$, and $H(t)$ are the known time-dependent matrices of dimensions $n \times n$, $n \times p$, and $m \times n$, respectively; $x_0$ is the Gaussian zero-mean vector of initial conditions with covariance matrix $P_0$; and $w(t)$ and $v(t)$ are the Gaussian zero-mean white noise with specified PSDs, that is,

$$E\{w(t)w^{\mathrm{T}}(\tau)\} = Q(t)\delta(t - \tau), \quad Q(t) \geq 0, \tag{8.26}$$

$$E\{v(t)v^{\mathrm{T}}(\tau)\} = R(t)\delta(t - \tau), \quad R(t) > 0. \tag{8.27}$$

As earlier, for simplification, suppose that the vectors $x_0$, $w_i$, $v_i$ are not correlated with each other:

$$E\{x_0 w^{\mathrm{T}}(t)\} = 0; \quad E\{w(t)v^{\mathrm{T}}(t)\} = 0; \quad E\{x_0 v^{\mathrm{T}}(t)\} = 0. \tag{8.28}$$

The aim of a filtering problem is to estimate $x(t)$ using measurements $Y(t) = \{y(\tau) : \tau \in [0, t]\}$ accumulated over an interval $[0, t]$, based on minimization of the criterion

$$P(t) = E\left\{\left(x(t) - \hat{x}(t)\right)\left(x(t) - \hat{x}(t)\right)^{\mathrm{T}}\right\}. \tag{8.29}$$

In joint research by R. Kalman and R. Bucy and works by R. Stratonovich, it was shown that the estimate and its error covariance matrix are determined using the following equations (Stratonovich, 1959a, b, 1966; Kalman and Bucy, 1961):

$$\dot{\hat{x}}(t) = F(t)\hat{x}(t) + K(t)\left(y(t) - H(t)\hat{x}(t)\right), \tag{8.30}$$

$$K(t) = P(t)H(t)^{\mathrm{T}} R^{-1}(t), \tag{8.31}$$

$$\dot{P}(t) = P(t)F(t)^{\mathrm{T}} + F(t)P(t) - P(t)H(t)^{\mathrm{T}} R^{-1}(t)H(t)P(t) + G(t)Q(t)G^{\mathrm{T}}(t). \tag{8.32}$$

From (8.30), it follows that the equation for the estimate can be written as

$$\dot{\hat{x}}(t) = \left(F(t) - K(t)H(t)\right)\hat{x}(t) + K(t)y(t), \tag{8.33}$$

that is, the Kalman filter is a linear time-varying system with dynamic properties determined by the matrix $(F(t) - K(t) H(t))$ and the input measurements $y(t)$ weighted by matrix $K(t)$.

It should be noted that in practice the filtering problem is formulated in continuous form at the statement step. However, when filtering algorithms are implemented using computing aids, a discrete analog of the problem is needed, which requires correct transformation from one system to the other (Gelb *et al.*, 1974; Grewal and Andrews, 1993; Brown and Hwang, 1997; Gibbs, 2011).

The algorithms minimizing the criteria (8.18) are usually called *optimal minimum variance estimators* (filters), or *optimal minimum variance filtering algorithms*. It is important to note that if the condition for the Gaussian character of sequences and processes is omitted, the algorithms and estimates given before minimize these criteria for the class of linear algorithms. It is this circumstance, that determines the effective use of the resultant algorithms in various applied estimation and filtering problems. Sometimes, such algorithms are called *linear optimal minimum variance estimators* or filters (Li and Jilkov, 2004; Stepanov, 2006).

After Kalman's first paper was published, filtering theory based on the state-space approach and resultant methods began progressing rapidly. The use of the Kalman filter in applied problems was associated with a number of difficulties in choosing models adequately describing the state vector and measurements; with the sensitivity of algorithms to the selected models; with the reduction of computational burdens in the development of suboptimal filtering algorithms due to decreased dimensionality in the estimated state vector; with simplified descriptions of the state vector and measurements; and with the computational stability of the proposed procedures, etc. Abundant publications and books are devoted to the development of Kalman's idea (Battin, 1964; Bucy and Joseph, 1968; Medich, 1969; Jazwinski, 1970; Sage and Melsa, 1971; Mehra, 1972; Gelb *et al.*, 1974; Kurzhanski, 1977; Anderson, 1979; Maybeck, 1979; Fomin, 1984; Pugachev and Sinitsyn, 1985; Åström, 1991; Kailath, 1991; Mathematical System Theory, 1991; Grewal and Andrews, 1993; Yarlykov and Mironov, 1993; Brown and Hwang, 1997; Bar-Shalom *et al.*, 2001, 2011, and many others). Much attention was given to various modifications of Kalman filters used to solve nonlinear problems (Jazwinski, 1970; Sage and Melsa, 1971; Gelb *et al.*, 1974; Lainiotis, 1976; Schmidt, 1981). In navigational applications, nonlinearities are often associated with the nonlinearity of functions in measurement models (Stepanov, 1998; Bergman, 1999). The design of the simplest variants of filters for nonlinear problems is based on the linearized representation of these functions.

For example, for nonlinear measurements

$$y_i = s_i(x_i) + v_i,$$

the following can be written:

$$s_i(x_i) \approx s_i(x_i^n) + \frac{ds_i}{dx_i^T}\bigg|_{x_i = x^n} (x_i - x_i^n) = s_i(x_i^n) + H_i(x_i^n)(x_i - x_i^n).$$

Here, $x_i^l$ is the linearization point, $H_i(x_i^l) = ds_i/dx_i^m\big|_{x=x^l}$. To obtain a linearized suboptimal Kalman filter, some fixed values such as $x_i^l = \overline{x}_i$, where $\overline{x}_i$ denotes *a priori* mathematical expectations, can be used. Note that in this case

$$\overline{y}_i^{\text{lin}} = s_i(\overline{x}_i) + H_i(\overline{x}_i)(\hat{x}_{i/i-1} - \overline{x}_i),$$

where $\bar{y}_i^{lin} = y_i - s_i\left(x_i^l\right)$, and covariance matrices $P_i^{lin}$ and gain factor $K_i^{lin}$ can be calculated using equations of type (8.22) and (8.23), where $H_i$ should be substituted with $H_i\left(\bar{x}_i\right) = \dfrac{ds_i}{dx_i^m}\bigg|_{x=\bar{x}_i}$ .

The extended Kalman filter is the most widely used suboptimal algorithm nowadays (Jazwinski, 1970; Gelb et al., 1974; Schmidt, 1981). Here, prediction estimate $x_i^l = \hat{x}_{i/i-1}$ is used as a linearization point in processing the measurements, and in the covariance equation, matrix $H_i$ should be substituted with $H_i(\hat{x}_{i/i-1}) = ds_i/dx_i^T\big|_{x=\hat{x}_{i/i-1}}$ . It should be noted that the extended Kalman filter is a nonlinear one, as the matrix $H_i(\hat{x}_{i/i-1})$ is dependent on measurements.

Later, so-called iterative filters were proposed. In these, the current measurement is repeatedly processed using the following algorithm (Jazwinski, 1970):

$$\hat{x}_i^{(\gamma+1)} = \hat{x}_{i/i-1} + K_i\left(\hat{x}_i^{(\gamma)}\right)\left(y_i - s_i\left(\hat{x}_i^{(\gamma)}\right) - H_i^{(\gamma)}\left(\hat{x}_i^{(\gamma)}\right)\left(\hat{x}_{i/i-1} - \hat{x}_i^{(\gamma)}\right)\right),$$

$$K_i\left(\hat{x}_i^{(\gamma)}\right) = P_i\left(\hat{x}_i^{(\gamma)}\right)H_i^T\left(\hat{x}_i^{(\gamma)}\right)R_i^{-1},$$

$$P_i\left(\hat{x}_i^{(\gamma)}\right) = \left(\left(P_{i/i-1}\right)^{-1} + H_i^T\left(\hat{x}_i^{(\gamma)}\right)R_i^{-1}H_i\left(\hat{x}_i^{(\gamma)}\right)\right)^{-1},$$

where

$$H_i^{(\gamma)}\left(\hat{x}_i^{(\gamma)}\right) = \dfrac{ds_i\left(x_i\right)}{dx_i^T}\bigg|_{x_i=\hat{x}_i^{(\gamma)}} \quad ; \quad \gamma = 1,2\dots, \quad \hat{x}_i^{(1)} = \hat{x}_{i/i-1}.$$

Suboptimal algorithms with structures determined by (8.21) and (8.30) are usually referred to as Kalman-type algorithms. Higher-order filters such as the second-order filters introduced later can also be treated as various modifications of Kalman-type algorithms (Jazwinski, 1970).

There are many applications where filtering algorithms are widely used. One of them is the development of radio engineering systems, including radionavigation systems. Usually, the problems considered here were for continuous time (Sage and Melsa, 1971; Sorenson, 1985; Yarlykov and Mironov, 1993; Shakhtarin, 2008). Another application is connected with navigation, guidance, and tracking, which deal with both continuous-time problems and their discrete variants (Chelpanov, 1967; Boguslavski, 1970; Gelb et al., 1974; Rivkin et al., 1976; Grewal and Andrews, 1993; Bar-Shalom et al., 2001, 2011). It should be emphasized that in the collection of papers published on the occasion of Kalman's 60th birthday, the section devoted to applications of the Kalman filter was illustrated by navigation problems (Mathematical System Theory, 1991).

### 8.2.4 Comparison of Kalman and Wiener Approaches

For time-invariant systems, when the matrices used in the filtering problem (8.24) and (8.25) do not depend on time, the gain factor $K(t)$ in the Kalman filter may have a steady-state value $K_\infty$. This is found from (8.32) with $\dot{P}(t) = 0$. In this case the Kalman filter described by the equation

$\dot{\hat{x}}(t) = (F - K_\infty H)\hat{x}(t) + K_\infty y(t)$ presents a time-invariant system in the steady-state mode. If the same problem is solved by this Kalman filter and the Wiener filter described by (8.14), they will obviously coincide. In transient mode, the gain factor $K(t)$ in the Kalman filter changes, and the filter can be represented by two blocks for steady-state and transient modes (Pamyati professora Nesenyuka, 2010; Gibbs, 2011).

Generally, comparing the Kalman and Wiener approaches, the following evident advantages of the Kalman approach can be distinguished (Anderson and Moore, 1991; Ivanovski, 2012; Stepanov, 2012):

- It provides filtering for both time-invariant and time-varying processes and minimizes the estimation error variance for any time. This feature guarantees minimum time to reach the steady-state mode if it does exist.
- Along with optimal estimates, the Kalman filter generates a covariance matrix characterizing the estimation accuracy, which is extremely important for navigation data processing.
- The Kalman filter is intended to solve multidimensional problems with multiple inputs and outputs so that as the problem is solved, optimal estimates for all components of state vectors, including those not measured directly, are generated.
- The Kalman filter can be conveniently used to design recurrence processing algorithms.

Other merits of the Kalman filter include its applicability to a wide range of problems such as adaptive estimation problems with augmented state vector, nonlinear estimation problems using extended Kalman filters or a bank of filters, and simultaneous solution of estimation problems and hypothesis testing.

The following limitations of the approach should be noted: the need to present sequences and processes in the form (8.16) or (8.24), presuming their Markov nature, and computational problems in algorithm design, particularly for a high-dimensional state vector.

For the Wiener approach, the following should be noted:

- It is focused on the filtering of time-invariant processes, and filtering error variance is minimized only in the steady-state mode.
- In the transient mode, the errors may reach large values, and the mode duration can be rather long.
- Descriptions of estimated processes and their measurement errors in terms of correlation functions or PSD are required.
- As the filtering problem is solved, only the measured process is estimated and no accuracy characteristics are generated: such characteristics can be obtained only for the steady-state mode.
- The approach is intended to solve one-dimensional problems with one input and one output.

Strong points of the Wiener approach include the following:

- Generally, the estimated processes need not be of Markov nature.
- The resultant estimator is time-invariant and, therefore, simpler to implement.
- Intuitively understandable tools can be used to design the filters and select an appropriate description for the signal and its measurement errors.

Note here the common features of the two approaches:

- In the steady-state mode of filtering the time-invariant processes, the Wiener and Kalman filters coincide.
- Minimum estimation error variance is achieved only for Gaussian processes.
- For non-Gaussian processes, minimum filtering error variance is achieved in the class of linear algorithms.
- The resultant algorithms are sensitive to the models used.
- Computational problems occur in filter design in the absence of white noise components in measurement errors.

It follows from the text that Kalman algorithms offer broader opportunities than Wiener filters: Kalman algorithms are able to solve time-varying problems, obtain optimal estimates for any point of time, and generate accuracy characteristics along with the estimates. However, despite clear advantages and wide application of Kalman-type algorithms, frequency methods based on the Wiener approach still find uses in some problems of navigation data processing, which allow the use of time-varying sequences and processes (Chelpanov, 1967; Chelpanov et al., 1978; Nebylov, 2004, Pamyati professora Nesenyuka, 2010; Loparev et al., 2014). In engineering practice, approximate (suboptimal) methods applied to solve this type of problem are the most popular. In navigation applications, the researchers widely use an approach proposed in references Chelpanov (1967), Chelpanov et al. (1978), and Zinenko (2012) and based on the method of rectified logarithmic curves, later called the method of local approximations of signal and measurement error PSDs (Loparev et al., 2014). This method can be applied if PSD $S_x(\omega)$ and $S_n(\omega)$ are used to describe the valid signal $x(t)$ and noise $n(t)$. To approximately describe them in logarithmic scale, straight-line approximations are used, corresponding to the so-called conditional PSD of the signal and noise $S_x(\omega) \approx a^2 (\rho/\omega)^{2p}$ and $S_n(\omega) \approx a^2 (\rho/\omega)^{2q}$, where $p$ and $q$ are integers. This method is based on the assumption that the filter properties are determined by the behavior of these PSD in crossing points. Equations for filter transfer functions with different combinations of $p$ and $q$ are presented in Loparev et al. (2014). A major advantage of the method is that the solution can be obtained using conditional PSD in rather narrow frequency bands, which facilitates the selection and correct simplification of signal models. Then, generally, the obtained filter transfer functions take the form well suited for hardware realization. Later, a connection was found between this class of Wiener filtering problems and the Kalman filtering problem in estimating multiply integrated white noise when the measurement error is white noise too. It predetermined a method of developing algorithms based on a combination of Kalman and Wiener approaches. In this case, research on the filtering problem is implemented and sometimes a corresponding statement is formulated within the Wiener approach using the method of local approximations, and the relevant Kalman filter is used at the final stage of the filter design. It eliminates the main disadvantage of Wiener filters—that during the transient process the error variances can significantly exceed their steady-state values (Loparev et al., 2014).

### 8.2.5  Beyond the Kalman Filter

The further development of filtering methods gained renewed impetus in the late 1990s. Among other factors, it was conditioned by the need to solve complicated navigation problems in robotics, and by significant progress in computing technology. While solving these problems, researchers often have to deal with problems involving the nonlinear estimation of $n$-dimensional state vectors described by the equation

$$x_i = g_i\left(x_{i-1}\right) + \Gamma_i w_i, \qquad (8.34)$$

using $m$-dimensional measurements

$$y_i = s_i\left(x_i\right) + v_i, \quad i = 1, 2, \ldots, \qquad (8.35)$$

where $i$ is the discrete time; $g_i\left(x_{i-1}\right)$ and $s_i(x_i)$ are the known nonlinear $n$- and $m$-dimensional vector functions; $\Gamma_i$ is the known matrix; and $w_i$ and $v_i$ are discrete zero-mean white noise of relevant dimensionality with known Probability Density Function (PDF). In this problem, the following recurrence equations can be written for *a posteriori* PDF $f(x_i/Y_i)$ (Yarlykov and Mironov, 1993; Stepanov, 1998; Bergman, 1999; Ristic *et al.*, 2004):

$$f\left(x_i / Y_i\right) = \frac{f\left(y_i / x_i\right) f\left(x_i / Y_{i-1}\right)}{\int f\left(y_i / x_i\right) f\left(x_i / Y_{i-1}\right) dx_i} = c_i^{-1} f\left(y_i / x_i\right) f\left(x_i / Y_{i-1}\right) \qquad (8.36)$$

Here, $f(x_i/Y_{i-1}) = \int f(x_i/x_{i-1}) f(x_{i-1}/Y_{i-1}) dx_{i-1}$ is the prediction PDF; $c_i = f(y_i/Y_{i-1}) = \int f(y_i/x_i) f(x_i/Y_{i-1}) dx_i$ is the normalizing coefficient; $f(y_i/x_i)$ and $f(x_i/x_{i-1})$ are the PDFs defined using (8.34) and (8.35).It should be noted that the problem of calculating the estimate

$$\hat{x}_i\left(Y_i\right) = \int x_i f\left(x_i / Y_i\right) dx_i \qquad (8.37)$$

and the corresponding covariance matrix

$$P_i\left(Y_i\right) = \int \left(x_i - \hat{x}_i\left(Y_i\right)\right)\left(x_i - \hat{x}_i\left(Y_i\right)\right)^{\mathrm{T}} f\left(x_i / Y_i\right) dx_i \qquad (8.38)$$

actually consists in the calculation of multiple integrals, which often determines the method used to find them.

Currently, two major trends in the development of suboptimal filtering algorithms can be distinguished.

The first is based on the assumption of the Gaussian nature of *a posteriori* PDF. For this case, Kalman-type algorithms are designed where two first moments of PDF are calculated at each step. Apart from the extended and iterative Kalman filters and the second-order filters mentioned earlier, a number of new filters were proposed: regression filters, sigma-point filters or Unscented Kalman Filters (UKF), cubature filters, etc. (Li and Jilkov, 2004). A large group of filters gained popularity, based on a rather simple idea where the procedure of taking the derivatives to get a linear function representation is substituted with a procedure close to statistical linearization. Such algorithms were proposed as early as the beginning of 1970s. Among the other publications, they are described in Gelb *et al.*, 1974, citing the first publications on this method. However, at that time they did not find real applications because of poorly developed computing technology. Ideologically, these algorithms are connected with the problem of designing linear optimal algorithms for nonlinear non-Gaussian systems.

The fundamental idea behind the algorithm can be exemplified by solving the problem of estimating the time-invariant vector using the measurements,

$$y = s(x) + v. \tag{8.39}$$

A linear optimal algorithm is found by minimizing the criterion (8.18) in the class of estimates depending linearly on the measurements. Consider a composite vector $\tilde{z} = (x^{\mathrm{T}}, y^{\mathrm{T}})^{\mathrm{T}}$. It is known that linear optimal estimate $x$ using the measurements (8.39), minimizing the criterion (8.11) in the class of linear estimates, and its *a posteriori* error covariance matrix can be found using the following equations (Li and Jilkov, 2004; Stepanov and Toropov, 2010):

$$\hat{x}(y) = \bar{x} + K^{\mathrm{lin}} (y - \bar{y}); \quad K^{\mathrm{lin}} = P^{xy} (P^{y})^{-1},$$

$$P^{\mathrm{lin}} = P^x - P^{xy} (P^{y})^{-1} P^{yx} = P^x - K^{\mathrm{lin}} P^{yx}, \tag{8.40}$$

where $\bar{x}$, $\bar{y}$ and $P^{xy}$, $P^y$ are the relevant mathematical expectations and covariance matrices. Therefore, to calculate linear optimal estimates and error covariance matrix with the known two first moments for vector $\tilde{z} = (x^{\mathrm{T}}, v^{\mathrm{T}})^{\mathrm{T}}$ using (8.40), it is necessary to find mathematical expectation $\bar{y}$, cross-covariance matrix $P^{xy}$, and covariance matrix $P^y$ for the measurement vector $y$. It can be seen that Equation (8.40) is similar to (8.8) and (8.10). Assuming for simplification that vectors $x$ and $v$ are independent of each other, and $v$ is zero-mean, the following formulas can be written:

$$\bar{y} = E_y (y) = E_{x,v} \{ s(x) + v \} = \int s(x) f(x) dx, \tag{8.41}$$

$$P^{xy} = E_{x,y} \left\{ (x - \bar{x})(y - \bar{y})^{\mathrm{T}} \right\} = E_{x,v} \left\{ (x - \bar{x})(s(x) + v - \bar{y})^{\mathrm{T}} \right\}$$

$$= \int (x - \bar{x})(s(x) - \bar{y})^{\mathrm{T}} f(x) dx, \tag{8.42}$$

$$P^y = E_{x,y} \left\{ (y - \bar{y})(y - \bar{y})^{\mathrm{T}} \right\} = \iint (s(x) + v - \bar{y})(s(x) + v - \bar{y})^{\mathrm{T}} f(x,v) dx dv$$

$$= \int (s(x) - \bar{y})(s(x) - \bar{y})^{\mathrm{T}} f(x) dx + R. \tag{8.43}$$

Note that obtaining a linear optimal algorithm in a linear problem for the non-Gaussian case is reduced to a standard Kalman filter. But implementing this algorithm in a nonlinear problem requires numerical integration in Equations (8.41)–(8.43) to determine the two statistical moments for the measurements and the mutual moments for the measurements and estimated parameters. These can be calculated using the Monte-Carlo method, for example. In estimating the random sequences, this procedure can be implemented recurrently, for example, by substituting *a posteriori* PDF at each step for Gaussian approximation of this PSD, using two of its statistical moments. Various modifications of algorithms are reduced to different methods of simplifying the integration required to find these moments. The most popular algorithms applied in this case are based on unscented transformation (Van der Merwe and Wan, 2001; Juiler and Uhlmann, 2004; Lefebvre *et al.*, 2005).

It should be emphasized that Kalman-type filtering algorithms based on Gaussian approximation of *a posteriori* PDF effectively operate if PDF has only one extremum. However, in applied problems, PDF is often multiextremal. In this case, more complicated suboptimal methods should be applied, such as the point mass method (Bucy and Senne, 1971); Gaussian sum approximation (Alspach and Sorenson, 1972; Daum, 2005); a partitioning method; and various others (Lainiotis, 1976). Currently, sequential Monte-Carlo algorithms are the most popular (Zaritsky *et al.*, 1975; Liu and Chen, 1998; Doucet *et al.*, 2001; Gustafsson *et al.*, 2002; Shön *et al.*, 2005; Doucet and Johansen, 2011).

In the sequential Monte-Carlo method, *a posteriori* PDF at each step is approximated as

$$f\left(x_i/Y_i\right) \approx \sum_{j=1}^{L} \tilde{\omega}_i^j \delta\left(x_i - x_i^j\right), \quad \sum_{j=1}^{L} \tilde{\omega}_i^j = 1, \tag{8.44}$$

where $\delta(\bullet)$ is the multidimensional delta-function; and $\tilde{\omega}_i^j$, $x_i^j$, $j = \overline{1.L}$ are the set of so-called normalized weights and samples of some independent random vectors. Formula 8.44 is considered as an approximation of $f(x_i/Y_i)$ if it can be used to calculate the PDF moments to the required accuracy. Particularly, from (8.44), it follows that the estimates and covariance matrices in the filtering problem can be calculated as follows:

$$\hat{x}_i\left(Y_i\right) \approx \sum_{j=1}^{L} \tilde{\omega}_i^j x_i^j, \quad P_i\left(Y_i\right) \approx \sum_{j=1}^{L} \tilde{\omega}_i^j x_i^j \left(x_i^j\right)^{\mathrm{T}} - \hat{x}_i\left(Y_i\right)\left(\hat{x}_i\left(Y_i\right)\right)^{\mathrm{T}}.$$

Consider briefly some of the main techniques used in the Monte-Carlo method that improve its performance (Zaritsky *et al.*, 1975; Smith and Gelfand, 1992; Gordon *et al.*, 1993; Liu and Chen, 1998; Ivanov *et al.*, 2000; Doucet *et al.*, 2001; Gustafsson *et al.*, 2002; Bolić *et al.*, 2004; Ristic *et al.*, 2004; Shön *et al.*, 2005).

If an approximation is constructed in the form (8.44), the selection of the set of samples $\left(X_i^j\right)^{\mathrm{T}} = \left[\left(x_1^j\right)^{\mathrm{T}} \quad \cdots \quad \left(x_i^j\right)^{\mathrm{T}}\right]$ becomes important. Samples should be generated in the area where *a posteriori* PDFs are much different from zero, which is why the importance sampling methods became popular in practice.

The primary merit of the Monte-Carlo method is that Approximation (8.44) can be implemented recurrently, which means that the approximation at the *i*-th step is generated using sampling $X_i^j$ by adding sample $x_i^j$ to sample $X_{i-1}^j = \left[\left(x_1^j\right)^{\mathrm{T}} \quad \cdots \quad \left(x_{i-1}^j\right)^{\mathrm{T}}\right]^{\mathrm{T}}$, that is, $X_i^j = \left[\left(X_{i-1}^j\right)^{\mathrm{T}}, \left(x_i^j\right)^{\mathrm{T}}\right]^{\mathrm{T}}$, and non-normalized weights $\omega_i^j$ are calculated using $\omega_{i-1}^j$.

The technique of using the importance sampling and recurrence procedures for generating the samples and calculating the weights is referred to as *Sequential Importance Sampling* in the context of Monte-Carlo methods (Zaritsky *et al.*, 1975; Liu and Chen, 1998; Doucet *et al.*, 2001).

In implementing Monte-Carlo methods, a problem of algorithm degeneracy occurs: with time only one weight will have a value close to 1, and all the others will be close to 0. This drawback is partly eliminated through the use of *Sequential Importance Resampling* or the *bootstrap procedure* (Smith and Gelfand, 1992; Gordon *et al.*, 1993; Doucet *et al.*, 2001). As applied to this problem, the bootstrap procedure consists in that new samples $\tilde{X}_i^j$, $j = \overline{1.L}$ are generated using the set $X_i^j$, $j = \overline{1.L}$, according to discrete distribution (8.44). This allows the use of the following approximation instead of (8.44):

$$f\left(x_i/Y_i\right) \approx \frac{1}{L}\sum_{j=1}^{L}\delta\left(x_i - \tilde{x}_i^j\right).\tag{8.45}$$

Various algorithms for generating the samples $\tilde{X}_i^j$, $j = \overline{1.L}$, corresponding to discrete distribution are discussed in Bolić *et al.* (2004). Note that the values of $\tilde{X}_i^j$ are selected from the earlier generated set $X_i^j$, $j = \overline{1.L}$, which makes it possible to recurrently generate the samples $X_i^j = \left(\tilde{X}_{i-1}^j, x_i^j\right)$. The relevant contribution of samples included in (8.45) is achieved due to their more frequent reproduction (cloning) as compared with those with lower weights, which is important because the introduction of the resampling procedure does not affect the design of the recurrence algorithm. Various methods used to determine if resampling is required are discussed in Doucet and Johansen (2011).

Finally, one more technique that significantly improves the performance of Monte-Carlo methods is the so-called Rao–Blackwellization method, which is based on the analytical integration of a part of variables. This can be done if a subvector can be distinguished in the state vector, the fixation of which reduces the nonlinear problem to a linear one. For instance, it can be made in the following problem: estimate a composite vector $x_i^{\mathrm{T}} = \left(\left(\theta_i\right)^{\mathrm{T}}, \left(x_i^l\right)^{\mathrm{T}}\right)$ described by the equations (Ivanov *et al.*, 2000; Gustafsson *et al.*, 2002; Shön *et al.*, 2005)

$$\theta_i = \Phi_i^\theta \theta_{i-1} + \Gamma_i^\theta w_i^\theta,$$

$$x_i^l = \Phi_i^l\left(\theta_{i-1}\right)x_{i-1}^l + \Gamma_i^l w_i^l,$$

using the measurements

$$y_i = s_i\left(\theta_i\right) + H_i\left(\theta_i\right)x_i^l + v_i, \quad i = 1,2,\ldots$$

It can be seen that if the vector $\Theta_i = \left(\theta_1^{\mathrm{T}}, \theta_2^{\mathrm{T}}, \ldots, \theta_i^{\mathrm{T}}\right)^{\mathrm{T}}$ is fixed, the considered problem becomes linear. Then, relevant Kalman filters can be used to calculate the parameters of PDFs $f\left(X_i^l/\Theta_i, Y_i\right)$, and the Monte-Carlo method is applied only to approximate PDF $f(\theta/Y_i)$ for the vector $\theta_i$, whose dimensionality is smaller than that of vector $x_i$, which is decisive for designing effective algorithms.

## 8.3   Filtering Problems for Integrated Navigation Systems

The proper statement of filtering problems is very important for their effective solutions and some possible statements commonly encountered in the processing of navigation data are given below (Brown, 1972–1973; Stepanov, 2012).

### 8.3.1   Filtering Problems Encountered in the Processing of Data from Systems Directly Measuring the Parameters to be Estimated

Assume measurements from two systems or sensors in the following form:

$$y^{I}(t) = X(t) + \Delta y^{I}(t), \tag{8.46}$$

$$y^{II}(t) = X(t) + \Delta y^{II}(t), \tag{8.47}$$

where $X(t) = (X_1(t), X_2(t), \ldots, X_n(t))^T$ is an $n$-dimensional vector of the parameters to be estimated, $y^j(t)$ and $\Delta y^j(t)$ are $n$-dimensional measurement vectors from the systems or sensors and their errors, and $j = I, II$ is the index for the system or sensor number.

Problem statement: estimate unknown vector $X(t)$ using measurements (8.46) and (8.47) accumulated by time $t$.

Consider two possible statements of the filtering problem and initially assume that *a priori* statistical information is only available for the errors of the two systems, whereas there is no information on vector $X(t)$ at all. In this case, it is appropriate to use the scheme of the so-called complementary filter (Brown, 1972–1973). According to this, the solution of the filtering problem for the vector $X(t)$ is reduced to estimating errors of one system against the background of the errors of the other system, using difference measurements of the form:

$$y(t) = y^{I}(t) - y^{II}(t) = \Delta y^{I}(t) - \Delta y^{II}(t). \tag{8.48}$$

Figure 8.1 provides an explanation for the basis of such a scheme in the scalar case.

From Figure 8.1, it follows that the measurements in the filtering problem are represented by difference measurements independent of the parameter $X(t)$, the estimate of which is formed by compensation for the measurement errors of one of the sensors. It is this feature that has motivated another name for this scheme: *the invariant ($X(t)$-independent) scheme of processing*

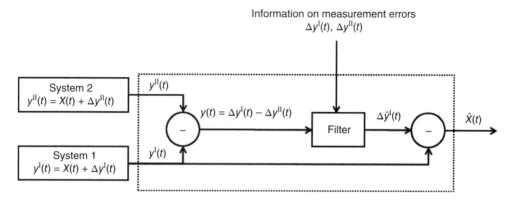

**Figure 8.1**   Invariant scheme for the processing

(Dmitriev *et al.*, 2000). To formulate the corresponding Kalman filtering problem statement for the vector case, it is necessary to form the following:

- A model for the state vector (i.e., describe it with a shaping filter (8.24))
- A model for measurements in the form of (8.25), relating these measurements to the state vector
- Measurements used in solving the filtering problem.

Assume that in the general case, the errors of the systems can be described as a sum of two components:

$$\Delta y^j(t) = \varepsilon^j(t) + v^j(t), \quad j = \mathrm{I}, \mathrm{II}, \tag{8.49}$$

where $v^{\mathrm{I}}(t)$ and $v^{\mathrm{II}}(t)$ are high-frequency error components; and $\varepsilon^{\mathrm{I}}(t)$ and $\varepsilon^{\mathrm{II}}(t)$ are error components that can be described by the shaping filters for vectors $x^{\mathrm{I}}(t)$ and $x^{\mathrm{II}}(t)$ of $n^{\mathrm{I}}$ and $n^{\mathrm{II}}$ dimensions, respectively, that is,

$$\dot{x}^j(t) = F^j(t) x^j(t) + G^j(t) w^j(t), \tag{8.50}$$

$$\varepsilon^j(t) = H^j x^j(t), \quad j = \mathrm{I}, \mathrm{II}, \tag{8.51}$$

where $F^j(t)$, $H^j(t)$, and $G^j(t)$ are the known matrices of $n^j \times n^j$, $n \times n^j$, and $n^j \times p^j$ dimensions, respectively; $v^j(t)$ and $w^j(t)$ are $n$- and $p^j$-dimensional zero-mean white noise, that is, $E\{w^j(t+\tau)w^j(t)\} = Q^j(t)\delta(\tau)$, $E\{v^j(t+\tau)v^j(t)\} = R^j(t)\delta(\tau)$ with matrices $Q^j(t)$ and $R^j(t)$ describing the noise PSD. For simplification, vectors $v^j(t)$, $w^j(t)$ and $x^j(0)$, $j = \mathrm{I}, \mathrm{II}$ are assumed to be uncorrelated with each other.

Introducing a composite $(n^{\mathrm{I}} + n^{\mathrm{II}})$-dimensional state vector $x^{\mathrm{I}}(t) = ((x^{\mathrm{I}}(t))^{\mathrm{T}}, (x^{\mathrm{II}}(t))^{\mathrm{T}})^{\mathrm{T}}$, it is possible to form the following model:

$$\dot{x}(t) = F(t) x(t) + G(t) w(t). \tag{8.52}$$

Here,

$$F(t) = \begin{bmatrix} F^{\mathrm{I}}(t) & 0 \\ 0 & F^{\mathrm{II}}(t) \end{bmatrix},$$

$$G(t) = \begin{bmatrix} G^{\mathrm{I}}(t) & 0 \\ 0 & G^{\mathrm{II}}(t) \end{bmatrix},$$

$$Q = \begin{bmatrix} Q^{\mathrm{I}}(t) & 0 \\ 0 & Q^{\mathrm{II}}(t) \end{bmatrix},$$

$$w^{\mathrm{T}} = \left(w^{\mathrm{I}}(t)\right)^{\mathrm{T}}, \left(w^{\mathrm{II}}(t)\right)^{\mathrm{T}}.$$

The $n$-dimensional measurements for the filtering problem are formed by using (8.48). Thus, the model for these measurements is determined as

$$y(t) = H(t)x(t) + v(t),$$                                    (8.53)

where $H(t) = [H^{\text{I}}(t), \; -H^{\text{II}}(t)]$, $v(t) = v^{\text{I}}(t) - v^{\text{II}}(t)$, and $R(t) = R^{\text{I}}(t) + R^{\text{II}}(t)$. Here, $H^{\text{I}}(t)$ and $H^{\text{II}}(t)$ are the $n \times n^{\text{I}}$ and $n \times n^{\text{II}}$ matrices, respectively.

Finally, it is necessary to specify the covariance matrix $P_0$, characterizing the state vector uncertainty at the initial time. Usually, it is a diagonal matrix with elements determining the level of *a priori* uncertainty in the knowledge of the relevant state vector components. Thus, the problem is reduced to the standard filtering problem of the state vector (8.52) with the use of measurements (8.53). This is solved by the standard relations of the Kalman filter given in Section 8.2. As a result, the optimal estimate of vector $x(t) = ((x^{\text{I}}(t))^{\text{T}}, (x^{\text{II}}(t))^{\text{T}})^{\text{T}}$ is obtained. Using this estimate and following the invariant scheme of processing, it is possible to form the required estimate $X(t)$ as $\hat{X}(t) = y^{\text{I}}(t) - \hat{\varepsilon}^{\text{I}}(t)$, where $\hat{\varepsilon}^{\text{I}}(t) = H^{\text{I}}(t)\hat{x}^{\text{I}}(t)$. A block diagram explaining the procedure for this processing scheme is given in Figure 8.2.

The problem statements described earlier can be illustrated by a simple example involving the processing of data on an aircraft altitude measured by a barometric altimeter and a satellite receiver.

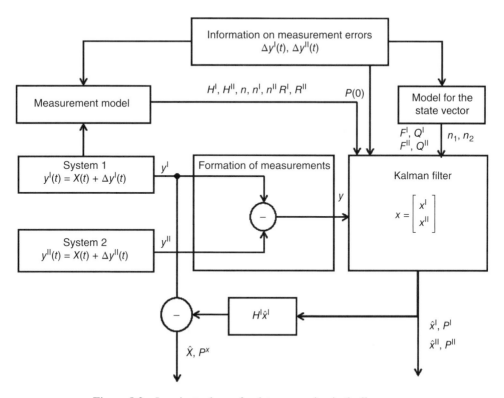

**Figure 8.2**   Invariant scheme for data processing in the linear case

## Example 8.1   Invariant Scheme

Assume scalar measurements of the form (8.46) and (8.47). Using the accepted notation and taking into consideration the scalar character of the measurements, these can be written as follows:

$$y^{I}(t) = h(t) + \Delta y^{I}(t), \qquad (8.54)$$

$$y^{II}(t) = h(t) + \Delta y^{II}(t). \qquad (8.55)$$

Here, $h(t)$, $y^{I}(t)$, $y^{II}(t)$, $\Delta y^{I}(t)$, and $\Delta y^{II}(t)$ are the scalar processes describing the true altitude, the measurements of the barometric altimeter (I = BAR) and the satellite receiver (II = SNS), and their errors. Assume that the barometric altimeter errors $\Delta y^{BAR}(t) = \varepsilon^{BAR}(t) + v^{BAR}(t)$ contain white noise $v^{BAR}(t)$ and the first-order Markov process, that is, $\dot{\varepsilon}^{BAR}(t) = -\alpha_B \varepsilon^{BAR}(t) + \sqrt{2\sigma_B^2 \alpha_B} w^{BAR}(t)$. The satellite receiver errors contain only white-noise components, which means that $x^{II}(t) = \varepsilon^{SNS}(t)$ is absent. In this case, the state vector $x(t) = x^{I}(t) = x^{BAR}(t) = \varepsilon^{BAR}(t)$, whence:

$$F = F^{BAR} = -\alpha_B, \quad G = G^{BAR} = \sqrt{2\sigma_B^2 \alpha_B}, \quad H = H^{BAR} = 1, \quad v(t) = v^{BAR}(t) - v^{SNS}(t),$$
$$w(t) = w^{BAR}(t), \quad Q(t) = Q^{BAR}(t) = 0, \quad R(t) = R^{BAR}(t) + R^{SNS}(t), \quad P_0 = \sigma_B^2.$$

Thus, Relations (8.52) and (8.53) can be written as

$$\dot{x}(t) = -\alpha_B x(t) + \sqrt{2\sigma_B^2 \alpha_B} w(t), \quad y(t) = x(t) + v(t),$$

and the problem is reduced to filtering of the Markov process against the background of white noise. The estimated altitude is formed as $\hat{h}(t) = y^{BAR}(t) - \hat{x}(t)$.

In the case where both a model in the form of (8.52) for the vector of parameters $X(t)$ being estimated and measurements of the form (8.46) and (8.47) are available, the filtering problem can be formulated based on the noninvariant scheme (Dmitriev et al., 2000; Mikhailov and Koshaev, 2014). A schematic of this procedure for the scalar case is given in Figure 8.3. Its essence is that the original measurements not subject to preliminary transformation are used as the filter input. The filtering problem is then solved directly for the vector of parameters $X(t)$ under estimation.

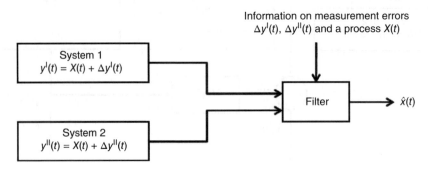

**Figure 8.3**   Noninvariant scheme for data processing

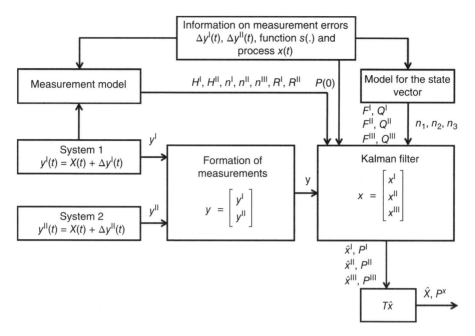

**Figure 8.4**  Noninvariant scheme for data processing in the linear case

For the Kalman statement of the problem in the vector case, it is necessary to enter an additional subvector $x^{III}(t)$ into the state vector, by which $X(t)$ is described as $X(t) = H^{III}(t)x^{III}(t)$. The filtering problem can then be solved for the $(n^{I} + n^{II} + n^{III})$-dimensional state vector $x(t) = ((x^{I}(t))^{T}, (x^{II}(t))^{T}, (x^{III}(t))^{T})^{T}$. Such a problem can be formulated by analogy with what has been done earlier, that is, by entering matrices $F^{III}(t)$, $G^{III}(t)$, and $Q^{III}(t)$. Measurements for the filtering problem are formed as $y(t) = ((y^{I}(t))^{T}, (y^{II}(t))^{T})^{T}$, and the model for these measurements is determined by Equation 8.53, where $H(t) = \begin{bmatrix} H^{I}(t) & 0_{n \times n^{II}} & H^{III}(t) \\ 0_{n \times n^{I}} & H^{II}(t) & H^{III}(t) \end{bmatrix}$,

$v(t) = ((v^{I}(t))^{T}, (v^{II}(t))^{T})^{T}$, and $R(t) = \begin{bmatrix} R^{I}(t) & 0 \\ 0 & R^{II}(t) \end{bmatrix}$.

Here, $H^{I}(t)$, $H^{II}(t)$, and $H^{III}(t)$ are $n \times n^{I}$, $n \times n^{III}$, and $n \times n^{II}$ matrices, respectively; and $0_{n \times n^{I}}$ and $0_{n \times n^{II}}$ are $n \times n^{I}$ and $n \times n^{II}$ zero matrices, respectively. As a result, estimates $\hat{x}(t) = ((\hat{x}^{I}(t))^{T}, (\hat{x}^{II}(t))^{T}, (\hat{x}^{III}(t))^{T})^{T}$ are obtained along with the corresponding covariance matrices $P^{I}(t)$, $P^{II}(t)$, and $P^{III}(t)$.

Estimate $X(t)$ can be calculated as $\hat{X}(t) = H^{III}(t)\hat{x}^{III}(t)$ or $\hat{X}(t) = T(t)\hat{x}(t)$, where $T(t) = (0_{n \times n_{1}}, \quad 0_{n \times n_{2}}, \quad H^{III}(t))$. A block diagram of this scheme is given in Figure 8.4.

### Example 8.2    Noninvariant Scheme

Assume that using measurements of the form (8.54) and (8.55), the vehicle's path in the vertical plane can be described as a first-degree polynomial $h(t) = x_{0} + Vt$, where $x_{0}$, $V$ are the initial altitude and the vertical velocity, both being time-invariant values; and $t$ being the time from the beginning of the observation. In this case, assuming that the errors in the satellite receiver and the barometric altimeter are described as in Example 8.1, and introducing the state vector $x(t) = (x_1(t), x_2(t), x_1(t))^T = ((x^{BAR}(t))^T, (x^{III}(t))^T)^T$, where $x^{BAR}(t) = \varepsilon^{BAR}(t)$, and $x^{III}(t) = (h(t), V)^T$, it is possible to formulate the filtering problem in which the matrices and vectors included in (8.52) and (8.53) are defined as

$$
F = \begin{bmatrix} -\alpha_B & 0 & 0 \\ 0 & 0 & 1 \\ 0 & 0 & 0 \end{bmatrix}, \quad G = \begin{bmatrix} \sqrt{2\sigma_B^2 \alpha_B} \\ 0 \\ 0 \end{bmatrix}, \quad w(t) = w^{BAR}(t), \quad P_0 = \begin{bmatrix} \sigma_B^2 & 0 & 0 \\ 0 & \sigma_0^2 & 0 \\ 0 & 0 & \sigma_V^2 \end{bmatrix},
$$

$$
H(t) = \begin{bmatrix} H^I(t) & H^{III}(t) \\ 0 & H^{III}(t) \end{bmatrix} = \begin{bmatrix} 1 & 1 & 0 \\ 0 & 1 & 0 \end{bmatrix}, \quad \text{and } v(t) = \begin{bmatrix} v^{BAR}(t) \\ v^{SNS}(t) \end{bmatrix}, \quad R = \begin{bmatrix} R^{BAR} & 0 \\ 0 & R^{SNS} \end{bmatrix}.
$$

That is, in order to obtain estimates of the altitude, it is necessary to solve the filtering problem of a three-dimensional state vector ($n^I = 1$, $n^{II} = 0$, and $n^{III} = 2$) described as

$$
\begin{bmatrix} \dot{x}_1 \\ \dot{x}_2 \\ \dot{x}_3 \end{bmatrix} = \begin{bmatrix} -\alpha_B & 0 & 0 \\ 0 & 0 & 1 \\ 0 & 0 & 0 \end{bmatrix} \begin{bmatrix} x_1 \\ x_2 \\ x_3 \end{bmatrix} + \begin{bmatrix} \sqrt{2\sigma_B^2 \alpha_B} \\ 0 \\ 0 \end{bmatrix} w^{BAR}(t)
$$

by using measurements

$$
y_1(t) = x_1(t) + x_2(t) + v^{BAR}(t);
$$
$$
y_2 = x_2(t) + v^{SNS}(t).
$$

The result of the formulated filtering problem is the estimated altitude, which is the optimal estimate of the state vector second component, that is, $\hat{h}(t) = T(t)\hat{x} = \hat{x}_2(t)$, where $T(t) = (0, 1, 0)$.

## 8.3.2 Filtering Problems in Aiding a Navigation System (Linearized Case)

A special feature of measurements of the form (8.46) and (8.47) is that both systems provide for the direct measurement of parameters to be estimated. A more common case for navigation data processing, where only one of the systems directly measures parameters to be estimated, is when the other one provides measurements of a known specified function of these parameters. Moreover, this function is usually nonlinear.

Consider the filtering problem in accordance with the invariant scheme, and assume the linearized description of the nonlinear function to be valid. It is supposed that the navigation system being aided generates measurements in the form of (8.46)

$$
y^I(t) = X(t) + \Delta y^I(t) \tag{8.56}
$$

along with additional aiding measurements in the following form:

$$
y^{II}(t) = s(X(t)) + \Delta y^{II}(t). \tag{8.57}
$$

For simplification, assume that it is required to estimate three-dimensional coordinates for a vehicle, that is, $X(t) = (X_1(t), X_2(t), X_3(t))^T$, and that $y^I(t)$ and $\Delta y^I(t)$ are the measurements of the navigation system and its errors, $y^{II}(t)$ and $\Delta y^{II}(t)$ are the $m$-dimensional vector of additional measurements and their errors, and $s(X(t))$ is the known $m$-dimensional vector function defined by the form of aiding measurements. As in the previous section, assume that errors $\Delta y^I(t)$ and $\Delta y^{II}(t)$ can be described by relations (8.49)–(8.51).

To obtain the invariant processing scheme, form the difference measurement

$$y(t) = y^{II}(t) - s\left(y^I(t)\right) \tag{8.58}$$

and then assume that the linearized description is valid for function $s(y^I(t))$ and write

$$s\left(y^I(t)\right) \approx s\left(X(t)\right) + \frac{ds\left(X(t)\right)}{dX^T}\bigg|_{X(t)=y^I(t)} \quad \Delta y^I(t) = s\left(X(t)\right) + H^*\left(y^I(t)\right)\Delta y^I(t), \tag{8.59}$$

where matrix $H^*\left(y^I(t)\right) = \dfrac{ds\left(X(t)\right)}{dX^T}\bigg|_{X(t)=y^I(t)}$ is defined as follows:

$$H^*\left(y^I\right) = \begin{bmatrix} \dfrac{\partial s_1\left(X(t)\right)}{\partial X_1} & \dfrac{\partial s_1\left(X(t)\right)}{\partial X_2} & \dfrac{\partial s_1\left(X(t)\right)}{\partial X_3} \\[2ex] \dfrac{\partial s_2\left(X(t)\right)}{\partial X_1} & \dfrac{\partial s_2\left(X(t)\right)}{\partial X_2} & \dfrac{\partial s_2\left(X(t)\right)}{\partial X_3} \\[2ex] \cdot & & \cdot \\[1ex] \dfrac{\partial s_m\left(X_i X(t)\right)}{\partial X_1} & \dfrac{\partial s_m\left(X(t)\right)}{\partial X_2} & \dfrac{\partial s_m\left(X(t)\right)}{\partial X_3} \end{bmatrix}_{X(t)=y^I(t)}. \tag{8.60}$$

Substituting (8.59) into (8.58), taking account of (8.57), the following formula for the difference measurements (8.58) can be easily obtained:

$$y(t) = -H^*\left(y^I(t)\right)\Delta y^I(t) + \Delta y^{II}(t). \tag{8.61}$$

Taking into account the fact that the description of errors $\Delta y^I(t)$ and $\Delta y^{II}(t)$ by Relations (8.49)–(8.51) is valid, it is also easy to formulate the filtering problem for the state vector $x(t) = ((x^I(t))^T, (x^{II}(t))^T)^T$, described by relations of the form (8.52) and by using difference measurements (8.58). In this case, the required matrices for the state vector model and measurement model are defined as

$$F = \begin{bmatrix} F^I & 0 \\ 0 & F^{II} \end{bmatrix}, \quad G = \begin{bmatrix} G^I & 0 \\ 0 & G^{II} \end{bmatrix}, \quad H = \begin{bmatrix} -H^* H^I, H^{II} \end{bmatrix},$$

$$v(t) = v^I(t) + H^* v^{II}(t), \quad w(t) = \left(w^I(t)\right)^T, \left(w^I(t)\right)^T, \quad Q_i = \begin{bmatrix} Q_i^I & 0 \\ 0 & Q_i^{II} \end{bmatrix},$$

$$R = R^{II} + H^* R^I \left(H^*\right)^T.$$

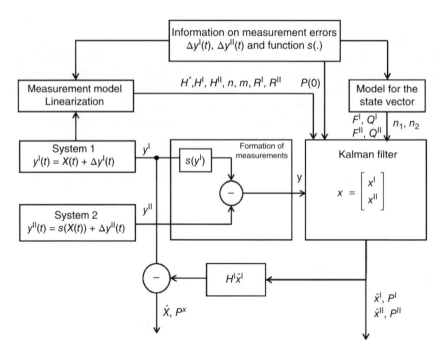

**Figure 8.5** Invariant scheme for processing data with the use of linearized measurements

A special feature of this scheme is that in the formation of $m$-dimensional difference measurements for the filtering problem, measurements $y^I(t)$ from one of the systems undergo nonlinear transformation $s(y^I(t))$. In this case, summand $s(X(t))$, which depends on unknown vector $X(t)$, is eliminated from the difference measurements, which virtually determines invariance (independence) of the estimate properties with regard to $X(t)$. This scheme is shown in Figure 8.5.

## 8.3.3 Filtering Problems in Aiding a Navigation System (Nonlinear Case)

Consider the features of the filtering problem statement in aiding the navigation system by using measurements of (8.57) when it is impossible to use a linearized description of function $s(\bullet)$ (Stepanov, 1998; Bergman, 1999). If linearization is unacceptable, summand $s(X(t))$ cannot be eliminated because the measurements are formed as follows:

$$y(t) = y^{II}(t) - s\left(y^I(t)\right) = s\left(X(t)\right) + \Delta y^{II}(t) - s\left(X(t) + \Delta y^I(t)\right).$$

For this reason, in order to formulate the filtering problem without a model for vector $X(t)$, it is necessary to rely on the principle of *distribution of navigational information* (Krasovski *et al.*, 1979). According to this principle, some navigational measurements to be processed can be considered as input signals in the processing algorithm. In the problem under consideration, measurements of the form (8.56) can be considered as an input signal. Therefore, measurement model (8.57) may be written as follows:

$$y^{II}(t) = s\left(y^I(t) - \Delta y^I(t)\right) + \Delta y^{II}(t) = \tilde{s}\left(\Delta y^I(t), t\right) + \Delta y^{II}(t). \qquad (8.62)$$

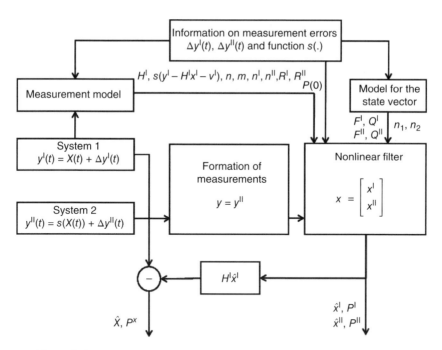

**Figure 8.6**   Scheme for the processing of data using nonlinear measurements

In the filtering problem statement, the state vector is the same as in Section 8.3.2. Introducing the measurements $y \equiv y^{II}(t)$, the measurement model used in the filtering problem may be written as follows:

$$y = \tilde{s}\left(\Delta y^I(t), t\right) + \Delta y^{II}(t) = s\left(y^I(t) - H^I(t)x(t) + v^I(t)\right) + H^{II}(t)x(t) + v^{II}(t). \quad (8.63)$$

In this case, $y^I(t)$ can be considered as a known deterministic input signal. It should be noted that although this statement is similar to the invariant one, in essence it is not the same. The fact is that the implicitly right-hand side of (8.63) depends on $y^I(t)$ because these measurements determine the function $\tilde{s}(\Delta y^I(t), t)$. The assumption that $y^I(t)$ is a known deterministic signal, alongside the assumption that vector $\Delta y^I(t)$ is a random one and that (8.56) is valid, virtually means the introduction of an assumption about the random nature of vector $X(t)$. In this case, the mathematical expectation of $X(t)$ is $y^I(t)$, and its statistical properties are determined by the properties of $\Delta y^I(t)$ since the PDF for $X(t)$ is determined by the PDF for $\Delta y^I(t)$ as $f(X(t)) = f_{\Delta y^I}(y^I(t) - \Delta y^I(t))$, where $f_{\Delta y^I}(\bullet)$ is the PDF for $\Delta y^I(t)$. Hence it follows that the algorithm will no longer possess the properties of invariance with respect to vector $X(t)$. However, the similarity to the invariant scheme is evident since in this case the problem of filtering (estimating) errors $\Delta y^I(t)$ of one system against the background of errors $\Delta y^{II}(t)$ of another system is also solved. Figure 8.6 shows a block diagram explaining the procedures described earlier.

A special feature of this procedure is that in the filtering problem, the measurements $y(t) = y^{II}(t)$, nonlinearly related to the vector of parameters to be estimated, are used, whereas the measurement model $\tilde{s}(\Delta y^I(t), t) = s(y^I(t) - \Delta y^I(t)) + \Delta y^{II}(t)$ is formed using measurements $y^I(t)$ and the information about function $s(\bullet)$. Note that during the linearization of (8.63), one arrives exactly at the statement of the problem that was considered earlier. Indeed, assuming that the derivative of function $s(\bullet)$ does not change in the *a priori* uncertainty domain, it is possible to write

$$ y^{II}(t) = s\left(y^I(t) - \Delta y^I(t)\right) + \Delta y^{II}(t) \approx s\left(y^I(t)\right) - \frac{ds\left(y^I(t)\right)}{d\left(y^I(t)\right)^T}\Delta y^I(t) + \Delta y^{II}(t). $$

Transporting the known value $s(y^I(t))$ into the left-hand side, an equation is obtained that coincides with (8.61).

The noninvariant statement of the filtering problem can be formulated when a model for vector $X(t) = H^{III}(t)x^{III}(t)$ is introduced. Then the model for $(n^I + n^{II} + n^{III})$-dimensional state vector $x(t) = ((x^I(t))^T, (x^{II}(t))^T, (x^{III}(t))^T)^T$ can be determined by Equation 8.52, where

$$ F(t) = \begin{bmatrix} F^I(t) & 0 & 0 \\ 0 & F^{II}(t) & 0 \\ 0 & 0 & F^{III}(t) \end{bmatrix}, \quad G(t) = \begin{bmatrix} G^I(t) & 0 & 0 \\ 0 & G^{II}(t) & 0 \\ 0 & 0 & G^{III}(t) \end{bmatrix}, $$

$$ Q(t) = \begin{bmatrix} Q^I(t) & 0 & 0 \\ 0 & Q^{II}(t) & 0 \\ 0 & 0 & Q^{III}(t) \end{bmatrix}. $$

A special feature of the filtering problem is the fact that composite $n + m$-dimensional measurements are used,

$$ y(t) = \begin{bmatrix} y^I(t) \\ y^{II}(t) \end{bmatrix} = \begin{matrix} X(t) + \Delta y^I(t), \\ s(X(t)) + \Delta y^{II}(t), \end{matrix} $$

including, among the others, nonlinear measurements. The model for these measurements is determined as follows:

$$ y^I(t) = H^{III}(t)x(t) + H^I(t)x(t) + v^I(t), $$

$$ y^{II}(t) = s\left(H^{III}(t)x(t)\right) + H^{II}(t)x(t) + v^{II}(t). $$

From this, the following estimates are obtained: $\hat{x}(t) = ((\hat{x}^I(t))^T, (\hat{x}^{II}(t))^T, (\hat{x}^{III}(t))^T)^T$. A block diagram explaining this scheme is shown in Figure 8.7.

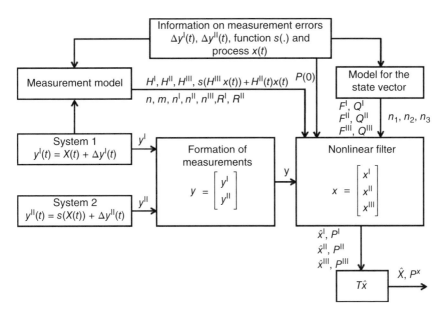

**Figure 8.7**   Noninvariant scheme for data processing with the use of nonlinear measurements

## Example 8.3    The problem of aiding a navigation system using terrain measurements.

The problem of aiding the navigation system using terrain profiles is a particular case of aiding using maps. The basic idea of this navigation method is to estimate a vehicle's position based on the comparison of measured data with reference data from a map. This method is known as *map-matching navigation, map-aided navigation,* or *correlation-extremal navigation* (Krasovski *et al.*, 1979; Stepanov, 1998; Bergman, 1999). The latter is commonly applied to systems using information in the form of images. This name is due to the fact that the design of processing algorithms frequently involves estimation of navigation parameters based on finding the extremum of the mutually correlated function between the measured and the reference images. The design of such systems is discussed in Chapter 6. At the same time, navigation parameters in such systems can be estimated within the context of filtering theory. This may be illustrated by a simple example of problem solution for aircraft navigation using a terrain map. For simplification, eliminate the problem of determining the altitude (assuming that the aircraft is moving at a constant altitude) and estimate coordinates $X_1(t), X_2(t)$ in the horizontal plane. In this case, the scalar measurements of the terrain can be represented as

$$y^{TR}(t) = \Psi\left(X_1(t), X_2(t)\right) + \Delta y^{TR}(t), \tag{8.64}$$

where function $\Psi(X_1(t), X_2(t))$ describes the dependence of terrain on the vehicle's coordinates and is specified using a map; and $\Delta y^{TR}(t)$ are the measurement errors, which are the sum of the map and sensor errors. Assuming that there is a navigation system to be aided, and that this system measures the vehicle's coordinates in the form of (8.56), that is, $y_1^{NS}(t) = X_1(t) + \Delta y_1^{NS}(t)$, $y_2^{NS}(t) = X_2(t) + \Delta y_2^{NS}(t)$, interpreting these measurements as input signals, it is possible to write:

$$\Psi(X_1, X_2) = \Psi\left(y_1^{NS}(t) - \Delta y_1^{NS}(t), y_2^{NS}(t) - \Delta y_2^{NS}(t)\right) = \tilde{s}\left(\Delta y_1^{NS}(t), \Delta y_2^{NS}(t), t\right).$$

Therefore, measurements of the form (8.64) can be represented as follows:

$$y^{TR}(t) = \tilde{s}\left(\Delta y_1^{NS}(t), \Delta y_2^{NS}(t), t\right) + \Delta y^{TR}(t). \tag{8.65}$$

Specifying error models for $\Delta y^{NS}(t)$ and $\Delta y^{TR}(t)$, a terrain-aided navigation problem as a problem of filtering using measurements of the form (8.65) may be formulated.

If an assumption of uniform motion of the vehicle during observation is introduced, the problem can be formulated in the context of noninvariant statements as follows: to estimate vector $x^{III}(t) = (X_1, V_1, X_2, V_2)^T$ described as

$$\dot{x}_1(t) = x_2(t),$$
$$\dot{x}_2 = 0,$$
$$\dot{x}_3(t) = x_4(t),$$
$$\dot{x}_4 = 0,$$

by using measurements

$$y_1^{NS}(t) = x_1(t) + \Delta y_1^{NS}(t),$$
$$y_2^{NS}(t) = x_2(t) + \Delta y_2^{NS}(t),$$
$$y^{TR}(t) = \Psi\left(x_1(t), x_2(t)\right) + \Delta y^{TR}(t).$$

Consider a simple case of a possible statement in the first variant in more detail. Assume that the errors in the navigation system during the observation do not change, that is, $\Delta y_1^{NS}(t) = x_1$, $\Delta y_2^{NS}(t) = x_2$ are random biases, and errors $\Delta y^{TR}(t)$ only contain a white-noise component. In this case, the problem is reduced to the vector filtering:

$$\dot{x}_1 = 0;$$
$$\dot{x}_2 = 0$$

by using measurements $y^{TR}(t) = \tilde{s}(x_1, x_2, t) + v^{TR}(t)$, where $\tilde{s}(x_1(t), x_2(t), t) = \Psi\left(y_1^{NS}(t) - x_1, y_2^{NS}(t) - x_2\right)$.

Suppose that measurements are taken at discrete points in time. Then, assuming that the estimated vector and the measurement errors are independent of each other and that they are Gaussian with similar variance $r^2$, then for *a posteriori* PDF, the following equation may be written (Stepanov, 2012):

$$f(x/Y_m) = \frac{\exp\{-J(x,Y_m)f(x)\}}{\int \exp\{-J(x,Y_m)f(x)\}dx},$$

where $J(x,Y_m) = \dfrac{1}{2r^2}\sum_{i=1}^{m}(y_i - \tilde{s}(x_1, x_2, t_i))^2$, $f(x) = N(x;0,\sigma_0^2 I_2)$, $I_2$ is a $2 \times 2$ unit matrix.

Figure 8.8 shows two graphs illustrating a possible behavior of *a posteriori* PDF, which is generally multiextremal.

A considerable number of various algorithms are known to have been designed to solve map-matching problems. The formulation of this problem as a filtering problem opens up the possibility of using the whole arsenal of available methods for solving nonlinear filtering

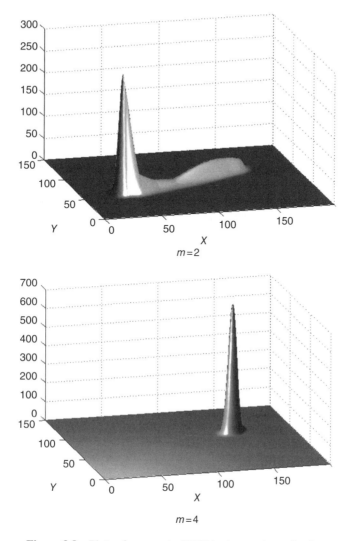

**Figure 8.8**   Plots of *a posteriori* PDF in the terrain navigation

problems in the process of algorithm design and their accuracy analysis. In particular, the best-suited variant of the algorithm depends significantly on the field properties, the possible levels of *a priori* position uncertainty, and many other factors.

## 8.4   Filtering Algorithms for Processing Data from Inertial and Satellite Systems

One of the main trends in the development of modern navigation systems is the integration of data from an INS and an SNS (Grewal and Andrews, 1993; Parkinson and Spiller, 1996; Barbour and Schmidt, 2001; Schmidt and Phillips, 2003). This section shows how the problem statements formulated in Section 8.3 can be applied to the processing of data from such inertial and satellite systems.

## 8.4.1   Inertial System Error Models

Equations describing INS errors are very important in the processing of data from integrated INS/SNS systems. To be specific, it is assumed that a strapdown INS (Anuchin and Emel'yantsev, 1999) is employed.

Following from the previous section, in order to formulate a filtering problem, it is necessary to have descriptions of INS errors. To describe INS errors, it is convenient to write the state vector in the form of two subvectors $x(t) = ((x^{l_1}(t))^T, (x^{l_2}(t))^T)^T$. One of them, subvector $x^{l_1}(t)$, describes the errors of the main navigation parameters: coordinates, velocities, and attitude angles. The other subvector $x^{l_2}(t)$ describes the errors of the INS sensors, namely, Angular Rate Sensors (ARSs) and accelerometers. Equations of shaping filters for the above two subvectors, based on the results of Anuchin and Emel'yantsev (1999) may then be written. To do this, consider the equation for subvector $x^{l_1}(t)$, assuming that it comprises the subvector of geographical coordinate errors $\Delta S(t) = (\Delta\varphi, \Delta\lambda, \Delta h)^T$, the subvector of velocity component errors $\Delta V(t) = (\Delta V_N, \Delta V_E, \Delta V_H)^T$ given in the axes of the geographical trihedron, and subvector $\Delta a(t) = (\alpha, \beta, \gamma)^T$, where $\alpha$ is the heading error and $\beta, \gamma$ are vertical errors (Figure 8.9).

Using the accepted notation, it is possible to write the following differential equation for the components of vector $x^{l_1}(t)$, including the error components in determining the coordinates, velocity, and attitude angles:

$$
\left.
\begin{aligned}
\Delta\dot{\varphi} &= \frac{\Delta V_N}{R}, \\[2mm]
\Delta\dot{\lambda} &= \frac{\Delta V_E}{R\cos\varphi} + \frac{V_E \cdot \sin\varphi}{R\cos^2\varphi}\Delta\varphi, \\[2mm]
\Delta\dot{h} &= \Delta V_H, \\[2mm]
\Delta\dot{V}_E &= n_N\alpha - n_H\gamma + \Delta a_E - \delta a_{BE}, \\[2mm]
\Delta\dot{V}_N &= -n_E\alpha + n_H\beta + \Delta a_N - \delta a_{BN}, \\[2mm]
\Delta\dot{V}_H &= n_E\gamma - n_N\beta + \Delta a_H - \delta a_{BH}, \\[2mm]
\dot{\alpha} &= \omega_N\beta - \omega_E\gamma + tg\varphi\frac{\Delta V_E}{R} + \left(\Omega\cos\varphi + \frac{V_E}{R\cos^2\varphi}\right)\Delta\varphi - \Delta\omega_H, \\[2mm]
\dot{\beta} &= -\omega_N\alpha + \omega_H\gamma - \frac{\Delta V_N}{R} - \Delta\omega_E, \\[2mm]
\dot{\gamma} &= \omega_E\alpha - \omega_H\beta + \frac{\Delta V_E}{R} - \Omega\sin\varphi\Delta\varphi - \Delta\omega_N,
\end{aligned}
\right\} \qquad (8.66)
$$

where $\omega_E, \omega_N, \omega_H$ are angular rate vector components of the geographical trihedron; $n_E, n_N, n_H$ are the projections of the specific force on the axes of the geographical trihedron; $\Delta\omega(t) = (\Delta\omega_N, \Delta\omega_E, \Delta\omega_H)^T$, $\Delta a(t) = (\Delta a_N, \Delta a_E, \Delta a_H)^T$ are the subvectors of the ARS (gyros)

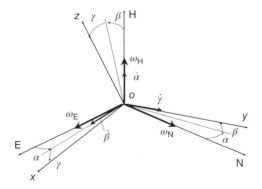

**Figure 8.9**   Errors of the geographical trihedron

and accelerometer errors in the projections on the geographical trihedron; $\delta a_{\mathrm{BE}}, \delta a_{\mathrm{BN}}, \delta a_{\mathrm{BH}}$ are the error components described as

$$
\left.\begin{aligned}
\delta a_{\mathrm{BE}} &= -\Delta V_{\mathrm{H}}\left(2\Omega + \dot{\lambda}\right)\cos\varphi + \Delta V_{\mathrm{N}}\left(2\Omega + \dot{\lambda}\right)\sin\varphi + \\
&\quad +\Delta\varphi[V_{\mathrm{H}}\left(2\Omega + \dot{\lambda}\right)\sin\varphi + \\
&\quad +V_{\mathrm{N}}\left(2\Omega + \dot{\lambda}\right)\cos\varphi] - \Delta\dot{\lambda}\left(V_{\mathrm{H}}\cos\varphi - V_{\mathrm{N}}\sin\varphi\right); \\
\delta a_{\mathrm{BN}} &= -\Delta V_{\mathrm{E}}\left(2\Omega + \dot{\lambda}\right)\sin\varphi - \Delta V_{\mathrm{H}}\dot{\varphi} - \Delta\varphi V_{\mathrm{E}}\left(2\Omega + \dot{\lambda}\right)\cos\varphi - \\
&\quad -\Delta\dot{\varphi}V_{\mathrm{H}} - \Delta\dot{\lambda}V_{\mathrm{E}}\sin\varphi; \\
\delta a_{\mathrm{BH}} &= \Delta V_{\mathrm{E}}\left(2\Omega + \dot{\lambda}\right)\cos\varphi + \Delta V_{\mathrm{N}}\dot{\varphi} - \Delta\varphi V_{\mathrm{E}}\left(2\Omega + \dot{\lambda}\right)\sin\varphi + \\
&\quad +\Delta\dot{\lambda}V_{\mathrm{E}}\cos\varphi + \Delta\dot{\varphi}V_{\mathrm{N}},
\end{aligned}\right\}
\tag{8.67}
$$

where $\varphi$ and $\lambda$ are the latitude and longitude, respectively; $V_{\mathrm{N}}, V_{\mathrm{E}}, V_{\mathrm{H}}$ are the velocity components; and $\Omega$ is the angular velocity of the Earth.

Note that the sensors' errors are given in the body trihedron axes $x_{\mathrm{b}} y_{\mathrm{b}} z_{\mathrm{b}}$. To determine the projections of these errors on the geographic trihedron $\Delta\omega_{\mathrm{E}}, \Delta\omega_{\mathrm{N}}, \Delta\omega_{\mathrm{H}}, \Delta a_{\mathrm{E}}, \Delta a_{\mathrm{N}}, \Delta a_{\mathrm{H}}$, the following relations are used:

$$
\begin{bmatrix} \Delta\omega_{\mathrm{E}} \\ \Delta\omega_{\mathrm{N}} \\ \Delta\omega_{\mathrm{H}} \end{bmatrix} = C\left(t\right) \begin{bmatrix} \Delta\omega_{bx} \\ \Delta\omega_{by} \\ \Delta\omega_{bz} \end{bmatrix}, \quad
\begin{bmatrix} \Delta a_{\mathrm{E}} \\ \Delta a_{\mathrm{N}} \\ \Delta a_{\mathrm{H}} \end{bmatrix} = C\left(t\right) \begin{bmatrix} \Delta a_{bx} \\ \Delta a_{by} \\ \Delta a_{bz} \end{bmatrix}.
\tag{8.68}
$$

Here, $\Delta\omega_{bi}, \Delta a_{bi}$, and $i = x, y, z$ are the gyro and accelerometer errors specified in the body trihedron $x_{\mathrm{b}} y_{\mathrm{b}} z_{\mathrm{b}}$; $C(t)$ is the direction cosine matrix, which determines the attitude of the geographical trihedron relative to the body trihedron:

$$
C = \begin{bmatrix}
\cos K \cos\theta + \sin K \sin\psi \sin\theta & \sin K \cos\psi & \cos K \sin\theta - \sin K \sin\psi \cos\theta \\
-\sin K \cos\theta + \cos K \sin\psi \sin\theta & \cos K \cos\psi & -\left(\sin K \sin\theta + \cos K \sin\psi \cos\theta\right) \\
-\cos\psi \sin\theta & \sin\psi & \cos\psi \cos\theta
\end{bmatrix},
$$

where $K$ is a heading and $\psi$ and $\theta$ are the roll and pitch, respectively.

Taking into account the introduced vector $x^{l_1}(t) = ((\Delta S(t))^T, (\Delta V(t))^T, (\Delta \alpha(t))^T)^T$ and Equations 8.66–8.68, the vector–matrix equation may be written:

$$\dot{x}^{l_1}(t) = F^{l_1} x^{l_1}(t) + G^{l_1} u(t),\qquad(8.69)$$

where $u(t) = (\Delta \omega_{bx}, \Delta \omega_{by}, \Delta \omega_{bz}, \Delta a_{bx}, \Delta a_{by}, \Delta a_{bz})^T$ is the vector of the total error components of the sensors, and the matrices included in this equation are defined as follows:

$$G^{l_1} = \begin{bmatrix} 0_{3\times3} & 0_{3\times3} \\ 0_{3\times3} & C \\ C & 0_{3\times3} \end{bmatrix}\qquad(8.70)$$

$$F^{l_1} = \begin{bmatrix}
0 & 0 & 0 & 0 & \dfrac{1}{R} & 0 & 0 & 0 & 0 \\
\dfrac{V_E \sin\varphi}{R\cos^2\varphi} & 0 & 0 & \dfrac{\cos\varphi}{R} & 0 & 0 & 0 & 0 & 0 \\
0 & 0 & 0 & 0 & 0 & 1 & 0 & 0 & 0 \\
f[4,1] & 0 & 0 & f[4,4] & f[4,5] & f[4,6] & n_N & 0 & -n_H \\
f[5,1] & 0 & 0 & f[5,4] & f[5,5] & f[5,6] & -n_E & n_H & 0 \\
f[6,1] & 0 & 0 & f[6,4] & f[6,5] & 0 & 0 & -n_N & n_E \\
\Omega\cos\varphi + \dfrac{V_E}{R\cos^2\varphi} & 0 & 0 & \dfrac{tg\varphi}{R} & 0 & 0 & 0 & \omega_N & -\omega_E \\
0 & 0 & 0 & 0 & -\dfrac{1}{R} & 0 & -\omega_N & 0 & \omega_H \\
-\Omega\sin\varphi & 0 & 0 & \dfrac{1}{R} & 0 & 0 & \omega_E & -\omega_H & 0
\end{bmatrix}\qquad(8.71)$$

The elements of matrix (8.71) denoted as $f[i,j]$ are defined as follows:

$$f[4,1] = \frac{V_E}{R}\left(V_H tg\varphi - V_N tg^2\varphi\right) - \left(2\Omega + \dot{\lambda}\right)\left(V_H \sin\varphi + V_N \cos\varphi\right);$$

$$f[4,4] = \frac{V_H}{R} + \frac{V_N}{R} tg\varphi;$$

$$f[4,5] = -\left(2\Omega + \dot{\lambda}\right)\sin\varphi;\quad f[4,6] = \left(2\Omega + \dot{\lambda}\right)\cos\varphi;$$

$$f[5,1] = \left(2\Omega + \dot{\lambda}\right)V_E \cos\varphi + \frac{V_E^2}{R} tg^2\varphi;$$

$$f[5,4] = \left(2\Omega + \dot{\lambda}\right)\sin\varphi + \frac{V_E}{R} tg\varphi;\quad f[5,5] = \frac{V_H}{R};$$

$$f[5,6] = \dot{\varphi}; \quad f[6,1] = \left(2\Omega + \dot{\lambda}\right)V_E \sin\varphi - \frac{V_E^2}{R}\mathrm{tg}^2\varphi;$$

$$f[6,4] = -\left(2\Omega + \dot{\lambda}\right)\cos\varphi - \frac{V_E}{R}; f[6,5] = -\dot{\varphi} - \frac{V_N}{R}.$$

Assume that the errors of the sensors can be represented as a sum of four independent components (Anuchin and Emel'yantsev, 1999), that is,

$$\Delta\varepsilon_{bi} = \Delta\bar{\varepsilon}_{bi} + \Delta\varepsilon_{bi}^m + \varepsilon_{bi}\Delta M_{\varepsilon i} + \tilde{w}_{bi}^{\varepsilon}, \quad \varepsilon = \omega, a; \quad i = x, y, z. \tag{8.72}$$

Here $\Delta\bar{\varepsilon}_{bi}$ are the drift components of the gyros ($\varepsilon = \omega$) and accelerometers ($\varepsilon = a$) that characterize zero shift from start to start as random bias, that is,

$$\Delta\dot{\bar{\varepsilon}}_{bi} = 0,$$

$\Delta\varepsilon_{bi}^m$ are the error components characterizing "zero" instability at start-up, described as first-order exponentially correlated Markov processes,

$$\Delta\dot{\varepsilon}_{bi}^m = -\alpha_i^{\varepsilon}\Delta\varepsilon_{bi}^m + \sqrt{2\alpha_i^{\varepsilon}\left(\sigma_i^{\varepsilon}\right)^2}w_{bi}^{\varepsilon},$$

where $w_{bi}^{\varepsilon}$ are zero-mean white noise with unity PSD, uncorrelated for various $i$ and with the initial state vector; $\left(\sigma_i^{\varepsilon}\right)^2$ and $\alpha_i^{\varepsilon}$ being the variances and reciprocals, respectively, of the correlation interval; $\Delta M_{\varepsilon i}$ are the scale factor errors described by the random bias $\Delta\dot{M}_{\varepsilon i} = 0$; $\tilde{w}_{bi}^{\varepsilon}$ are zero-mean white-noise error components with known PSD, uncorrelated for various $i$ and with the initial state vector; $\varepsilon_{bi}$ are the components of the corresponding parameter being measured, namely, $\varepsilon = \omega$ for the gyros and $\varepsilon = a$ for the accelerometers, $\varepsilon = \omega, a, i = x, y, z$.

Taking these into consideration, it is possible to introduce subvector $x^{l_2}$:

$$x^{l_2} = \left(\left(x^g\right)^{\mathrm{T}}, \left(x^a\right)^{\mathrm{T}}\right)^{\mathrm{T}},$$

where

$$x^g = \left(\Delta\bar{\omega}_{bx} \quad \Delta\bar{\omega}_{by} \quad \Delta\bar{\omega}_{bz} \quad \Delta\omega_{bx}^m \quad \Delta\omega_{by}^m \quad \Delta\omega_{bz}^m \quad \Delta M_{\omega x} \quad \Delta M_{\omega y} \quad \Delta M_{\omega z}\right)^{\mathrm{T}},$$

$$x^a = \left(\Delta\bar{a}_{bx} \quad \Delta\bar{a}_{by} \quad \Delta\bar{a}_{bz} \quad \Delta a_{bx}^m \quad \Delta a_{by}^m \quad \Delta a_{bz}^m \quad \Delta M_{ax} \quad \Delta M_{ay} \quad \Delta M_{az}\right)^{\mathrm{T}}$$

are the subvectors describing the error components of the sensor in the axes of the body trihedron.

From the aforementioned, it follows that the INS error equations can now be written as

$$\dot{x}^{l_1} = F^{l_1}x^{l_1} + F_{12}x^{l_2} + G^{l_1}w^{l_1}, \tag{8.73}$$

$$\dot{x}^{l_2} = F^{l_2}x^{l_2} + G^{l_2}w^{l_2}, \tag{8.74}$$

where $F^{l_1}$ is a nine-dimensional square matrix, defined by (8.71); $G^{l_1}$ is a matrix of $9 \times 6$ dimension, defined by (8.70); $F^{l_2}$ is an 18-dimensional block-diagonal matrix with blocks of $3 \times 3$ dimension, wherein all the blocks are zero ones except the second diagonal block $\left( F_\omega = \mathrm{diag}\{-\alpha_i^\omega\} \right)$ and the fifth diagonal block $\left( F_a = \mathrm{diag}\{-\alpha_i^a\} \right)$, $i = x, y, z$; $F_{12}$ is determined as

$$F_{12} = \begin{bmatrix} 0_{3\times9} & 0_{3\times9} & 0_{3\times9} & 0_{3\times9} & 0_{3\times9} & 0_{3\times9} \\ 0_{3\times9} & 0_{3\times9} & 0_{3\times9} & C & C & CF_a \\ C & C & CF_\omega & 0_{3\times9} & 0_{3\times9} & 0_{3\times9} \end{bmatrix}, \tag{8.75}$$

where $F_\varepsilon = \mathrm{diag}\{\varepsilon_{bi}\}$, $i = x, y, z$, $\varepsilon = \omega, a$,

$$\left( G^{l_2} \right)^{\mathrm{T}} = \begin{bmatrix} 0_{3\times3} & 0_{3\times3} & 0_{3\times3} & 0_{3\times3} & G_\omega^{l_2} & 0_{3\times3} \\ 0_{3\times3} & G_a^{l_2} & 0_{3\times3} & 0_{3\times3} & 0_{3\times3} & 0_{3\times3} \end{bmatrix}, \tag{8.76}$$

$$G_\varepsilon^{l_2} = \mathrm{diag}\left\{ \sqrt{2\alpha_i^\varepsilon \left( \sigma_i^\varepsilon \right)^2} \right\}, \quad i = x, y, z, \quad \varepsilon = \omega, a,$$

$w^{l_1} = \left( \tilde{w}_{bx}^\omega, \tilde{w}_{by}^\omega, \tilde{w}_{bz}^\omega, \tilde{w}_{bx}^a, \tilde{w}_{by}^a, \tilde{w}_{bz}^a \right)^{\mathrm{T}}$ is a six-dimensional vector of the sensor white-noise components in the axes of the body trihedron with specified PSDs $q_{\varepsilon i}^2$, $i = x, y, z$, $\varepsilon = \omega, a$; and $w^{l_2} = \left( w_{bx}^\omega, w_{by}^\omega, w_{bz}^\omega, w_{bx}^a, w_{by}^a, w_{bz}^a \right)^{\mathrm{T}}$ is a six-dimensional vector of white noise with the unity PSDs for exponentially correlated error components. Here, $w^{l_1}$ and $w^{l_2}$ are assumed uncorrelated with each other and with the initial conditions.

Analyzing Equations 8.73 and 8.74, it is necessary to pay attention to the following fact. In these equations, the matrices depend on the geographical coordinates, velocity components, angular velocities, and specific force as well as the attitude angles that are included in the matrix of the direction cosines $C(t)$. In other words, these matrices depend, among the others, on the parameters that are to be estimated. To overcome this contradiction in the calculation of the matrix elements in (8.73) and (8.74), measurements directly from the INS being aided are used. The possibility of such replacement for the example of replacing $\varphi^{\mathrm{INS}}$ by $\varphi$ may be explained as follows. Using the obvious relation $\varphi = \varphi^{\mathrm{INS}} - \Delta\varphi$ in the formation of matrix $F^{l_1}$, the products of $\varphi^{\mathrm{INS}} - \Delta\varphi$ by the other components of the state vector are obtained. It therefore follows that, for example, $\varphi\Delta\varphi = (\varphi^{\mathrm{INS}} - \Delta\varphi)\Delta\varphi \approx \varphi^{\mathrm{INS}}\Delta\varphi$ with an accuracy of infinitely small second-order summands. However, an important point is that the description of INS errors by linear equations is only valid when these errors vary within certain limits. It is this fact that determines the need for a feedback, that is, taking into account the corrections obtained as a result of the filtering problem solution in the formation of matrices in Equation 8.73.

It should also be noted that Equations 8.70, 8.75, and 8.76 for matrices $G^{l_1}$, $F_{12}$, and $G^{l_2}$, respectively, are dictated mostly by the necessity of transforming the sensor errors specified in the axes of the body trihedron on the axes of the geographical trihedron. Note also that it is the need for such transformations that determines the nonstationary (time-varying) character of Equation 8.73 for strapdown INSs.

### 8.4.2 The Filtering Problem in Loosely Coupled INS/SNS

Consider the filtering problem for the fusion of INS and SNS data in a loosely coupled processing scheme (Schmidt and Phillips, 2003). Proceeding from an invariant statement of the problem, the previously considered processing scheme (Figure 8.1) can be presented as in Figure 8.10 and in more detail in Figure 8.2.

To solve the problem, we assume that the aircraft altitude is determined in a separate filter, the so-called vertical channel filter. In this case, the measurements from the SNS are the components of the coordinates and velocity in the horizontal plane. In accordance with the invariant scheme, the measurements for the filtering problem can be written as follows:

$$
\left.
\begin{aligned}
y_1(t) &= \varphi^{INS}(t) - \varphi^{SNS}(t) = \Delta\varphi(t) - \Delta\varphi^{SNS}(t); \\
y_2(t) &= \lambda^{INS}(t) - \lambda^{SNS}(t) = \Delta\lambda(t) - \Delta\lambda^{SNS}(t); \\
y_3(t) &= V_E^{INS}(t) - V_E^{SNS}(t) = \Delta V_E(t) - \Delta V_E^{SNS}(t); \\
y_4(t) &= V_N^{INS}(t) - V_N^{SNS}(t) = \Delta V_N(t) - \Delta V_N^{SNS}(t).
\end{aligned}
\right\}
$$

Here, $\varphi^{INS}$, $\varphi^{SNS}$ and $\lambda^{INS}$, $\lambda^{SNS}$ are the latitude and longitude and $V_i^{INS}$, $V_i^{SNS}$, $i = E, N$ are the velocity components from the INS and SNS, respectively; $\Delta\varphi(t)$, $\Delta\lambda(t)$, $\Delta V_E(t)$, $\Delta V_N(t)$ and $\Delta\varphi^{SNS}(t)$, $\Delta\lambda^{SNS}(t)$, $\Delta V_E^{SNS}(t)$, $\Delta V_N^{SNS}(t)$ are the INS and SNS errors. Assume for simplification that the SNS errors contain only white-noise error components, that is, $x^{II}(t) = x^{SNS}(t)$ are absent, $\Delta y^{II}(t) = \Delta y^{SNS}(t) = v^{SNS}(t)$, and $R^{SNS} = \mathrm{diag}\{R[i,i]\}$, $i = \overline{1.4}$. In this case, the equations for the state vector are completely defined by Equations 8.73 and 8.74 for the 27-dimensional vector $x(t) = x^{INS}(t) = ((x^{I_1}(t))^T, (x^{I_2}(t))^T)^T$ of the SINS errors; $w(t) = ((w^{I_1})^T, (w^{I_2})^T)^T$ is the 12-dimensional vector of system white noise; and $v = -v^{SNS}(t)$ is the four-dimensional vector of measurement white noise. The $27 \times 27$, $27 \times 12$ and $4 \times 27$-dimensional matrices $F$, $G$ and $H$ in (8.52) and (8.53) are defined as follows:

$$
F = \begin{bmatrix} F^{I_1} & F_{12} \\ 0 & F^{I_2} \end{bmatrix}, \quad
G = \begin{bmatrix} G^{I_1} & 0 \\ 0 & G^{I_2} \end{bmatrix}, \quad
H = \begin{bmatrix} H_1, 0_{4\times22} \end{bmatrix},
$$

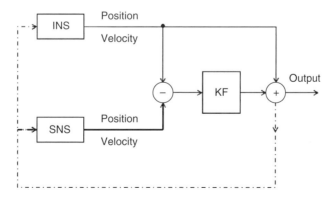

**Figure 8.10** Block diagram for an invariant loosely coupled scheme for INS and SNS data fusion

$$
H_1 = \begin{bmatrix} 1 & 0 & 0 & 0 & 0 \\ 0 & 1 & 0 & 0 & 0 \\ 0 & 0 & 0 & 1 & 0 \\ 0 & 0 & 0 & 0 & 1 \end{bmatrix}.
$$

Here $F^{l_1}$, $F^{l_2}$, $F_{12}$, $G^{l_1}$, and $G^{l_2}$ are determined by (8.70), (8.71), (8.75), and (8.76). After the estimates $\hat{x}(t)$ have been obtained for the formulated filtering problem, the coordinates from the integrated navigation system are determined as follows:

$$
\hat{\varphi}(t) = \varphi^{\text{INS}}(t) - \hat{\Delta}\varphi(t), \quad \hat{\lambda}(t) = \lambda^{\text{INS}}(t) - \hat{\Delta}\lambda(t), \quad \hat{V}_{\text{N}}(t) = V_{\text{N}}^{\text{INS}}(t) - \hat{\Delta}V_{\text{N}}(t),
$$
$$
\hat{V}_{\text{E}}(t) = V_{\text{E}}^{\text{INS}}(t) - \hat{\Delta}V_{\text{E}}(t),
$$

Here, $\hat{\Delta}\varphi(t)$, $\hat{\Delta}\lambda(t)$, $\hat{\Delta}V_{\text{E}}(t)$, and $\hat{\Delta}V_{\text{N}}(t)$ are the corresponding components of $\hat{x}(t)$.

It should be noted that the state vector includes sensor errors and, as a result of the filtering problem solution, these errors are also estimated. In this regard, note that the dotted lines in Figure 8.10 imply that the estimates generated in the algorithm can be used not only to aid the output data but also to estimate the errors of the INS sensors and improve the quality of SNS signal tracking in satellite-user equipment. It should also be noted that, despite the fact that one of the goals of the filtering problem is to increase the accuracy of coordinate determination, it is not the main goal in INS aiding. The point is that usually the coordinate accuracy for the integrated system with the availability of SNS data does not differ much from that of the SNS. However, aiding makes it possible to estimate the errors of INS sensors, which provides a substantial increase in the accuracy of navigation parameters and coordinates in the case when the satellite signals are missing and the Kalman filter operates in prediction mode, that is, when only INS data are used.

### 8.4.3   The Filtering Problem in Tightly Coupled INS/SNS

Consider the formulation of the filtering problem relevant to the fusion of INS and SNS data in the case of tightly coupled processing (Schmidt and Phillips, 2003). The invariant structure of a tightly coupled scheme of INS and SNS data processing is depicted in Figure 8.11 and in more detail in Figure 8.5.

As can be seen from Figure 8.11, when data are processed in accordance with the tightly coupled scheme, measurements of coordinates and velocities from the inertial system (I = INS) can be used as measurements of the form (8.56), whereas measurements of the form (8.57) are represented by pseudoranges $\rho^{\text{SNS}}(t)$ and Doppler shifts $\dot{\rho}^{\text{SNS}}(t)$ from the SNS (II = SNS) for each of the satellites, which can be written by the following equations:

$$
\rho(t) = \sqrt{\left(X_1^i - X_1\right)^2 + \left(X_2^i - X_2\right)^2 + \left(X_3^i - X_3\right)^2} + c\Delta t + \varepsilon(t), \tag{8.77}
$$

$$
\dot{\rho}_i = \frac{\left(X_1^i - X_1\right)\left(\dot{X}_1^i - \dot{X}_1\right) + \left(X_2^i - X_2\right)\left(\dot{X}_2^i - \dot{X}_2\right) + \left(X_3^i - X_3\right)\left(\dot{X}_3^i - \dot{X}_3\right)}{\sqrt{\left(X_1^i - X_1\right)^2 + \left(X_2^i - X_2\right)^2 + \left(X_3^i - X_3\right)^2}} + c\dot{\Delta}t + \tilde{\varepsilon}_i,
$$

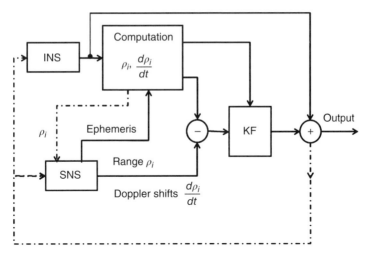

**Figure 8.11** Block diagram of an invariant tightly coupled scheme for INS and SNS data fusion

where $X_1, X_2, X_3$ are the user's unknown coordinates in the orthogonal coordinate system; $X_j^i$, $j = 1, 2, 3$ are the $i$-th satellite coordinates in the same coordinate system transmitted to the user in the navigation message; $\dot{X}_j$; $\dot{X}_j^i$, $j = 1, 2, 3$ are the components of the user's velocity and the $i$-th satellite, respectively; $\Delta t$ is the user's clock error; $\varepsilon_i$ is the total measurement error; $c$ is the velocity of light; $\Delta \dot{t}$ is the error due to user's drift; and $\tilde{\varepsilon}_i$ is the total Doppler measurement error. The values of the pseudoranges and Doppler shifts are calculated and the difference measurements of the (8.58) type are formed. Taking these into account , it is possible to write (8.58) as follows:

$$y(t) = \rho^{\text{SNS}}(t) - \rho^{\text{INS}}\left(X^{\text{INS}}(t)\right) = \Delta\rho^{\text{SNS}}(t) - \Delta\rho^{\text{INS}}(t), \qquad (8.78)$$

where $\rho^{\text{INS}}(X^{\text{INS}}(t))$ are the pseudoranges calculated with the use of INS coordinates and the ephemerides received from the satellite; $\Delta\rho^{\text{INS}}$ are calculation errors of these pseudoranges; and $\rho^{\text{SNS}}(t)$ and $\Delta\rho^{\text{SNS}}(t)$ are the measured values of pseudoranges and the corresponding errors.

To describe measurements of the form (8.78), it is necessary to introduce the state vector $x(t)$ with two subvectors, $x^{\text{I}}(t) = x^{\text{INS}}(t)$ and $x^{\text{II}}(t) = x^{\text{SNS}}(t)$ of $n^{\text{INS}}$, $n^{\text{SNS}}$ dimensions, by which the INS and SNS errors are described. Subvector $x^{\text{I}}(t) = x^{\text{INS}}(t)$ coincides with a similar subvector for the loosely coupled scheme considered in the previous section, whereas subvector $x^{\text{I}}(t) = x^{\text{SNS}}(t)$ has a different structure. When data from $m$ satellites are used, this vector can be written as

$$x^{\text{SNS}}(t) = \left(\Delta(t), \left(x_\rho^1(t)\right)^{\text{T}}, \left(x_\rho^2(t)\right)^{\text{T}}, \dots, \left(x_\rho^m(t)\right)^{\text{T}}\right)^{\text{T}},$$

where $\Delta(t) = \Delta t$ is the user's clock error and subvectors $x_\rho^j(t)$, $j = \overline{1.m}$ are used to describe the error components $\varepsilon_\rho^j(t)$ for each satellite.

In a simple case, with the presence of only white-noise error components, that is, $\varepsilon_p^j(t) = 0$, $j = \overline{1.m}$, subvector $x^{SNS}(t) = \Delta(t)$, and its dimension is unity. If, in addition to white-noise errors, there are slowly varying error components $\varepsilon^j(t)$ and they are described, for example, by random walk or exponentially correlated processes, then subvector $x^{SNS}(t) = \left(\Delta(t), x_p^1(t), x_p^2(t), \ldots, x_p^m(t)\right)^T$, and its dimension is $m+1$.

With the accepted notation, $\Delta\rho^{SNS}$ and $\Delta\rho^{INS}$ take the form:

$$\Delta\rho^{SNS}(t) = H_p^{SNS} x^{SNS}(t) + v_p^{SNS}(t),$$
$$\Delta\rho^{INS}(t) = H_p^{INS} x^{INS}(t),$$

where $H_p^{SNS}$ and $H_p^{INS}$ are the matrices providing formation of the corresponding range measurement error components from the components of subvectors $x^{SNS}(t)$ and $x^{INS}(t)$; and $v_p^{SNS}(t)$ are the white-noise range measurement error components (Parkinson and Spiller, 1996; Grewall et al., 2013).

Defining the vectors and matrices used to describe SNS errors for the case of four satellites, the slowly varying components of the pseudomeasurement errors $\varepsilon^j(t)$ are described by random walk processes. In this case, the dimension of vector $x^{SNS}(t)$ is five, and the dimension of vectors $w^{SNS}(t)$ and $v^{SNS}(t)$ is four:

$$F^{SNS}(t) = 0_5, \quad Q^{SNS}(t) = q^2\delta(t)E_4, \quad R^{SNS}(t) = r^2\delta(t)E_4,$$

$$G^{SNS}(t) = \begin{bmatrix} 0 & 0 & 0 & 0 \\ 1 & 0 & 0 & 0 \\ 0 & 1 & 0 & 0 \\ 0 & 0 & 1 & 0 \\ 0 & 0 & 0 & 1 \end{bmatrix}, \quad H_p^{SNS}(t) = \begin{bmatrix} 1 & 1 & 0 & 0 & 0 \\ 1 & 0 & 1 & 0 & 0 \\ 1 & 0 & 0 & 1 & 0 \\ 1 & 0 & 0 & 0 & 1 \end{bmatrix},$$

$$P_0^{SNS} = \begin{bmatrix} \sigma_\Delta^2 & 0 & 0 & 0 & 0 \\ 0 & \sigma^2 & 0 & 0 & 0 \\ 0 & 0 & \sigma^2 & 0 & 0 \\ 0 & 0 & 0 & \sigma^2 & 0 \\ 0 & 0 & 0 & 0 & \sigma^2 \end{bmatrix},$$

where $q^2$ and $r^2$ are the PSDs for system and measurement noise, $\sigma_\Delta^2$ determines the level of uncertainty in the knowledge of the user's clock error, and $\sigma^2$ is the RMS value for the random walk component at the initial time. For simplification, these quantities are assumed to be the same for different satellites.

For the tightly coupled processing scheme, as in the case of the loosely coupled variant, the dotted lines in block diagram of Figure 8.11 show the feedback that allows the results of the filtering problem to be used for estimating the data from the INS sensors and for improving the accuracy of SNS signal tracking.

Note that in the case of the tightly coupled scheme, the state vector can be used without the user clock error component, which is the same for all satellites. For this purpose, difference pseudorange measurements from different satellites should be formed beforehand and one of the satellites should be taken as a reference. In so doing, the SNS measurement dimension becomes $m-1$, that is, reduced by unity. When formulating the corresponding filtering problem, it is necessary to take into account the fact that the covariance matrix $R^{\text{SNS}}(t)$ will no longer be a diagonal one (Parkinson and Spiller, 1996; Grewall *et al.*, 2013).

### 8.4.4   *Example of Filtering Algorithms for an Integrated INS/SNS*

Consider the solution examples for a filtering problem formulated already, namely, filtering for a loosely coupled INS/SNS system. Here, matrices $F$, $G$, and $H$ are determined as described in Section 8.4.2 and the calculations involve a specially developed program for designing linear filtering algorithms for integrated navigation systems (Koshaev and Stepanov, 2011). Then, the following input data are required:

- A diagonal initial covariance matrix specifying the initial covariance for all 27 components of the state vector
- $q_{\varepsilon i}^2, i = x, y, z, \varepsilon = \omega, a$, specifying the PSDs for the white-noise components of sensor errors
- $\left(\sigma_i^\varepsilon\right)^2, \alpha_i^\varepsilon$, specifying the variances and correlation intervals of exponentially correlated processes describing the sensor "zero" instability at start-up
- $R[i,i], i = 1.4$, specifying levels of the measurement errors
- The vehicle trajectory, which determines the elements of dynamics matrix
- A time diagram for using the satellite measurements.

For purposes of a simulation, assume that the vehicle is moving rectilinearly with a speed of $V = 50$ m/s and a heading $K=30°$ with initial $\varphi=60°$. Also assume that the system uses microelectromechanical (MEMS) angular rate sensors and accelerometers with the performance characteristics of Table 8.1. The parameters for all three axes are supposed to be identical.

To determine matrix $F^{12}$ and covariance matrix $P_0$ at the initial time, data from Table 8.1 must be used. It may also be assumed that $P_0[i,i] = 0, i = 1.6$, these values being taken to better observe how position and velocity errors increase; $P_0[i,i] = (0.5°)^2$ for heading errors; and $P_0[i,i] = (0.3°)^2$ for vertical misalignment.

To initially estimate how different components of sensor errors contribute to the position and velocity errors, it can be shown that in INS autonomous mode (without SNS aiding) the changes in position and velocity errors are approximately determined by the following simple relationships (Stepanov, 2012):

$$\sigma_{\Delta\phi} = g\sigma_{\Delta\omega}\frac{t^3}{6}, \quad \sigma_{\Delta V_N} = g\sigma_{\Delta\omega}\frac{t^2}{2}.$$

Position and velocity errors calculated by these formulas for $t = 100$ s are given in Table 8.2.

From Table 8.2, it can be seen that ARS bias is the main contributor to the error in navigation parameters.

**Table 8.1**  Accuracy performance of angular rate sensors and accelerometers

| Parameter | ARS | Accelerometers |
|---|---|---|
| Bias from start to start, RMS | 30°/h | 0.01 m/s² |
| Scale factor error, RMS | 1% | 1% |
| Sensor "zero" instability at start-up: RMS, | $\sigma^\omega = 10°/h$ | $\sigma^a = 0.01$ m/s² |
| correlation intervals $\tau^\varepsilon = 1/\alpha^\varepsilon$, $\varepsilon = \omega, a$ | $\tau^\omega = 180$ s | $\tau^a = 180$ s |
| PSDs of white noise | $q_\omega^2 = (30°/h)^2$ s | $q_a^2 = (0.03 \text{ m}/s^2)^2$ s |

**Table 8.2**  The contributions of sensor error components to position and velocity errors (RMS)

| Error source | | Coordinates (m) | Velocity (m/s) |
|---|---|---|---|
| ARS | Bias | 237 | 7.1 |
| | White noise | 32 | 0.8 |
| | Markov component | 74.7 | 2.2 |
| Accelerometer | Bias | 50 | 1 |
| | White noise | 17.3 | 0.3 |
| | Markov component | 46.5 | 0.9 |
| Total (without Markov component) | | 245.5 | 7.3 |
| Total (with Markov component) | | 262.8 | 7.6 |

The results obtained using approximate relationships mostly agree with similar results obtained using the complete model (8.71), which is confirmed by the plots in Figure 8.12.

Consider now the results with satellite aiding. The levels of the satellite measurement error set by the relevant PSD were taken to be $R[i,i] = (5\text{m})^2$ s, $i = 1,2$ (position) and $R[i,i] = (0.1\text{m}/\text{s})^2$ s, $i = 3,4$ (velocity). These values correspond to RMS errors of 5 m (position) and 0.1 m/s (velocity) with white-noise averaging interval of 1 s.

Assuming that satellite measurements were not available for the first and last 100 s over total 400 s, the results for position and velocity errors will be as in Figure 8.13.

The main conclusion from the plots is that without SNS the errors increase significantly slower at the second interval, starting from the 300th second, in comparison with the first interval of 0–100 s. It completely agrees with the expected results: with SNS, the ARS bias values are estimated over 100–300 s interval (Figure 8.14a), which slows down the error accumulation after SNS signal outage. This is actually a major objective of INS aiding: estimating the sensor errors to improve the system performance during SNS outage. At the same time, the results reveal that during the aiding, accelerometer bias errors and ARS instabilities described by exponentially correlated Markov processes remain at *a priori* level (Figure 8.14b). This creates prerequisites for designing suboptimally reduced filters with the reduced state vector (Rivkin *et al.*, 1976).

Here, calculations for such a filter with a state vector with excluded Markov error components are provided. In these calculations it was taken that INS errors are described with a 27-dimensional vector, including exponentially correlated components of sensor errors. However, the Kalman filter used was assumed not to include such components, that is, the dimensionality of the state vector in the Kalman filter was reduced by six components. In this

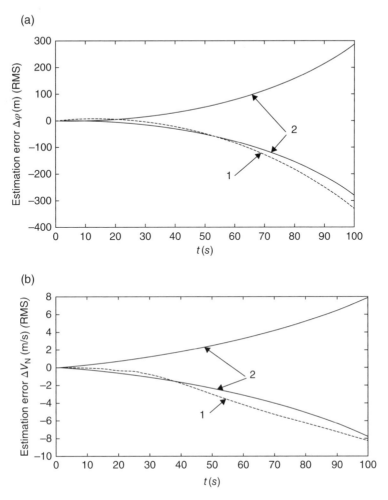

(a)

(b)

**Figure 8.12** Sample of errors (1) and RMS errors (2) in latitude (a) and velocity northern component (b) Autonomous mode

case, it should be remembered that the covariance matrix generated in the Kalman filter (calculation matrix) differs from the actual (real) covariance matrix of the filter. Figures 8.15 and 8.16 present the results of calculations made using the aforementioned program (Koshaev and Stepanov, 2011). It helps to study the filter sensitivity in case of discrepancy between the actual model for the state vector and the assumed model used in the Kalman filter. In the calculations, the satellite measurements were taken to be available within the period of 25–75 s.

Analysis of the plots reveals that Root-Mean-Square (RMS) estimation errors for optimal and suboptimal filters differ slightly. At the same time, RMS errors calculated in the filter (assumed RMS) are significantly lower than the actual errors—that is, the assumed RMS errors are more optimistic. Therefore, simplifying the algorithm by

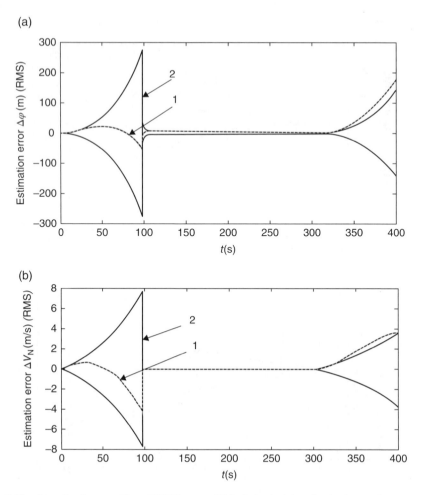

**Figure 8.13** Sample of errors (1) and RMS errors (2) in latitude (a) and velocity northern component (b) Satellite aiding mode

reducing the dimensionality of the state vector brings about the problem that the covariance matrix generated in the Kalman filter is inadequate for the actual covariance matrix. This problem can be solved by using the so-called filters with guaranteed estimation quality, wherein an adequate analysis covariance matrix is provided. A method for designing such filters is considered, for example, in Litvinenko *et al.* (2009) and Kulakova and Nebylov (2008).

In closing Section 8.4 that was concerned with the application of the filtering approach for integrated INS/SNS systems, it should be noted that this approach can also be used for autonomous satellite orbit determination using SNS receivers (Mikhailov *et al.*, 2010; Mikhailov and Mikhailov, 2013a). This is also referred to in Mikhailov and Mikhailov (2013b) for discussion of the filtering results on SNS data.

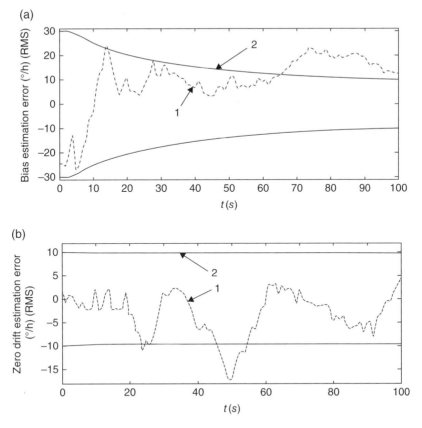

**Figure 8.14**   Sample of errors (1) and ±RMS errors (2) for bias (a) and bias instabilities (b) for ARS with horizontal sensitivity axes in aided mode

## 8.5   Filtering and Smoothing Problems Based on the Combined Use of Kalman and Wiener Approaches for Aviation Gravimetry

The estimation of Gravity Anomalies (GAs) using gravimeters on board aircraft (aviation gravimetry) is now a widely used technology (Koshaev and Stepanov, 2010; Kulakova *et al.*, 2010; Krasnov *et al.*, 2011a, 2011b, 2014a, 2014b).

Data about GA are very important for high precision navigation, and progress in aviation gravimetry is largely dependent on the ability to determine aircraft position and velocity in the vertical plane with high accuracy. These data are used in order to exclude unknown vertical acceleration components of the aircraft from gravimeter readings. Such information can be obtained from SNS operating in the differential mode with phase measurements. At the same time, the difference between the second integral of gravimeter readings and altitude measurement from SNS can be formed. Hence, filtering and smoothing problems can be formulated for processing all data. Kalman and Wiener approaches can be jointly used to solve relevant optimal filtering and smoothing problems and are considered below (Stepanov, 2004; Koshaev and Stepanov, 2010).

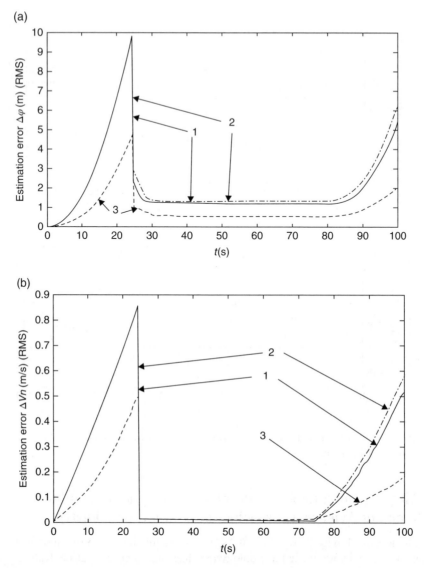

**Figure 8.15** RMS errors in latitude (a) and velocity northern component (b) for optimal (1) and sub-optimal Kalman filter (2) and assumed RMS error (3) generated in a reduced Kalman filter

### 8.5.1   Statement of the Optimal Filtering and Smoothing Problems in the Processing of Gravimeter and Satellite Measurements

Formulate the statement of the filtering and smoothing problems encountered in processing gravimeter data, satellite measurements of the vertical components of coordinates (altitude),

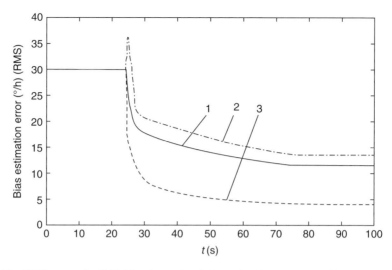

**Figure 8.16** RMS errors in ARS bias for optimal (1), suboptimal Kalman filter (2), and estimated RMS error (3)

and velocity (vertical velocity) (Koshaev and Stepanov, 2010). In this problem, the following measurements are used:

$$h^{\text{SNS}}(t) = h(t) + \delta h^{\text{SNS}}(t),\tag{8.79}$$

$$V^{\text{SNS}}(t) = V(t) + \delta V^{\text{SNS}}(t),\tag{8.80}$$

$$\tilde{g}^{\text{GR}}(t) = \dot{V}(t) + \tilde{g}(t) + \delta g(t),\tag{8.81}$$

where $h(t)$, $V(t)$, and $\dot{V}(t)$ are the altitude, vertical velocity, and acceleration, respectively; $\delta h^{\text{SNS}}(t)$, $\delta V^{\text{SNS}}(t)$ are the errors in altitude and vertical velocity obtained using SNS data; $\tilde{g}(t)$ are the GAs; $\delta g(t)$ are the gravimeter errors. For simplification, assume that the initial values of altitude and vertical velocity are known. By successively integrating the gravimeter data, the altitude and vertical velocity can be calculated. Their errors are given by

$$\begin{cases} \Delta \dot{h}^{\text{GR}}(t) = \Delta V^{\text{GR}}(t); \\ \Delta \dot{V}^{\text{GR}}(t) = \tilde{g}(t) + \delta g(t), \end{cases}\tag{8.82}$$

where $\Delta h^{\text{GR}}(t)$ and $\Delta V^{\text{GR}}(t)$ are the errors in altitude and vertical velocity, $\tilde{g}(t)$ are the GAs, and $\delta g(t)$ are the gravimeter errors.

Using the invariant scheme of processing, the measurements for the estimation problem can be formed as a difference between the first and the second integrals of gravimeter data (8.81), on the one hand, and satellite measurements of altitude and vertical velocity (8.79) and (8.80), on the other. In an operator form, these differences can be presented as follows:

$$y^h = \frac{\tilde{g}^{\text{GR}}}{p^2} - h^{\text{SNS}} = \frac{\tilde{g} + \delta g}{p^2} - \delta h^{\text{SNS}},$$

$$y^V = \frac{\tilde{g}^{GR}}{p} - V^{SNS} = \frac{\tilde{g} + \delta g}{p} - \delta V^{SNS}.$$

These measurements may be written as follows:

$$y^h = \Delta h^{GR} - \delta h^{SNS},$$  (8.83)

$$y^V = \Delta V^{GR} - \delta V^{SNS}.$$  (8.84)

Taking (8.79)–(8.82) into consideration and specifying the stochastic models for $\tilde{g}$, $\delta h^{SNS}$, $\delta V^{SNS}$, and $\delta g$, it becomes possible to formulate the relevant estimation problem: filtering and/or smoothing (Koshaev and Stepanov, 2010). Its objective is to estimate the GA $\tilde{g}$, altitude, and vertical velocity using measurements (8.83) and (8.84).

It should be noted that due to the specific form of difference measurements for the estimation problem, the considered processing scheme is invariant with respect to altitude and noninvariant with respect to GA. It conditions the need to introduce a statistical model to describe the GA.

In what follows, the ways of how the GA estimation problem can be solved are analyzed.

### 8.5.2  Problem Statement and Solution within the Kalman Approach

Assume that $\tilde{g}$ can be presented as a component of a Markov process described by a relevant shaping filter. Then, introducing the state vector $x = (\Delta h^{GR}, \Delta V^{GR}, \tilde{g}, \ldots)^T$, it is possible to specify the estimation problem within the Kalman approach taking account of the models of altitude errors $\delta h^{SNS}$, vertical velocity errors $\delta V^{SNS}$, and gravimeter errors $\delta g$. The problem may then be formulated following (Koshaev and Stepanov, 2010), and assuming that the GA can be described, for example, by using the so-called Jordan model. According to this model, the correlation function of GA along a rectilinear trajectory can be represented as in Jordan (1972):

$$K_{\tilde{g}}(\rho) = \sigma_{\tilde{g}}^2 \left( 1 + \alpha\rho - \frac{(\alpha\rho)^2}{2} \right) e^{-\alpha\rho},$$  (8.85)

where $\sigma_{\tilde{g}}^2$ is the variance, $\alpha$ is the parameter determining the spatial variability of GA, and $\rho$ is the path length along the rectilinear trajectory. The relevant PSD $\tilde{g}$ is given by

$$S_{\tilde{g}}(\omega) = 2\alpha^3 \sigma_{\tilde{g}}^2 \frac{5\omega^2 + \alpha^2}{(\omega^2 + \alpha^2)^3},$$  (8.86)

where $\omega$ is the analog of the angular frequency for the process depending on the length.

To transform (8.85) to the time domain, $\rho = Vt$ should be used instead of $\rho$, where $V$ is the speed. The process with correlation function (8.85) is a differentiable one, and the variance of its derivative can be determined as follows:

$$\sigma_{\partial\tilde{g}/\partial\rho}^2 = -\frac{d^2}{d\rho^2} K_{\tilde{g}}(\rho)\bigg|_{\rho=0} = 2\alpha^2\sigma_{\tilde{g}}^2.$$  (8.87)

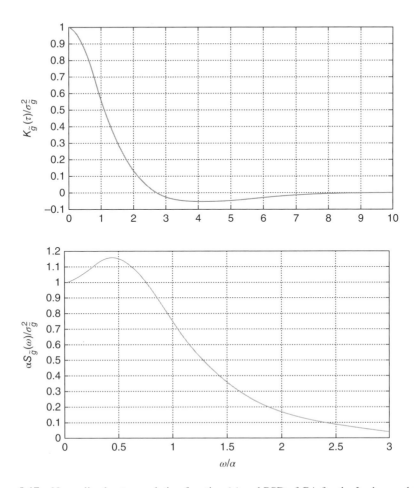

**Figure 8.17**  Normalized autocorrelation function (a) and PSD of GA for the Jordan model (b)

Plots of the normalized correlation function $K_{\tilde{g}}(\tau)/\sigma_{\tilde{g}}^2$ versus $\rho\alpha$ (a) and of $\alpha S_{\tilde{g}}(\omega)/\sigma_{\tilde{g}}^2$ versus $\omega/\alpha$ (b) are presented in Figure 8.17.

Using the known procedure for designing a shaping filter for processes with a specified PSD, a shaping filter for the process with PSD (8.86) can be obtained. This PSD can be presented as

$$S_{\tilde{g}}(\omega) = 2\alpha^3 \sigma_{\tilde{g}}^2 \frac{\left(\alpha + \sqrt{5}\,j\omega\right)\left(\alpha - \sqrt{5}\,j\omega\right)}{\left(\alpha + j\omega\right)^3 \left(\alpha - j\omega\right)^3},$$

and it can be shown that the shaping filter for $\tilde{g}$ can be presented as

$$\begin{cases} \dot{g}_1(t) = -\beta g_1(t) + g_2(t), \\ \dot{g}_2(t) = -\beta g_2(t) + g_3(t), \\ \dot{g}_3(t) = -\beta g_3(t) + q_w w(t), \end{cases} \tag{8.88}$$

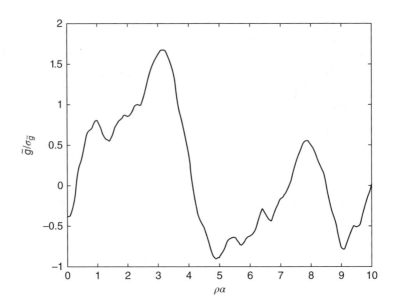

**Figure 8.18**   GA normalized sample for the Jordan model

$$\tilde{g}(t) = \dot{\tilde{g}}_1(t) + \tilde{g}_1(t)\frac{\beta}{\sqrt{5}} = -\beta\frac{\sqrt{5}-1}{\sqrt{5}}\tilde{g}_1(t) + \tilde{g}_2(t), \qquad (8.89)$$

where $\beta = V\alpha$ is the parameter determining the field variability in the time domain, and $w(t)$ is the white noise with unity PSD and $q_w = \sqrt{10\beta^3\sigma_{\tilde{g}}^2}$.

A normalized sample $\tilde{g}(t)$, obtained by simulation using Equations 8.88 and 8.89, is shown in Figure 8.18.

Assume that errors in altitude, vertical velocity, and gravimeter data are described by uncorrelated white noise with PSDs $R_h$, $R_V$, and $R_{GR}$. With these assumptions the problem can be reduced to estimating a five-dimensional vector $x(t)$ with components $x(t) = (\Delta h^{GR}(t), \Delta V^{GR}(t), x_3(t), x_4(t), x_5(t))^T$, where the first two components are described by Equation 8.82, and $(x_3, x_4, x_5)^T \equiv (g_1, g_2, g_3)^T$. Thus, the following can be written for the state vector components

$$\dot{x}_1(t) = x_2(t),$$
$$\dot{x}_2(t) = -\beta\zeta x_3(t) + x_4(t) + \sqrt{R_{GR}}\, w_{GR}(t),$$
$$\dot{x}_3(t) = -\beta x_3(t) + x_4(t),$$
$$\dot{x}_4(t) = -\beta x_4(t) + x_5(t),$$
$$\dot{x}_5(t) = -\beta x_5(t) + q_w w(t),$$

where $\zeta = (\sqrt{5}-1)/\sqrt{5}$ is a dimensionless coefficient, and $w_{GR}(t)$ is the white noise with unity PSD. Hence, a problem may be obtained in the form (8.24) and (8.25) with $n = 5, m = 2, p = 2$ and with the matrices:

$$F = \begin{bmatrix} 0 & 1 & 0 & 0 & 0 \\ 0 & 0 & -\beta\zeta & 1 & 0 \\ 0 & 0 & -\beta & 1 & 0 \\ 0 & 0 & 0 & -\beta & 1 \\ 0 & 0 & 0 & 0 & -\beta \end{bmatrix}, \quad G = \begin{bmatrix} 0 & 0 \\ \sqrt{R_{GR}} & 0 \\ 0 & 0 \\ 0 & 0 \\ 0 & q_w \end{bmatrix}, \tag{8.90}$$

$$H = \begin{bmatrix} I_2 & 0_{2\times3} \end{bmatrix}, \quad Q = I_2, \quad R = \begin{bmatrix} R_h & 0 \\ 0 & R_V \end{bmatrix}, \tag{8.91}$$

where $I_2, 0_{2\times3}$ are the $2\times2$ unit matrix and $2\times3$ zero matrix. For simplification, assume that the PSD of the white noise is $q_w^2$ and the corresponding coefficient in Equation 8.90 is 1. The problem of designing the filtering algorithm is reduced to implementation of the Kalman filter for the models specified by (8.90) and (8.91). The algorithm for solving the smoothing problem can be implemented in turn using the methods considered in Koshaev and Stepanov (2010) and Stepanov (2004, 2012). To analyze the accuracy of optimal filtering and smoothing, their error covariance matrices should be calculated by solving the relevant covariance equations and further analyzing their elements. However, it should be noted that this accuracy analysis is not quite convenient because a huge volume of computations is needed to analyze the accuracy, depending on the vehicle's velocity, altitude, velocity accuracy, and GA variability. The aforementioned method of PSD local approximation (Loparev et al., 2014) can also be used to solve the problem, which mainly overcomes this difficulty.

### 8.5.3 Solution Using the Method of PSD Local Approximations

In the method of PSD local approximations, the transfer function of an optimal filter in the steady-state mode is mainly determined by the PSD crossing point for the valid signal and noise and their inclinations at this point. Suppose that the gravitational acceleration is determined in airborne conditions with $V = 50$ m/s. According to the specifications of modern satellite receivers, take $R_V = 0.01^2 \,(\text{m/s})^2\,\text{s}$, $R_h = 0.005^2\,\text{m}^2\,\text{s}$, and gravimeter errors $R_{GR} = 5^2$ mGal²s (Krasnov et al., 2014a, b). Preliminary investigations have shown that in these conditions gravimeter instrumental errors can be neglected since they barely affect the accuracy of GA estimations. Let us construct the plots for GA PSD and conditional PSD $\omega^2 R_V$, $\omega^4 R_h$ for vertical velocity and altitude errors on logarithmic scales. Analysis of (8.86) reveals that in these conditions the PSD of GA in the vicinity of the crossing points with $\omega^2 R_V$, $\omega^4 R_h$ can be rather accurately approximated with $\omega \gg \alpha$ by

$$S_g(\omega) \approx \frac{q_w^2}{\omega^4}. \tag{8.92}$$

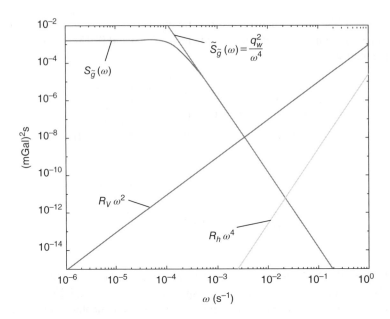

**Figure 8.19**  PSD of GA with its approximation and the conventional PSD of altitude and vertical velocity errors

This corresponds to the description of GA as a second integral of the white noise with PSD, that is, $q_w^2 = 10\beta^3\sigma_{\tilde{g}}^2$. The curves of all the aforementioned PSDs are given in Figure 8.19.

In state space, model (8.92) for GAs can be written as

$$\begin{cases} \dot{g}_1(t) = g_2(t), \\ \dot{g}_2(t) = q_w w(t), \end{cases} \tag{8.93}$$

$$\tilde{g}(t) = g_1(t),$$

where $w(t)$ is the white noise of unity PSD; $q_w = \sqrt{10\beta^3\sigma_{\tilde{g}}^2}$.

Given the earlier account, a simpler model can be applied instead of (8.90) and (8.91). As the model for GA is the second integral of the white noise, the state vector is given by $x(t) = (\Delta h^{\mathrm{GR}}(t), \Delta V^{\mathrm{GR}}(t), g_1(t), g_2(t))^{\mathrm{T}}$ with the same measurements and matrices:

$$F = \begin{bmatrix} 0 & 1 & 0 & 0 \\ 0 & 0 & 1 & 0 \\ 0 & 0 & 0 & 1 \\ 0 & 0 & 0 & 0 \end{bmatrix}, \quad Q = 1, \quad G = \begin{bmatrix} 0 \\ 0 \\ 0 \\ q_w \end{bmatrix}, \quad H = \begin{bmatrix} I_2 & 0_{2\times2} \end{bmatrix}. \tag{8.94}$$

Using the results obtained in Koshaev and Stepanov (2010) and Loparev *et al.* (2014), transfer functions $W^f(p)$, $W^s(p)$ for steady-state filtering and smoothing modes can easily be obtained using altitude measurements:

$$W^f(p) = W_B(p) \begin{bmatrix} \gamma \left( \dfrac{p}{\rho} \right)^3 + \dfrac{\gamma^2}{2} \left( \dfrac{p}{\rho} \right)^2 + \gamma \dfrac{p}{\rho} + 1 \\[2ex] p \left( \dfrac{\gamma^2}{2} \left( \dfrac{p}{\rho} \right)^2 + \gamma \dfrac{p}{\rho} + 1 \right) \\[2ex] p^2 \left( \gamma \dfrac{p}{\rho} + 1 \right) \\[2ex] p^3 \end{bmatrix}, \tag{8.95}$$

$$W^s(p) = W_B(p) W_B(-p) \begin{bmatrix} 1 \\ p \\ p^2 \\ p^3 \end{bmatrix}, \tag{8.96}$$

where

$$W_B(p) = \dfrac{\rho^4}{p^4 + \gamma p^3 \rho + \dfrac{\gamma^2}{2} p^2 \rho^2 + \gamma p \rho^3 + \rho^4} \tag{8.97}$$

and $\rho = \left( q_w / \sqrt{R_h} \right)^{1/4}$. $\gamma = \sqrt{2\left(2 + \sqrt{2}\right)}$ is the dimensionless coefficient.

It can also be shown that the filtering error covariance matrix in steady-state mode is given by

$$P_\infty^f = R_h \begin{pmatrix} \gamma\rho & \dfrac{\gamma^2 \rho^2}{2} & \gamma\rho^3 & \rho^4 \\[2ex] \dfrac{\gamma^2 \rho^2}{2} & \gamma\xi\rho^3 & \dfrac{\gamma^4 \rho^4}{8} & \gamma\rho^5 \\[2ex] \gamma\rho^3 & \dfrac{\gamma^4 \rho^4}{8} & \gamma\xi\rho^5 & \dfrac{\gamma^2 \rho^6}{2} \\[2ex] \rho^4 & \gamma\rho^5 & \dfrac{\gamma^2 \rho^6}{2} & \gamma\rho^7 \end{pmatrix}, \tag{8.98}$$

where $\rho = \left( q_w / \sqrt{R_h} \right)^{1/4}$, $\gamma = \sqrt{2\left(2 + \sqrt{2}\right)}$, $\xi = \sqrt{2} + 1$. Using the results obtained in Stepanov (2012) and Equation 8.98, the formula for smoothing error covariance matrix $P_\infty^s$ can also be found.

Analysis of the performances of the filtering and smoothing algorithms tuned to the simplified model (8.93) has revealed that they are rather effective. In particular, their accuracies prove to be close to that of the optimal filter if the real model complies with PSD (8.86). Also,

**Table 8.3** Formulas for error variance in gravity anomaly estimation

| Problem | SNS measurements | Estimation error variance |
| --- | --- | --- |
| Filtering | Vertical velocity (8.80) | $3R_V^{1/2}q_w$ |
| Smoothing | Altitude (8.79) | $6.31R_h^{3/8}q_w^{5/4}$ |
|  | Vertical velocity (8.80) | $0.17R_V^{1/2}q_w$ |
|  | Altitude (8.79) | $0.14R_h^{3/8}q_w^{5/4}$ |

**Table 8.4** GA RMS errors: filtering (numerator) and smoothing (denominator) (mGal)

| Measurements | Gradient $\partial\tilde{g}/\partial\rho$ (mGal/km) | | | |
| --- | --- | --- | --- | --- |
|  | 1 | 3 | 5 | 10 |
| $V_z$ | 3.4/0.8 | 7.7/1.8 | 11.2/2.7 | 18.8/4.5 |
| $h$ | 0.8/0.12 | 2.2/0.33 | 3.5/0.53 | 6.7/1.0 |
| $h+V_z$ | 0.8/0.12 | 2.2/0.33 | 3.5/0.53 | 6.7/1.0 |

the use of simplified models, apart from providing explicit transfer functions for filtering and smoothing problems, yields analytical formulas for estimation error variances for the sought components of the state vector. Table 8.3 contains formulas for error variances for GA filtering and smoothing using altitude and vertical velocity measurements specified by (8.83) and (8.84) if $R_{GR} = 0$. For example, these formulas can be obtained using the relevant diagonal elements of error covariance matrices for filtering and smoothing. Thus, error variance for GA filtering in the case of coordinate measurement is determined by the third diagonal element of matrix (8.98).

It should be noted that if velocity measurements are used to obtain a formula for the transfer function corresponding to GAs and a formula for estimated error variance, the state vector $x(t) = (\Delta V^{GR}(t), g_1(t), g_2(t))^T$ can be used.

The filtering and smoothing errors depend significantly on the changeability of GA along the trajectory. This changeability can be characterized by the derivative along the motion direction $\nabla\tilde{g} = \partial\tilde{g}/\partial\rho$.

Table 8.4 contains examples of calculating filtering and smoothing RMS errors for the accepted altitude and vertical velocity errors with various $\nabla\tilde{g}$. Then, the PSD $q_w^2$ of white noise $q_w w(t)$ should be set as $q_w^2 = \nabla\tilde{g}^2 3V^3/\rho$, where $\rho$ is the length of the path for which the GA increment is determined, $V$ being the speed. To be certain, the formula for the variance of the second integral of the white noise should be used, for which $\sigma_{\tilde{g}}^2(t) = q_w^2 t^3/3$ is true according to the results obtained in Stepanov (2012). Setting this value equal to the specified increment variance $(\nabla\tilde{g}\rho)^2$, it is possible to write $\sigma_{\tilde{g}}^2 = q_w^2 \dfrac{t^3}{3} = \left(\dfrac{\partial\tilde{g}}{\partial\rho}\rho\right)^2 = (\nabla\tilde{g}\rho)^2$. It then follows that

$$q_w^2 = \frac{3\sigma_{\tilde{g}}^2}{\rho^3}V^3 = \frac{3\left(\dfrac{\partial\tilde{g}}{\partial\rho}\rho\right)^2}{\rho^3}V^3 = 3(\nabla\tilde{g})^2 V^3/\rho.$$

From the presented results, the RMS error of GA estimation can be significantly reduced when a smoothing problem is solved instead of a filtering problem. If velocity measurements are used, the gain determined by the ratio between the filtering RMS and the smoothing RMS is 4.2, and 6.7 if altitude measurements are used.

The analysis of the results shows that with the current ratios between altitude and velocity accuracies of satellite measurements, velocity aiding in the presence of altitude data does not essentially affect the accuracy of gravity anomaly estimation.

## Acknowledgment

This work was supported by the Russian Science Foundation, project #14-29-00160.

## References

Alspach, D.L. and Sorenson, H.W. (1972) Nonlinear Bayesian Estimation Using Gaussian Sum Approximations, *IEEE Transactions on Aerospace and Electronic Systems*, AC-17(4), 439–448.

Anderson, B.D.O. (1979) *Optimal Filtering*. Englewood Cliffs: Prentice-Hall.

Anderson, B.D.O. and Moore, J.B. (1991) *Kalman Filtering: Whence, What, and Whither? Mathematical System Theory: The Influence of R.E. Kalman: A Festschrift in Honor of Professor R.E. Kalman on the Occasion of his 60th Birthday*, Antoulas, A.C., Ed., Berlin: Springer-Verlag.

Anuchin, O.N. and Emel'yantsev, G.I. (1999) *Integrated Attitude and Navigation Systems for Marine Vehicles*. St. Petersburg: Elektropribor.

Åström, K. J. (1991) *Adaptive Control, Mathematical System Theory: The Influence of R.E. Kalman: A Festschrift in Honor of Professor R.E. Kalman on the Occasion of his 60th Birthday*, Antoulas, A.C., Ed., Berlin: Springer Verlag.

Barbour, N. and Schmidt, G. (2001) Inertial Sensor Technology Trends. *IEEE Sensors Journal*, 1(4), 332–339.

Bar-Shalom, Y., Rong Li, X., Kirubarajan, T. (2001) *Estimation with Applications to Tracking and Navigation*. New York: John Wiley & Sons, Inc.

Bar-Shalom, Y., Willett, P.K, and Tian, X. (2011) *Tracking and Data Fusion. A Handbook of Algorithms*. Storrs: YBS Publishing.

Basar, T. Ed. (2001) *Control Theory. Twenty Five Seminal Papers (1932–1981)*. New York: IEEE Press.

Battin, R.H. (1964) *Astronautical Guidance*. New York: McGraw-Hill.

Bergman, N. (1999) Recursive Bayesian Estimation. Navigation and Tracking Applications. Ph.D. dissertation No. 579. Department of Electrical Engineering Linkoping University, Sweden.

Boguslavski, I.A. (1970) *Metody navigatsii i upravleniya po nepolnoi statisticheskoi informatsii [Methods for Navigation and Control Using Incomplete Statistical Information]*, Moscow: Mashinostroenie (in Russian).

Bolić, M., Djurić, S., and Hong, P.M. (2004) Resampling Algorithms for Particle Filters: A Computational Complexity Perspective, *EURASIP Journal on Applied Signal Processing* (15), 2267–2277.

Brown, R.G. (1972–1973) Integrated Navigation Systems and Kalman Filtering: A Perspective. *Navigation: Journal of the Institute of Navigation*, 19(4), 355–362.

Brown R.G. and Hwang, P.Y.C. (1997) *Introduction to Random Signals and Applied Kalman Filtering with Matlab Exercises and Solutions*. New York: John Wiley & Sons.

Bucy, R.S. and Joseph, P.D. (1968) *Filtering for Stochastic Processes with Applications to Guidance*. New York: Wiley-Interscience.

Bucy, R.S. and Senne, K.D. (1971) Digital Synthesis of Nonlinear Filters, *Automatica*, 7(3), 287–298.

Chelpanov, I.B. (1967) *Optimal'naya obrabotka signalov v navigatsionnykh sistemakh [Optimal Signal Processing in Navigation Systems]*, Moscow: Nauka (in Russian).

Chelpanov, I.B., Nesenyuk, L.P., and Braginskii, M.V. (1978) *Raschet kharaktiristik navigatsionnykh priborov [Calculation of Characteristics of Navigation Gyro Instruments]*, Leningrad: Sudostroenie (in Russian).

Daum, F. (2005) Nonlinear Filters: Beyond the Kalman Filter, *IEEE Aerospace and Electronic Systems Magazine*, 20(8), 57–71.

Dmitriev S.P., Stepanov, O.A. and Pelevin, A.E. (2000) Motion Control and Non-Invariant Algorithms Using the Vehicle Dynamics for Inertial-Satellite Integrated Navigation, Proceedings of the Seventh Saint Petersburg International Conference on Integrated Navigation Systems, St. Petersburg, Russia, May 29–31. IEE, pp. 75–82.

Doucet, A. and Johansen, A.M. (2011) A tutorial on particle filtering and smoothing: Fifteen years later. In: *The Oxford Handbook of Nonlinear Filtering. Oxford Handbooks in Mathematics*. Oxford: Oxford University Press, pp. 656–705.

Doucet, A., de Freitas, N., and Gordon, N.J. (2001) *Sequential Monte Carlo Methods in Practice*. New York: Springer-Verlag, p. 581.

Fisher, R.A. (1925) *Statistical Methods for Research Workers*. Edinburgh: Oliver and Boyd; republished by Oxford University Press, Oxford, 1990.

Fomin, V.N. (1984) *Rekurentnoye otsenivanie i adaptivnaya fil'tratsiya [Recurrent Estimation and Adaptive Filtering]*, Moscow: Nauka (in Russian).

Gelb, A., Kasper, J., Nash, R., Price, C., and Sutherland, A. (1974) *Applied Optimal Estimation*. Cambridge, MA; MIT Press.

Gibbs, B. P. (2011) *Advanced Kalman Filtering, Least-Squares and Modeling: A Practical Handbook*, Hoboken: John Wiley & Sons, Inc.

Gordon, N.J., Salmond, D.J., and Smith, A.F.M. (1993) Novel Approach to Nonlinear/Non-Gaussian Bayesian State Estimate, *IEEE Proceedings on Radar and Signal Processing*, 140(2), 107–113.

Grewal, M.S. and Andrews, A.P. (1993) *Kalman Filtering: Theory and Practice*. New Jersey: Prentice Hall.

Grewall M., Weill L.R., and Andrews A.P. (2013) *Global Navigation Satellite Systems, Inertial Navigation, and Integration*. 3rd Edition, New York: John Wiley & Sons, Inc.

Gustafsson, F., Gunnarsson, F., Bergman, N., Forssell, U., Jansson, J., Karlsson, R., and Nordlund, P.-J. (2002) Particle Filters for Positioning, Navigation and Tracking, *IEEE Transactions on Signal Processing*, 50(2), 425–437.

Ivanov, V.M., Stepanov, O.A., and Korenevski, M.L. (2000) Monte Carlo Methods for a Special Nonlinear Filtering Problem, 11th IFAC International Workshop on Control Applications of Optimization, vol. 1, pp. 347–353.

Ivanovski, R.I. (2012) Some Aspects of Development and Application of Stationary Filters to Navigation Systems, *Gyroscopy and Navigation*, 3(1), 1–8.

Jazwinski, A.H. (1970) *Stochastic Processes and Filtering Theory*. New York: Academic Press.

Jordan, S.K. (1972) Self-consistent Statistical Models for Gravity Anomaly and Undulation of the Geoid, *Journal of Geophysical Research*, 77(20), 3660–3670.

Juiler, S.J. and Uhlmann, J.K. (2004) Unscented Filtering and Nonlinear Estimation, *Proceedings of the IEEE*, 92(3), 401–422.

Kailath, T. (1974) View of Three Decades of Linear Filtering Theory. *IEEE Transactions on Information Theory*, IT–20, 146–181.

Kailath, T. (1980) *Linear Systems*. Prentice-Hall, Inc, Englewood Cliffs.

Kailath, T. (1991) *From Kalman Filtering to Innovations, Martingales, Scattering and Other Nice Things. Mathematical System Theory: The Influence of R.E. Kalman: A Festschrift in Honor of Professor R.E. Kalman on the Occasion of his 60th Birthday*, Antoulas, A.C., Ed., Berlin: Springer-Verlag.

Kailath, T., A. Sayed, and B. Hassibi (2000) *Linear Estimation*. Upper Saddle River: Prentice-Hall.

Kalman, R.E. (1960) A New Approach to Linear Filtering and Prediction Problems. *Transactions of the ASME—Journal of Basic Engineering*, 82 (Series D), pp. 35–45.

Kalman, R.E. and Bucy, R.S. (1961) New Results in Linear Filtering and Prediction Theory, *Transactions of the ASME—Journal of Basic Engineering*, 83, pp. 95–107.

Kolmogoroff, A. (1939) Sur l'interpolation et extrapolation des suites stationnaires, *Comptes Rundus de l'Acad. Sci. Paris*, 208, 2043–2045.

Kolmogorov, A.N. (1941) Interopolyatsiya i ekstrapolyatsiya statsionarnykh sluchainykh posledovatel'nostei, Izvestiya Akad. *Nauk SSSR, Seriya Matematika*, 5, 3–14 (in Russian); Kolmogorov, A.N. (1962) Interpolation and Extrapolation of Stationary Random Sequences, 3090-PR. Santa Monica: Rand Corporation.

Koshaev, D.A. and Stepanov, O.A. (2010) Analysis of Filtering and Smoothing Techniques as Applied to Aerogravimetry, *Gyroscopy and Navigation*, 1(1), pp. 19–25.

Koshaev, D.A. and Stepanov, O.A. (2011) A Program for Designing Linear Filtering Algorithms for Integrated Navigation Systems. *IFAC Proceedings Volumes (IFAC-PapersOnline)*, 18(Part 1), 4256–4259.

Krasnov, A.A., Nesenyuk, L.P., Peshekhonov, V.G., Sokolov, A.V., Elinson L.S. (2011a) Integrated Marine Gravimetric System. Development and Operation Results. *Gyroscopy and Navigation*, 2(2), 75–81.

Krasnov, A.A., Sokolov, A.V., and Usov, S.V. (2011b) Modern Equipment and Methods for Gravity Investigation in Hard-to-Reach Regions, *Gyroscopy and Navigation*, 2(3), 178–183.

Krasnov, A.A., Sokolov, A.V., and Elinson, L.S. (2014a) Operational Experience with the Chekan-AM Gravimeters, *Gyroscopy and Navigation*, 5(3), 181–185.

Krasnov A.A., Sokolov A.V., Elinson L.S. (2014b) A New Air-Sea Shelf Gravimeter of the Chekan Series, *Gyroscopy and Navigation*, 5(3), 131–137.

Krasovski, A.A., Beloglazov, E.N., and Chigin, G.P. (1979) *Teoriya korrelyatsionno-ekstremal'nykh navigatsionnikh sistem [Theory of Correlation Extremal Navigation Systems]*. Moscow: Nauka (in Russian).

Kulakova V.I. and Nebylov A.V. (2008) Guaranteed Estimation of Signals with Bounded Variances of Derivatives. *Automation and Remote Control*, 69(1), 76–88.

Kulakova, V.I., Nebylov, A.V., and Stepanov, O.A. (2010) Using the H 2/H ∞ Approach to Aviation Gravimetry Problems, *Gyroscopy and Navigation*, 1(2), 141–145.

Kurzhanski, A.B. (1977) *Upravlenie i nablyudenie v usloviyakh neopredelennosti [Control and Observation under Uncertainty]*, Moscow: Nauka (in Russian).

Kushner, H.J. (1964) On the Dynamical Equations of Conditional Probability Density Functions, with Applications to Optimal Stochastic Control Theory. *Journal of Mathematical Analysis and Applications*, 8, 332–344.

Lainiotis, D.G. (1974a) Estimation Algorithms, I: Nonlinear Estimation, *Information Sciences*, 7(3/4), 203–235, II: Linear Estimation, Information Sciences, 7(3/4), 317–340.

Lainiotis, D.G. (1974b) Estimation: Brief Survey, *Information Sciences*, 7, 191–202.

Lainiotis, D. G. (1976) Partitioning: A Unifying Framework for Adaptive Systems, 1: Estimation, *Proceedings of the IEEE* 64, 1126–1142.

Lefebvre, T., Bruyninckx, H., and De Schutter, J. (2005) *Nonlinear Kalman Filtering for Force-Controlled Robot Tasks*. Berlin: Springer.

Li, X.R. and Jilkov, V.P. (2004) A Survey of Maneuvering Target Tracking: Approximation Techniques for Nonlinear Filtering. Proceedings of the SPIE Conference on Signal and Data Processing of Small Targets, San Diego, CA, USA, pp. 537–535.

Liptser, R.Sh. and Shiryaev, A.N. (1974) *Statistika sluchainykh protsessov [Statistics of Random Processes]*, Moscow: Nauka (in Russian).

Litvinenko, J.A., Stepanov, O.A., and Tupysev, V.A. (2009) Guaranteed Estimation in the Problems of Navigation Information Processing, Proceedings of the IEEE Multi-Conference on Systems and Control. Conference on Control Applications (CCA'09), St. Petersburg, Russia, July.

Liu, J.S. and Chen, R. (1998) Sequential Monte Carlo Methods for Dynamic Systems, *Journal of the American Statistical Association*, 93(443), 1032–1044.

Loparev, A.V., Stepanov, O.A., and Chelpanov, I.B. (2014) Time-Frequency Approach to Navigation Information Processing. *Automation and Remote Control*, 75(6), 1091–1109.

Mathematical System Theory (1991) *The Influence of R.E. Kalman: A Festschrift in Honor of Professor R.E. Kalman on the Occasion of his 60th Birthday*, Antoulas, A.C., Ed., Berlin: Springer-Verlag.

Maybeck, P.S. (1979) *Stochastic Models, Estimation and Control*, vol. 1. New York: Academic Press.

Medich, J.S. (1969) *Stochastic Optimal Linear Estimation and Control*. New York: McGraw-Hill.

Mehra, R.K. (1972), Approaches to Adaptive Filtering, *IEEE Transactions on Automatic Control*, 17(5), pp. 693–698.

Mikhailov, N.V. and Koshaev, D.A. (2014) Positioning of a Spacecraft in Geostationary Orbit Using the Model of its Perturbed Motion and the Satellite Navigation Receiver, Proceedings of the 21st Saint Petersburg International Conference on Integrated Navigation System, St. Petersburg, Russia, May 26–28, pp. 416–425.

Mikhailov, N.V. and Mikhailov, V.F. (2013a) Determining the Coordinates and Velocities of a Geostationary Spacecraft Using Measurements of Satellite Navigation Systems, *Uspekhi sovremennoi radioelektroniki*, 2013(2), 113–121 (in Russian).

Mikhailov, N.V. and Mikhailov, V.F. (2013b) Determining the Parameters of a Geostationary Satellite Orbit Using a Satellite Navigation System, *Izvestiya vysshikh uchebnykh zavedenii Rossii, Radioelektronika*, 2013(2), 71–76 (in Russian).

Mikhailov N.V., Mikhailov, V.F., and Vasil'ev, M.V. (2010) Autonomous Orbit Determination of Artificial Earth Satellites Using Satellite Navigation Systems, *Giroskopiya i navigatsiya*, 2010(4), 41–52 (in Russian).

Nebylov, A.V. (2004) *Ensuring Control Accuracy. Lecture Notes in Control and Information Sciences*, vol. 305. Heidelberg: Springer-Verlag.

Pamyati professora Nesenyuka L.P. (2010) *Izbrannye trudy i vospominaniya [In Commemoration of Professor L.P. Nesenyuk. Selected Works and Recollections]*. St. Petersburg: OAO Kontsern Elektropribor (in Russian).

Parkinson, B.W. and Spiller, J.J., Eds. (1996). *Global Positioning System: Theory and Applications*, vol. I/II. Washington, DC: American Institute of Aeronautics and Astronautics.

Pugachev, V.S. and Sinitsyn I.N. (1985) *Stokhasticheskie differentsial'nye sistemy [Stochastic Differential Systems]*. Moscow: Nauka (in Russian).

Ristic, B., Arulampalam, S., and Gordon, N. (2004) *Beyond the Kalman Filter: Particle Filter for Tracking Applications*. Norwood, MA: Artech House Radar Library.

Rivkin, S.S., Ivanovsky, R.E., and Kostrov, A.V. (1976) *Statisticheskaya optimizatsiya navigatsionnykh sistem [Statistical Optimization of Navigation Systems]*. Leningrad: Sudostroenie (in Russian).

Rozov, A.K. (2002) *Nelineinaya fil'tratsiya signalov [Nonlinear Filtering of Signals]*. Saint Petersburg: Politekhnika, 372 pp. (in Russian).

Sage, A.R. and Melsa, J.L. (1971) *Estimation Theory with Applications to Communications and Control*. New York: McGraw-Hill.

Schmidt, S.F. (1981) The Kalman Filter: its Recognition and Development for Aerospace Applications, *AIAA Journal of Guidance and Control*, 4, 4–7.

Schmidt, G. and Phillips, R. (2003) INS/GPS Integration Architectures. *NATO RTO Lecture (NATO)*. Advances in Navigation Sensors and Integration Technology (232): 5-1–5-15.

Shakhtarin, B.E. (2008) *Fil'try Kalmana and Vinera [Kalman and Wiener filters]*. Moscow: Gelios, ARB (in Russian).

Shön, T., Gustaffson, F., and Nordlund, P.-J. (2005) Marginalized Particle Filters for Linear/Nonlinear State-space Models. *IEEE Transactions on Signal Processing*, 53(7), 2279–2289.

Sinitsyn, E.N. (2006) *Fil'try Kalmana i Pugacheva [Kalman and Pugachev Filters]*, Moscow: Logos (in Russian).

Smith, A.F.M. and Gelfand, A.E. (1992) Bayesian Statistics without Tears: A Sampling-Resampling Perspective, *The American Statistician*, 46, 84–88.

Sorenson, H.W. (1970), Least Square Estimation from Gauss to Kalman, *IEEE Spectrum*, 7(7), 63–68.

Sorenson, H.W. (1985) *Kalman Filtering: Theory and Application*. New York: IEEE Press.

Stepanov, O.A. (1998) *Primenenie teorii nelineinoi fil'tratsii v zadachakh obrabotki navigatsionnoi informatsii [Application of Nonlinear Filtering Theory for Processing Navigation Information]*. St. Petersburg: Elektropribor (in Russian).

Stepanov, O.A. (2004) An Efficient Unified Algorithm for Filtering and Smoothing Problems, Proceedings of the IFAC Workshop on Adaptation and Learning in Control and Signal Processing, ALCOSP 04, Yokohama, Japan, August 30–September 1, pp. 759–763.

Stepanov, O.A. (2006) Linear Optimal Algorithm for Navigation Problems, *Giroskopiya i navigatsiya*, 4, 11–20 (in Russian).

Stepanov, O.A. (2011) Kalman Filtering: Past and Present. An Outlook from Russia [On the Occasion of the 80th birthday of Rudolf Emil Kalman], *Gyroscopy and Navigation*, 2(2), pp. 99–110.

Stepanov, O.A. (2012) *Osnovy teorii otsenivaniya s prilozheniyami k zadacham obrabotki navigatsionnoi informatsii [Fundamentals of the Estimation Theory with Applications to the Problems of Navigation Information Processing], Part 1, Vvedenie v teoriyu otsenivaniya [Introduction to the Estimation Theory]*. St. Petersburg: TsNII Elektropribor, 2010 (in Russian); Part 2, Vvedenie v teoriyu filtrazii [Introduction to the Filtering Theory], St. Petersburg: Elektropribor, 2012 (in Russian).

Stepanov, O.A. and Toropov, A.B. (2010) A Comparison of Linear and Nonlinear Optimal Estimators in Nonlinear Navigation Problems. *Gyroscopy and Navigation*, 1(3), 183–190.

Stratonovich, R.L. (1959a) Optimum Nonlinear Systems which Bring about a Separation of a Signal with Constant Parameters from Noise, *Izvestiya vuzov SSSR, Seriya Radiofizika*, 2, 862–901 (in Russian).

Stratonovich, R.L. (1959b) K teorii optimal'noi nelineinoi fil'tratsii sluchainykh funktsii [Some Aspects of Optimal Nonlinear Filtering of Random Functions], *Teoriya veroyatnostei i eyo primeneniya*, 4(2), pp. 239–241 (in Russian).

Stratonovich, R. L. (1960) Conditional Markov Processes. *Theory Probability and its Applications*, 5, 156–178.

Stratonovich, R.L. (1966) *Uslovnye markovskie protsessy i ikh primenenie k teorii optimal'nogo upravleniya [Conditional Markov Processes and their Application to Optimal Control Theory]*. Moscow: MGU.

Tikhonov, V.I. (1983) Development of Optimal Filtering Theory in the USSR, *Radiotekhnika*, 11, 11–25 (in Russian).

Tikhonov, V.I. (1999) Development of Optimal Nonlinear Estimation of Random Processes and Fields in Russia, *Radiotekhnika*, 10, 4–20 (in Russian).

Van der Merwe, R. and Wan, E.A. (2001) The Unscented Kalman Filter. In: *Kalman Filtering and Neural Networks*, Haykin, S., Ed., New York: John Wiley & Sons, Inc., pp. 221–268.

Van Trees, H.L. (1968) *Detection, Estimation, and Modulation Theory*. New York: John Wiley & Sons, Inc.

Wiener, N. (1949) *Extrapolation, Interpolation and Smoothing of Stationary Time Series, with Engineering Applications*. John Wiley, New York, 1949 (Originally issued in February 1942, as a classified National Defense Research Council Report).

Yarlykov, M.S. and Mironov, M.A. (1993) *Markovskaya teoriya otsenivaniya sluchainykh protsessov*. Moscow: Radio i svyaz' (in Russian); (1996) The Markov Theory of Estimating Random Processes, Telecommunications and Radioengineering, vol. 50, no. 2–12. New York: Begell House.

Zaritsky, V.S., Svetnik, V.B., and Shimelevich, L.I. (1975) Metod Monte-Karlo v zadachakh optimal'noi obrabotki informatsii, *Avtomatika i telemekhanika*, 12, 95–103 (in Russian); (1975) Zaritsky, V.S., Svetnik, V.B., and Shimelevich, L.I. The Monte-Carlo Techniques in Problems of Optimal Information Processing. Automation and Remote Control, 36, 2015–2022.

Zinenko, V.M. (2012) Application of Suboptimal Time-Invariant Filters, *Gyroscopy and Navigation*, 3(4), 286–297.

# 9

# Navigational Displays

Ron T. Ogan

*Captain, U.S. Civil Air Patrol and Senior Member, IEEE, USA*

## 9.1 Introduction to Modern Aerospace Navigational Displays

Modern avionics systems for manned aircraft display navigational data for pilot and crew situational awareness and active control of aircraft flight systems. The aircraft involved include commercial, military, and general aviation (GA) categories with each having different constraints on the complexity of the cockpit avionics systems. Commercial aircraft systems have led navigational advancement via ease-of-use and safety margin innovations. Military navigational displays have serious space constraints for the accommodation of targeting, night vision, and infrared imaging systems. General Aviation, as its name implies, accommodates aircraft ranging from light sport classes to luxury 'bizjets' such as the Bombardier Learjets™.

Navigational displays thus have the common requirement of providing situational awareness for the pilot and crew, so making possible the control of the aircraft flight systems for safe delivery of passengers and cargo over varied terrain, and in weather from visual to instrument conditions.

The United States Federal Aviation Administration (FAA) specifies the aircraft equipage requirements for operation in National Airspace. Aircraft displays will continue to change to meet the requirements of the FAA NextGen flight control system, designated as Automatic Dependent Surveillance-Broadcast (ADS-B IN/OUT) with an implementation scheduled for January 1, 2020. These display systems will be linked to over 700 ground station in the continental United States to provide Traffic Information Service-Broadcast (TIS-B) and weather updates through (FIS-B) for improved safety and more direct routing for greater efficiency (Federal Register, 2010; Traffic Control, 2010).

*Aerospace Navigation Systems*, First Edition. Edited by Alexander V. Nebylov and Joseph Watson.
© 2016 John Wiley & Sons, Ltd. Published 2016 by John Wiley & Sons, Ltd.

### 9.1.1   The Human Interface for Display Control—Buttonology

Most of the older generations of general aviation (GA) aircraft have legacy cockpit arrangements as in Figure 9.1, including the popular single-engine Cessna 172. Also, Boeing 737 jet airliners built before the year 2000 also had traditional round-dial or "steam gauges" for navigation displays, primarily based on lower costs and a lack of stable electronic displays at the time. Like most light aircraft, the C-172 instruments are powered by the aircraft electrical system (the turn co-ordinator), the pitot-static system (the airspeed indicator, vertical speed indicator, and altimeter), or by the engine-driven vacuum system (the attitude indicator and directional gyroscope). These systems and instruments are more fully described in the companion volume "aerospace sensors." Furthermore, some systems utilized two methods of powering for safety reasons. Pilot adjustments were limited to setting the altimeter using the prevailing barometric pressure, zero-setting the attitude indicator, and calibration of the directional gyroscope to match the magnetic compass before take off and subsequent re-adjustment during level flight to correct for precession errors.

The average commercial transport aircraft cockpit in the mid-1970s had more than 100 cockpit instruments and controls, and the primary flight instruments were already crowded with indicators, crossbars, and other symbols. Thus, the growing number of cockpit elements was competing for cockpit space and pilot attention. Avionics designs varied widely, without common human interfaces, lacking efficient layouts. During this period, Light-Emitting Diode (LED) displays were becoming available but varying intensities made them difficult to read in varying ambient or artificial lighting. Some of the LEDs had custom laser-trimmed resistors to even out the emitted light intensities to provide more readable displays.

A Lockheed C-5A flight deck with control panels and analog instrument displays is shown in Figure 9.2, and an AMP-upgraded instrument panel for a C5A Galaxy aircraft is shown in Figure 9.3 for comparison. Pilot training for the analog instruments was difficult because of

**Figure 9.1**   Typical Cessna C-172 cockpit layout with round flight gauges

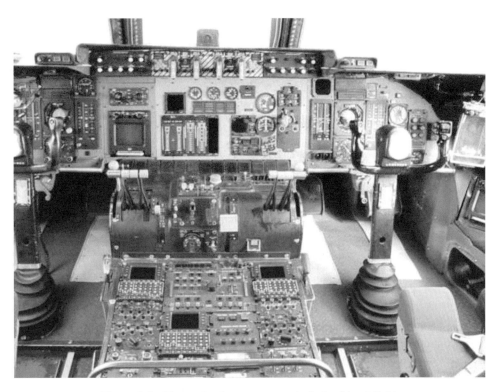

**Figure 9.2**    The analog instrument panel of a Lockheed C-5A

**Figure 9.3**    The AMP-upgraded instrument panel for a Lockheed C-5A

**Figure 9.4**    The Honeywell HUD 2020 Primus Epic® head-up display

the complexity, lack of standardization and number of instruments: during flights the pilots had to continually scan instruments to monitor and control aircraft performance over a large panel area and ergonomics had not been seen as a priority. Modern displays allow flexibility and customization of instruments for the most efficient use by the pilots.

Modern aerospace navigational displays cover three main types of equipment: radar displays, HUDs, and avionics display systems. Figure 9.4 shows the Honeywell HUD 2020 Head-Up-Display (HUD) that helps the pilot's visual focus to remain on the outside world. The compact HUD 2020 electro-optical overhead unit generates an image on a lightweight combiner that provides the pilot with real-time flight and aircraft performance data. The HUD 2020 was designed by pilots for pilots from a complete systems perspective to integrate fully with the entire suite of Honeywell standard avionics on board Gulfstream's stable of large business jets. The Honeywell-integrated Visual Guidance System™ (VGS) is a HUD designed specifically for the Primus Epic® system (Honeywell).

The US Air Force/Lockheed F-22 fighter aircraft accommodates an all-glass layout designed to provide more light, space, and visibility for improving the human habitability of the cockpit, as shown in Figure 9.5. Aviation performance and safety are optimized by display designs and built-in diagnostics that reduce maintenance and human error. The Hands on Throttle and Stick (HOTAS) and display controls increase safety and survivability for the pilot by decreasing the workload. Safety is further enhanced by providing better visibility through the canopy and optimizing the locations of panel edge lights, anti-collision lights, and strobe lights. A Standby Flight Group (SFG) remains operational as other systems fail, so providing critical aircraft information to the pilot and thereby increasing system and personnel survivability. The Integrated Caution, Advisory, and Warning (ICAW) system streamlines situational awareness by defining system failures and recommending appropriate responses, so reducing pilot fatigue and increasing mission effectiveness.

The HUD for the F-22 shown in Figure 9.5 is at an optimum eye level, approximately 4.5 in. (11.4 cm), and uses the standardized symbology developed by the US Air Force

**Figure 9.5**   F-22 HUD radar and LCD displays

Instrument Flight Center, which results in no new cognitive requirements for the aircrew. The tactical symbol set is the same as that which is used on the F-22 Head-Down-Display (HDD), which permits a crosscheck and reduces the chance of human error. Six Liquid Crystal Display (LCD) panels in the cockpit present information in full color and are fully readable in direct sunlight. The SFG is tied to the last source of electrical power in the aircraft, and will remain powered despite other electrical failures, so ensuring additional safety and survivability.

Pilot workload and fatigue in analyzing and reacting to changing flight activities is reduced by the uniquely designed ICAW. Messages appear on a display easily visible to the pilot, which identifies specific problems and provides an electronic checklist for addressing them.

The HOTAS switching design simplifies the otherwise more than 60 time-critical functions. The aircraft's primary flight controls, a side-stick controller and two throttles, have been re-located as part of the new design. The stick is located on the right console with a swing-out adjustable arm-rest for pilot comfort and stability, and the throttles are located on the left console. The HOTAS also incorporates buttons that control offensive and defensive weapons systems in addition to display management (US Air Force).

The Aspen Avionics Evolution 2500 is a state-of-the-art integrated glass cockpit display into which a pilot inputs data or commands using buttons and knobs located along the bottom and sides as shown in Figure 9.6. The Evolution 2500 combines the powerful EFD1000 Pro Primary Flight Display (PFD), the EFD1000 Multifunction Flight Display (MFD), and the EFD500 MFD to deliver Aspen's total glass cockpit solution. The Pro PFD and 1000 MFD provide the safety and confidence of DuoSafeTM PFD redundancy, and the 500 MFD expands display area to include more flight data precisely where required by the pilot. This makes the total glass cockpit experience for GA aircraft both easy-to-use and with patent-pending retrofit technology and superior compatibility for substantially lower installation and total ownership costs. Software menus compliant with RTCA DO-178B Level C provide control of the displays for traffic, weather, engine performance, and collision avoidance alerts in addition to the critical flight parameters required for safe flight (Aspen Avionics).

**Figure 9.6**   The Aspen Avionics Evolution 2500 display

## 9.1.2   *Rapidly Configurable Displays for Glass Cockpit Customization Purposes*

The term "glass cockpit" was thought to have originated during the time when NASA astronaut Fred Gregory learned that the cockpit in Shuttle Atlantis was due for a total technology update. He advised the Shuttle managers to talk with the agency's aeronautics experts, and this accounts for why the Atlantis' new cockpit resembles a future airliner cockpit, with colorful multifunction computer displays stretching from one side to the other. This radical new look is an accurate reflection of the cockpit's equally radical new capabilities. Gregory, originally from NASA's Langley Research Center who later became a NASA Associate Administrator for Safety and Mission Assurance, was aware that Langley Research Center had pioneered the glass cockpit concept in ground simulators and demonstration flights in the NASA 737 flying laboratory. Based on that work and via a favorable response from industry customers, Boeing had developed the very successful first glass cockpits for production airliners (NASA).

Currently, glass cockpits feature digital electronic instrument displays, typically via large LCD screens that replace the traditional analog dials and gauges. While a traditional cockpit relies on numerous mechanical gauges to display information, a glass cockpit uses several displays driven by Flight Management Systems (FMS) that can be adjusted to display flight information as needed. This simplifies aircraft operation and navigation and allows pilots to focus on the most pertinent information. They are also popular with airline companies because they usually eliminate the need for a flight engineer, so saving costs. In recent years this technology has also become widely available in small aircraft.

Modern aerospace navigational displays have a PFD that is located in front of the aircraft captain or Pilot-In-Command (PIC), and a secondary MFD that is located in the center panel or in front of the co-pilot or second-in-command of the aircraft. The MFD typically displays traffic, weather, and engine performance but may be reconfigured as a PFD or any other function needed in the event of an equipment failure.

As aircraft displays have modernized, so have the sensors that feed them. Traditional gyro-scopic flight instruments have been replaced by electronic Attitude and Heading Reference System (AHRS) and Air Data Computers (ADCs), so improving reliability and reducing the initial and maintenance costs. Furthermore, Global Positioning System (GPS) receivers are also usually integrated into glass cockpits.

Various technological developments have made possible the integration of glass cockpit avionics into modern aircraft, including the LCD that have replaced the older cathode ray tubes and segmented LED data displays. Flexibility has been the driving force in the evolution of the LCD screens that allow the pilot or crew to reconfigure the displays to show critical airspeed, attitude, and altitude in addition to terrain mapping, traffic awareness, and weather patterns in near real time.

The flat-panel LCD has a lifespan approaching 100000 h on average when used continu-ously under controlled conditions, for example, in an environment with "standard ambient" lighting conditions and 25°C (77°F) temperatures throughout.

The Society of Motion Picture and Television Engineers (SMPTE) recommends making use of a screen size that fills not less than a 30° field of view at the seating position. In the aircraft environment this implies that for best results the viewing distance from the pilot's eye position to the display should be such that the extreme ends of the display subtend a minimum angle of 30° (In TV terms, this corresponds to an LCD TV–viewing distance of 1.87 times the screen width).

Another design consideration is how the LCD display placement relates to the viewer's personal vision system or, more specifically, the pilot's visual acuity. The issue with visual acuity does not deal with identifying the best viewing position for the screen size, but to the maximum distance beyond which it will not be easy to see all the picture detail having regard to the limitations of the pilot's eyesight.

Visual acuity is by definition a measure of the eye's spatial resolving power, that is, it spec-ifies the *angular size* of the smallest resolvable detail (Figure 9.7).

A person with normal vision—often referred to as 20/20 (or 6/6 when expressed in meters)—can resolve a spatial pattern where each element within that pattern subtends an

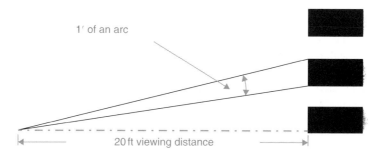

1′ of an arc

20 ft viewing distance

**Figure 9.7**   The Minimum Angle of Resolution (MAR)

angle of 1′ of arc at the eye (i.e., 1/60th of a degree) when viewed at a distance of 20 ft (6.1 m). This represents the Minimum Angle of Resolution (MAR), which means that a person with normal eyesight can resolve an object with a height of 1.77 mm at a distance of 6.1 m (20 ft).

From an LCD-viewing distance perspective, visual acuity represents the distance beyond which some of the picture detail will no longer be resolved because it will appear to blend in with adjacent picture information. In practice, this implies that the smallest image element— the pixel—should have a size that is not smaller than 0.0233-in. (0.59 mm) when viewed from a distance of 3 ft (0.915 m), which is the typical maximum range for viewing instruments in the aircraft cockpit.

## 9.2   A Global Positioning System Receiver and Map Display

The US GPS received an operational status in 1995 with a baseline configuration of 24 satellites in nearly circular orbits with a radius of 26 560 km (16 503 miles), a period of approximately 12 h, and stationary ground tracks. The satellites are arranged in six orbital planes inclined at 55° relative to the equatorial plane, with four primary satellite slots distributed unevenly in each orbit. Although GPS was designed originally for military projects, civilian maritime, land-based and airborne applications have developed rapidly (Garmin™).

Garmin International of Olathe, Kansas, is a leader in the development of aviation GPS navigation equipment, and one of its products will serve as a prime example of current approaches to electronic navigation. By combining visual clues and data readouts once scattered across numerous instruments, the Garmin G1000™ system makes flight information easier to scan so that pilots can respond more quickly, intuitively, and confidently. Typically configured as a two-or three-display system with flat screens ranging in size from 10 to 15 in., it is adaptable to a broad range of aircraft models from next-generation business jets to piston singles and forms a versatile, fully integrated glass flight deck. The system's large XGA high-resolution (1024×768) screens allow pilots to see at a glance the needed data without the necessity of sequencing through many pages of individual sensor readouts or navigation screens.

The following figures show the Garmin G1000 components. Figure 9.8 shows the Basic System Configuration, Figure 9.9 shows the PFD, Figure 9.10 shows the MFD, and Figure 9.11 shows the System Overview.

Each display is powered by an X-scale microprocessor along with a high-performance graphics accelerator for 3-D rendering. Offering wide side-to-side viewing angles and sharp, sunlight-readable TFT optics, these displays provide the pilot with very comprehensive flight-critical data and other information. Modular Line Replaceable Units (LRU) allow a convenient plug-and-play setup procedure for the hardware: all the components share a common language along with high-speed Ethernet connectivity in order to synchronize the integrated avionics and instrument and flight control functions. The centrally located MFD is used for engine and fuel system monitoring, plus detailed moving-map graphics. The map function is designed to interface with a variety of sensor inputs, so making it possible to overlay weather, lightning, traffic, terrain, and other avoidance system advisories as desired. Hence, these displays allow the pilot to add or deselect such overlays to customize the map view for any given phase of flight. Other graphical features, including Engine Indication System (EIS) and Crew Alerting System (CAS) advisories are accommodated via built-in system interfaces.

## G1000 Components

Primary Flight Display (PFD)        Multi-Function Display (MFD)

Attitude and Heading Reference          Engine interface unit
System (AHRS)
                                        Moving map + other displays
Air Data Computer (ADC)

Magnetometer

Mode S transponder

**Figure 9.8**    The Garmin G1000 basic configuration

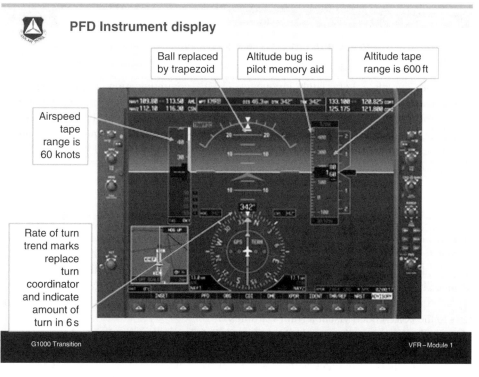

## PFD Instrument display

Ball replaced by trapezoid

Altitude bug is pilot memory aid

Altitude tape range is 600 ft

Airspeed tape range is 60 knots

Rate of turn trend marks replace turn coordinator and indicate amount of turn in 6 s

**Figure 9.9**    The Garmin G1000 primary flight display

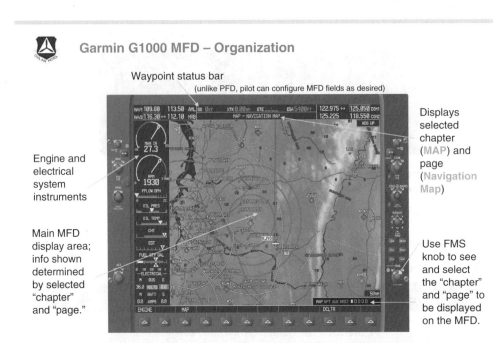

**Figure 9.10**   The Garmin G1000 multifunction display

In addition, complete Very High Frequency (VHF) communication, Class 3 Wide Area Augmentation System (WAAS)-certified GPS, VOR/ILS and transponder inputs are provided and controlled by knobs and function keys located on the PFD and MFD bezels. Onscreen navigation, along with communication and mapping functions, is supported by built-in database capabilities that may be easily updated by means of standard front-loading data subscription cards, typically updated at three monthly intervals.

## 9.2.1   Databases

A number of built-in terrain and mapping databases provide the Garmin G1000 with graphical references for navigation. At the most basic level, a worldwide base map helps to identify cities, roads, rivers, lakes, and other ground features for display on the MFD. Flight safety is enhanced by a built-in terrain elevation database that uses color coding to help alert pilots as they approach rising terrain. (Class B TAWS, optionally available on the G1000, adds voice alerts to the color-coded visual terrain warnings.)

For on-the-ground navigation, built-in Garmin SafeTaxi® diagrams help pilots to visualize their aircraft's exact location and direction of travel over numerous airports. For takeoffs and landings, standard Garmin FliteCharts® are available with electronic versions of NACO terminal procedure charts for US airports. As an alternative, pilots can opt for Garmin ChartView™ using

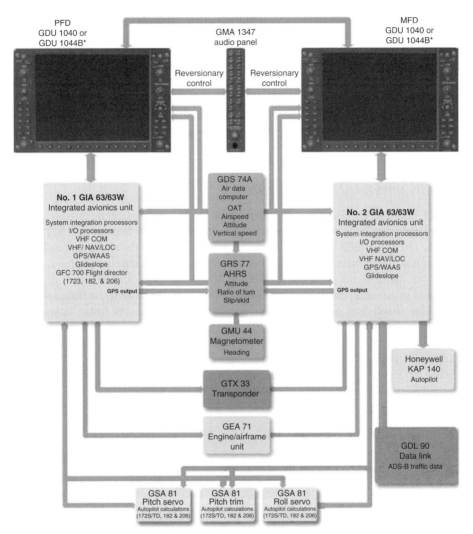

* The GDU 1040 is available in systems not using the GFC 700 Automatic Flight Control System.
  The GDU 1044B is available in systems using the Garmin GFC 700 Automatic Flight Control System.

**Figure 9.11**    Overview of the Garmin G1000 System

Jeppesen data and optional instrument approach plates along with surface charts in Jeppesen format. ChartView is unique in its ability to overlay a geo-referenced aircraft symbol on the electronic approach chart, so providing a visual crosscheck of inbound flight progress. Upon landing, the ChartView will automatically display the destination airport's surface diagram.

State-of-the-art datalink technology provides the G1000 system with an array of weather and inflight information sources. These include the GDL 69A™ datalink receiver with XM WX satellite weather capability providing graphical depictions of Next-Generation Radar (NEXRAD) weather, METARs, TAFs, Temporary Flight Restrictions (TFRs), and winds aloft, along with echo tops, surface precipitation, lightning strikes, and storm cell data from anywhere in the United States regardless of altitude. Users can also zoom out the NEXRAD screen range to 2500 nm for nationwide monitoring of weather patterns. For cabin entertainment en route, the G1000 also provides a user interface offering more than 170 channels of digital-quality audio programming via XM Satellite Radio. Garmin's GDL® 90 Universal Access Transceiver (UAT) datalink connectivity unit is certified for Automatic Dependent Surveillance-Broadcast (ADS-B) operation. This enables aircraft to automatically transmit position, velocity, and heading information for enhanced air traffic surveillance within the FAA's infrastructure of ADS-B ground stations. These will ultimately be used to provide interactive traffic separation services nationwide without the need for ground-based radar.

## 9.2.2   Fully Integrated Flight Control

As a further example, the Garmin GFC 700 digital flight control system provides AHRS-based situational reference and dual-channel self-monitoring safety features. Using prestored data from the aircraft's flight manual to optimize performance over the entire airspeed regime, the system can maintain precise lateral and vertical navigation guidance for all phases of flight along with the ability to maintain airspeed holds, VNAV profiles, vertical speed references, and automated go-around procedures. For climb and descent, the system's software modeling ensures smooth round-outs and vertical intercepts, while automatic nav-to-nav captures help to streamline *en route* and approach transitions in busy terminal areas. Half-bank, control wheel steering, soft ride, roll attitude, and hold facilities are also provided, and the system is fully enabled for WAAS GPS-only instrument approach operations. The WAAS is now bringing ILS-like glidepath approach guidance into numerous airports having no ground-based approach aids—which can be rigid and costly—of any kind.

## 9.2.3   Advanced AHRS Architecture

For the reliable output and referencing of aircraft position, rate, vector, and acceleration data, the G1000 uses the GRS77 AHRS. Able to properly reference itself even while the aircraft is moving, the AHRS also uses additional comparative inputs from GPS, magnetometer, and air data computer systems to achieve high levels of integrity, reliability, and precision.

## 9.2.4   Weather and Digital Audio Functions

The G1000 system provides weather and radio information via WxWorx's current weather information service, which is delivered through the continuous broadcasting of data from XM's Satellite Network. Weather data is retrieved and updated constantly from WxWorx sources located at several thousand points across the United States.

XM's satellite delivery system is driven by two commercial satellites that simultaneously transmit data and provide constant coverage over the whole of the continental United States. Given an XM service subscription, the GDL 69/69A datalink receiver can acquire weather data and display it as required on the PFD and MFD at any altitude or on the ground.

However, it is very important to note that the weather data received from XM Weather in the cockpit is not approved to satisfy FAR 91.103: approved weather data must be obtained from the National Weather Service/FSS. Hence, the use of datalink weather must be regarded as a supporting reference only, and XM Weather must not to be used as a main source for making decisions before or during flight.

The MFD accommodates TFRs on all maps, and the display of cell movement for all MFD pages is controlled through the Weather Datalink Page Setup Menu.

On the Navigation Map Page, NEXRAD and XM Lightning data can be displayed by pressing the MAP Softkey followed by the appropriately labeled softkey 4.

On the Weather Datalink Page, all available weather products can be displayed via softkeys. The Weather Information Page is displayed on the first Waypoint (WPT) page by pressing the WX Softkey to display METAR and TAF text information.

XM Weather Products and their corresponding update rates are as follows:

NEXRAD 5 min Cloud Tops (15 min)
Cell Movement (12 min County Warnings 5 min)
METARs (12 min AIRMETs 12 min)
TAFs (12 min SIGMETs 12 min)
Lightning (5 min Echo Tops 7.5 min)
TFRs (12 min, Winds Aloft 12 min)
Surface Analysis (12 min)

The reported/forecast products are marked with the date and Zulu time as reported by appropriate weather stations, not on when the data is received by the datalink receiver.

XM Weather provides only cloud-to-ground lightning. To receive cloud-to-cloud lightning strike information, the aircraft must be equipped with an additional lightning detection equipment, such as Stormscope™.

The NEXRAD is a Doppler radar system that has greatly improved the detection of meteorological events such as thunderstorms, tornadoes, and hurricanes.

## 9.2.5   Traffic Information Service

In the following sections, to conform with US commercial instrument and FAA conventions, distance (as from an intruding aircraft) is given in feet (ft), the conversion to meters (m) being 1 ft = 0.3048 m. Similarly, weights are given in pounds (lbs), where the conversion is 1 lb = 0.4536 kg. Similarly, the US nautical mile (nm) is 1.852 km.

To be visible on the TIS display, any intruder aircraft must have at minimum an operating transponder (Mode A, C, or S). For using the TIS, both aircraft must be equipped with the appropriate equipment and must be flying within the radar coverage of a Mode S radar site capable of providing that TIS. Typically, this will be within a 55 nm range of the site and in line-of-sight of the Mode S radar.

With TIS, GIFD can display up to eight intruder aircraft on the PFD/MFD for a TIS service volume defined as being within 7 nm and 3500 ft above and 3000 ft below any intruder aircraft. TIS information is broadcast every 5 s.

A Traffic Advisory (TA) is a level of alert concerning any aircraft within ±500 ft of altitude and within a distance of 0.5 nm. A TA is issued when the aircraft are estimated to be within 34 s of potential collision regardless of distance or altitude. When a TA is issued, a yellow Traffic Annunciation circle is displayed on the PFD and an audio alert, "TRAFFIC," is generated.

A Proximity Advisory (PA) is a level of alert given when an intruder aircraft approaches within ±1200 ft and 4 nm.

An FAA Air Traffic Controller (ATC) may issue a TA when a potential collision is detected, and the ATC instructions must be followed unless the (PIC) elects not to comply based upon an over-riding safety judgment. The traffic display is intended to assist the pilot in visual acquisition of these aircraft whilst flying in visual meteorological conditions (VMC) under visual flight rules (VFR). The TIS is not intended to be used as a collision-avoidance system and does not relieve the pilot of responsibility to "see and avoid" other aircraft. It must not be used for avoidance maneuvers during Instrument Meteorological Conditions (IMC) or at other times when there is no visual contact with an intruder aircraft.

The WAAS uses a network of ground stations to provide necessary corrections to received GPS SPS navigation signals. Precisely surveyed ground reference stations are strategically positioned across the United States (including Alaska, Hawaii, and Puerto Rico) to collect GPS satellite data. Using this information, messages are developed to correct any signal errors, and these are broadcast to on-board receivers on frequencies used by communication satellite GPS signals. The WAAS is designed to provide the additional accuracy, availability, and integrity necessary to enable users to rely on GPS for all phases of flight from *en route* through approaches to all qualified airports within the WAAS coverage areas.

The WAAS supplies two different sets of corrections: corrected GPS parameters (position, clock, etc.) and ionospheric parameters. The first set of corrections is user-position independent, that is, they apply to all users located within the WAAS service area. The second set is area specific: the WAAS supplies correction parameters for a number of points (organized in a grid pattern) across the WAAS service area. The on-board user receiver computes ionospheric corrections for the received GPS signals based on algorithms, which use appropriate grid points for the user location. However, the appropriate grid points may differ for each GPS satellite signal received and processed by the user receiver because GPS satellites are located at various positions in the sky relative to that user. The combination of these two sets of corrections allows for significantly increased user position accuracy and confidence anywhere within the WAAS service area.

Technical Standard Order (TSO)-C146a (GPS equipment) automatically utilizes WAAS error corrections for the safe navigation of LNAV/VNAV instrument approaches and provides both horizontal and vertical guidance. This equipment can also navigate the LPV approaches developed specifically for WAAS, so providing ILS-like performance.

The positional accuracy of TSO-C146a GPS equipment with WAAS is as follows:

- Horizontal position accuracy ≈ 50 ft (15 m)
- Vertical position accuracy ≈ 74 ft (23 m)
- GPS position update rate = 1 Hz

For TSO-C146a systems:

- Horizontal position accuracy < 10 ft (3 m)
- Vertical position accuracy < 15 ft (4.5 m)
- GPS position update rate = 5 Hz

(GIFD Pilot's Training Guide Instructor's Reference 190-00368-06 Rev. B)

## 9.3  Automatic Dependent Surveillance-Broadcast (ADS-B) System Displays

The FAA NextGen ADS-B system requires a Mode-S transponder, a datalink (either 1090 Extended Squinter or UAT at 978 MHz depending upon the altitude), and a multifunction display on board the aircraft. ADS-B provides a Traffic Information Service-Broadcast (TIS-B) and a Flight Information Service-Broadcast (FIS-B) for improved situational awareness by the pilot in terms of real-time aircraft traffic and weather data.

Unless otherwise authorized by ATC, no person may operate an aircraft within controlled Class A, B, C, D, and E airspace unless that aircraft is equipped with the applicable equipment specified in § 91.215, and after January 1, 2020, § 91.225 (FAA, 2010; Grappell & Wiken, 2006).

A Mode-S transponder broadcasts, automatically and independently of any radar interrogation, over the busy 1090-MHz frequency. Mode-S transponders will be required in all controlled airspace that currently require Mode-C, with the advantage that Mode-S will respond to interrogation from ground stations connected to ATC and can provide aircraft ID, encoded altitude, velocity or speed, and direction of flight. Other safety parameters may be added as needed within the limitations of messaging over the datalink in use.

The current radar control–based Air Traffic Control (ATC) system is being transformed over the period 2013–2020 to a global positioning satellite and performance-based system. The United States FAA and the European Union EUROCONTROL agencies are conducting this process through the ATS (NextGen) and the Single European Sky Air Traffic Management Research (SESAR) programs, respectively. This is expected to be the most significant and expensive improvement in aviation navigation and safety so far, and will cost over $250 billion in the United States and $350 billion in the rest of the world.

By 2018, NextGen could reduce total flight delays by about 21%, so providing $22 billion in cumulative benefits to the traveling public, aircraft operators, and the FAA. This requires early implementation before the 2020 requirement for ADS-B (out) to achieve reduced ground delays, reduced air miles via efficient routing, and increased safety margins via spacing and turbulence vortex and situational awareness.

ADS-B includes an FIS-B that is in real time with minimal latency to improve the safety of flight since weather is a major factor in aircraft accidents. Figure 9.12 shows an example of weather patterns as displayed in the cockpit.

The TIS-B provides ADS-B equipped aircraft with position reports from secondary surveillance radar on non-ADS-B equipped aircraft as shown in Figure 9.13.

The FAA NextGen ATC system scheduled for full implementation by January 1, 2020 replaces the current radar-based system with a new GPS-based system for improved accuracy

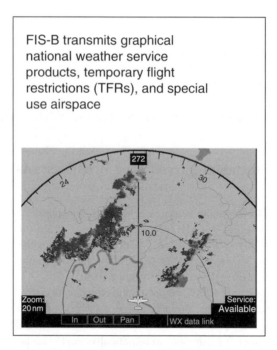

Figure 9.12   An ADS-B Flight Information Service-Broadcast (FIS-B) displaying current weather conditions

Figure 9.13   An ADS-B Traffic Information Service-Broadcast (TIS-B) display

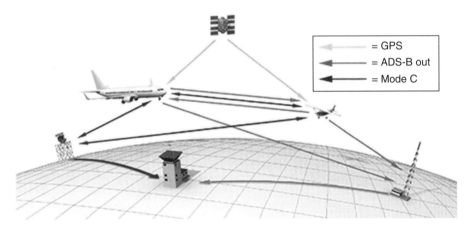

**Figure 9.14**  Operational Overview of the FAA NextGen ADS-B system

and routing efficiency. Properly equipped aircraft will be continuously monitored for vertical and lateral spacing with positions reported through datalinks to ground stations connected to the ATC facility as shown in Figure 9.14.

## 9.4  Collision Avoidance and Ground Warning Displays

TCAS II provides TAs and Resolution Advisories (RAs) consisting of recommended escape maneuvers, in the vertical dimension to either increase or maintain the existing vertical separation between aircraft. TCAS II is mandated in the United States for commercial aircraft, including regional airline aircraft with more than 30 seats or a maximum takeoff weight greater than 33 000 lbs (14 968 kg). Although not mandated for GA use, many turbine-powered GA aircraft and some helicopters are also equipped with TCAS II.

TCAS II was designed to operate in traffic densities of up to 0.3 aircraft per square nautical mile (nm), that is, 24 aircraft within a 5 nm radius, which was the highest traffic density envisioned through 2030. The TCAS computer processor performs airspace surveillance, intruder tracking, own aircraft altitude tracking, threat detection, RA maneuver determination and selection, and the generation of advisories. This processor uses pressure altitude, radar altitude, and discrete aircraft status inputs from own aircraft to control the collision avoidance logic parameters that determine the protection volume around the TCAS.

TCAS monitors the tracked aircraft to determine if an avoidance maneuver has been selected that will provide adequate vertical miss distance from the intruder while generally minimizing any perturbations to the existing flight path. If the threat aircraft is also equipped with TCAS II, the avoidance maneuver will be coordinated with the threat aircraft.

Figure 9.15 shows typical TCAS displays with the lower left image indicating symbology for own aircraft (white solid triangle), other aircraft (white open diamonds), proximal aircraft (white solid diamonds), TA (yellow solid circle) and RA data (red solid square). Figure 9.16 shows a Honeywell™ TCAS II System Display (FAA, 2011) and Figure 9.17 shows the relevant functional block diagram. The complete operational block diagram, including interactions with the pilot, is depicted in Figure 9.18.

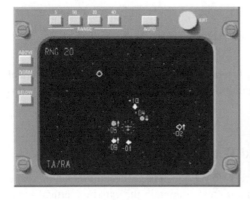

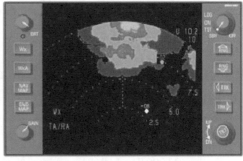

**Figure 9.15** Typical TCAS displays (*See insert for color representation of the figure.*)

**Figure 9.16** Honeywell™ TCAS II System display

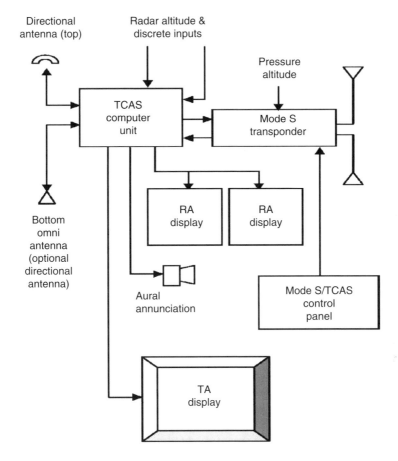

**Figure 9.17** TCAS II Functional Block Diagram

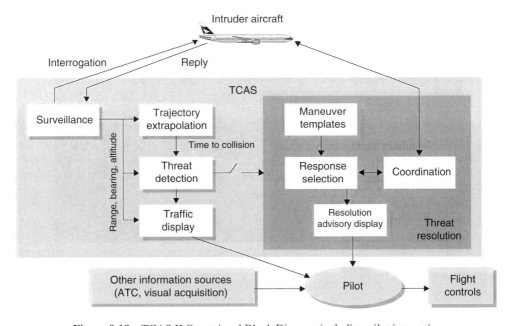

**Figure 9.18** TCAS II Operational Block Diagram including pilot interaction

## 9.4.1   Terrain Awareness Warning System (TAWS): Classes A and B

TAWS provides the highest level of protection against Controlled Flight into Terrain (CFIT) accidents. TAWS Class A provides all the functionality of the Class B system, plus a terrain awareness display to the aircraft's display system and a fully autonomous Ground Proximity Warning System (GPWS). It will also support smart bank angle alerts, minimum callouts, and altitude callouts at selected altitudes. Furthermore, it provides a class-specific (required RTCA DO-161A and TSO-C92c GPWS) warning of imminent contact with the ground.

Approved in accordance with TSO-C151b requirements, TAWS Class A and B systems both offer Forward Looking Terrain Avoidance (FLTA) based on terrain data and the aircraft's state and predicted flight path, Premature Descent Alerts Attention (the aural "five hundred" callout), and alerts based on temperature-compensated GPS altitudes that warn of imminent contact with the ground.

TAWS integrates with the FMS to provide an additional unique predictive alerting feature, based on information in the flight plan. It provides crisp and clear graphical depictions of actual terrain in three view formats, Map View, 3-D Perspective View, and Profile View, on the FMS CDUs or flight deck displays such as the MFD-640 or EFI-890R (Universal Avionics).

The high-resolution terrain database is stored in internal flash memory and updated using the Data Transfer Unit via a high-speed Ethernet bus. It features a data point approximately every 0.5 mile world-wide and up to 0.1 mile at mountainous airports. The terrain database also includes data for depicting oceans and large inland bodies of water.

Figure 9.19 shows the topology of a sample flight path and indicates, in red, any geographic or man-made object that is above the aircraft flight altitude.

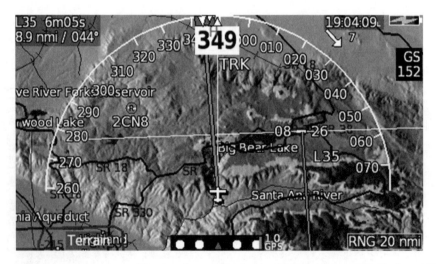

**Figure 9.19**   Terrain Awareness Warning System display by GRT Avionics EFIS Horizon HX 6.5″ Display Unit (GRT Avionics)

## Appendix: Terminology and Review of Some US Federal Aviation Regulations

APV: Approach with vertical guidance is an International Civil Aviation Organization (ICAO) term referring to specific ICAO criteria adopted in May 2000. This approach classification allows stabilized descent using vertical guidance without the accuracy required for traditional precision approach procedures. The United States has developed criteria for lateral/vertical navigation (LNAV/VNAV) and LPV approach procedures that meet this approach classification. LNAV/VNAV and LPV approaches provide guidance in both the lateral and vertical planes.

LNAV: Lateral Navigation provides lateral guidance for a profile or path. This terminology is used for GPS nonprecision approaches.

LNAV/VNAV: Lateral Navigation/Vertical Navigation describes an approach in which lateral guidance as well as vertical glide slope guidance is provided.

LPV: Localizer Performance with Vertical guidance is a new type of APV approach procedure in which the angular lateral precision is combined with an electronic glidepath. TERPS approach criteria are very similar to that used for ILS approaches.

RAIM: Receiver Autonomous Integrity Monitoring provides integrity check capability to ensure the safe use of GPS for IFR flight operations (see the AIM for drawing verification). The only time RAIM prediction is needed for TSO-C146a systems is when the WAAS is not available.

SBAS: Satellite-Based Augmentation Systems use satellites and networks of ground stations to provide improved accuracy for received GPS satellite signals. Internationally, many countries are working with the International Civil Aviation Organization (ICAO) to standardize SBAS globally. The WAAS is an SBAS currently being implemented in the United States.

WAAS: The WAAS is a satellite navigation system consisting of equipment and software, which augment the GPS Standard Positioning Service (SPS). It provides enhanced integrity, accuracy, availability, and continuity over and above GPS SPS. The differential correction function provides improved accuracy required for precision approach.

VNAV (or VNV): Vertical Navigation manages the altitude profile. The concept of the VNAV is to assign a target altitude to assist in controlled decent operations. Using the flight plan, a target altitude may be assigned to any waypoint. To reach the target altitude, the rate of descent is calculated and displayed, and vertical guidance is provided to reach the descent profile.

## References

Aspen Avionics. Available at: http://www.aspenavionics.com/products/general-aviation/evolution-2500 (accessed on December 18, 2015).

FAA. AC20-165A Airworthiness Approval of Automatic Dependent Surveillance-Broadcast (Out) (May 21, 2010).

FAA Publication. Introduction to TCASII (2011). FAA, Washington, DC.

Federal Register (2010-12645). FAA, 14 CFR Part 91 Automatic Dependent Surveillance—Broadcast (ADS–B) Out Performance Requirements.

Garmin™: G-1000™ Integrated Flight System. Available at: www.garmin.com; GPS www.gps.gov (accessed on December 18, 2015).

GIFD: Pilot's Training Guide Instructor's Reference 190-00368-06 Rev. B

GRT Avionics: Terrain Awareness Warning System display by EFIS Horizon HX 6.5″ Display Unit. Available at: www.planesfactory.it (accessed on December 18, 2015).

R D Grappel, R Wiken (April 3, 2006) "Compliance Verification for S-Mode Transponder Elementary Surveillance (ELS), Enhanced Surveillance (EHS) and Automatic Dependent Surveillance via Broadcast (ADS-B) Applications." Massachusetts Institute of Technology, Lincoln Laboratory, Lexington, MA. Project Memorandum Number 42PM-AFST-0002.

Honeywell. Available at: http://aerospace.honeywell.com/products/cockpit-displays/head-up-displays#sthash. 9SuO6fSs.dpuf (accessed on December 18, 2015).

NASA: FS-2000-06-43-LaRC, 'The Glass Cockpit'.

Universal Avionics: Terrain Awareness Warning Systems.

Traffic Control (ATC) Service Final Rule, May 28, 2010.

US Air Force IHSI I 311HSW/PA Case File #07-205 at http:www.globalsecurity.org/military/systems/aircraft/ f-22-cockpit.htm

# 10

# Unmanned Aerospace Vehicle Navigation

Vladimir Y. Raspopov[1], Alexander V. Nebylov[2], Sukrit Sharan[3] and Bijay Agarwal[3]
[1]Tula State University, Tula, Russia
[2]State University of Aerospace Instrumentation, Saint Petersburg, Russia
[3]Sattva E-Tech India Pvt Ltd., Bangalore, India

## 10.1   The Unmanned Aerospace Vehicle

The Unmanned Aerial Vehicles (UAVs), often called drones, are vehicles wherein the entire control and piloting is carried out remotely, with no pilot on board. Usually, it is an aircraft capable of flying in a totally automatic mode as part of a complex that includes a ground control station containing its central element—a human operator.

In US Department of Defense (DoD) documentation, the term "Unmanned Aircraft System," or UAS, is normally used. The UAV is more generic and may refer to an airplane, a helicopter, a lighter-than-air vehicle, or indeed any other unmanned flying device.

## 10.2   Small-Sized UAVs

The size and weight of the UAS for the control of different categories of UAV vary by several orders of magnitude along with those of the vehicles themselves. "SUAVs" are small UAVs that may have take-off weights from 12 g to 10 kg and these form close to 50% of the total of all categories (Austin Reg, 2010; Beard and McLain, 2012; Unmanned Aircraft Systems, 2008/2009; Valavanis and Vachtsevanos, 2015; Raspopov, 2010).

The aerodynamic design of a UAV is influenced by the internal layout of the engine, the control surfaces, and the drive equipment linking them with elements of the airframe. That is, the UAV shape is greatly dependent on the required control forces and control moments and hence on the mutual arrangement of the devices that create those forces and moments.

(a)                                                  (b)

(c)                                                  (d)

**Figure 10.1**   Four different high-wing SUAV designs. (a) "Orlan-10" (Russia, STC), (b) "Aladin" (Germany, EMT), (c) "Aerosonde" (USA, AAI), and (d) Indian NAL "Slybird" UAV (for civilian applications—Livefist, India)

UAV aerodynamic schemes divide into fixed-wing and helicopter types. Fixed-wing configurations include those using upper, lower, and middle-wing designs and also *canard* (duck) and flying-wing types. The most common design uses the high-wing configuration where the wings terminate in the upper part of the fuselage. High wings provide improved longitudinal stability at high angles of attack and also make wing-mounted engines possible. Their disadvantages include a decrease in the efficiency of a vertical tail at high angles of attack where the tail enters the slipstream from the wing. The four typical designs shown in Figure 10.1 illustrate variations in the locations and numbers of engines, wing dihedrals, and tail configurations.

Mid-wing and low-wing SUAVs are much less common because they result in reductions in static stability and exhibit difficulties in landing, solutions for which complicate the overall design.

The canard configuration, in which a longitudinal control surface is located in front of the wing, is so called because of the similarity of the first aircraft (the Wright Brothers "Flyer") to the eponymous bird. The main advantage of the scheme is that the resultant placement of the rear wing provides better aerodynamics at high speeds because the center of aerodynamic pressure is shifted backward. Also, it is possible to reduce the weight of the vehicle because

**Figure 10.2**   Canard SUAV configurations: Indian "Rustom" Class UAV (India, Livefist)

there is no need to mount the rudder even further back. Furthermore, the canard scheme increases the critical angle-of-attack and thereby improves maneuverability. As for the high-wing configuration, the arrangement of the individual structural elements may vary widely, as shown in Figure 10.2.

A common alternative aerodynamic scheme for UAVs is the flying-wing or tailless config-uration, which provides no separate control surface for the vertical plane, but uses only the ailerons mounted on the rear edge of the wing (Figure 10.3).

The advantages of this configuration are lesser airframe weight and drag; however, the aileron efficiency is much worsened, which leads to lesser efficient pitch control. However, the introduction of a fly-by-wire control system minimizes this shortcoming. Furthermore, the lower inertia of the vehicle leads to increased maneuverability. Also, for military use, this form of construction makes optimization for radar visibility reduction comparatively easy.

There are several ways of taking off and landing a drone, depending on its mass, size, strength, and the value of the initial speed. For light SUAVs, a suitable launch can be via a hand or an arm. For heavier SUAVs needing higher launch speeds, catapults are used (Figure 10.4). Also, some drones can take off on a runway in the same manner as a piloted airplane. Landing a drone using a parachute (as currently provided in some piloted light aircraft for an emergency operation) is also common. Landing using nonstandard technologies such as translating level flight into a flat spin, resulting in a "belly-flop" onto an inflatable container, is also possible.

Civilian SUAV applications include the following tasks:

1. Aerial photography (cartography); inspection of, and compliance with contractual "open sky" and hydrological obligations; observing the meteorological situation; and policing actively radiating objects.
2. Monitoring environmental conditions including natural radiation, gas and chemical concentrations, and surveying seismic sensors.
3. The development of regional and interregional telecommunication networks including mobile, television, radio broadcasting retransmission centers, and navigation systems.

(a)                                             (b)

(c)

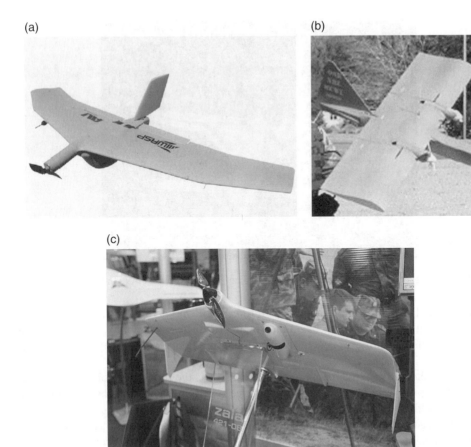

**Figure 10.3** Flying-wing SUAVs. (a) "Black Widow Wasp" (USA, AeroVironment), (b) RQ-14 "Dragon Eye" NRL/MCWL prototype (USA, AeroVironment), and (c) ZALA 421-08 (Russia, ZALA AERO)

4. The control of maritime traffic: the seeking and detection of vessels; assisting in the prevention of accidents in ports; monitoring sea borders; and checking conformity with fishing regulations.
5. Oceanography, including the exploration of ice conditions; monitoring sea conditions, and searching for schools of fish.
6. Assisting agricultural operations and geological prospecting; the characterization of soil; exploration of mineral deposits, and general subsurface sensing.

Military missions assisted by SUAVs can include the following:

1. Conducting reconnaissance flights in mountains or over water; determining the locations of small targets in an enemy territory; establishing the optimum deployment of military units to assist commanders of military operations; and conducting flight operations over positions of forces to obtain information about their deployment in otherwise inaccessible regions.

(a)

(b)

**Figure 10.4** SUAV take-off methods: (a) via a hand catapult and (b) pneumatic or mechanical catapult launchers

2. Reconnaissance and battlefield surveillance for the correction of artillery fire.
3. The laser sighting of target locations.
4. Conducting electronic warfare (electronic intelligence, electronic reconnaissance, exploration of communications; electronic countermeasures; and the suppression of radio-electronic countermeasures.
5. Providing microwave transmission facilities.

6. Participation in training operations as an airborne target.
7. The protection of important objects.
8. The destruction of ground targets in well-defended areas to avoid the loss of piloted aircraft.
9. The conduct of aerial warfare.

In many countries, such widespread use of SUAVs has caused great interest in new developments. Some of the technical characteristics of SUAVs from a selection of developers are shown in Table 10.1

Operational data for UAV systems should include the causes of any losses, the main ones being as follows (with typical results): engine problems (37%); control system errors (25%); operator errors (17%); communication problems (11%); and others (10%).

## 10.3 The UAV as a Controlled Object

The movement of a UAV as a rigid body can be described by differential equations in vector form:

$$m \left\{ \frac{d\mathbf{V}}{dt} + \omega \times \mathbf{V} \right\} = \mathbf{R};$$

$$\frac{d\mathbf{K}}{dt} + \omega \times \mathbf{K} = \mathbf{M}, \tag{10.1}$$

where $m$ is the UAV mass; $\mathbf{V}$ is the velocity vector of the mass center; $\mathbf{R}$ is an external force vector; $\mathbf{M}$ is the external force main moment; $\mathbf{K}$ is the UAV angular momentum; and $\omega$ is the angular velocity vector.

The use of equations (10.1) requires, above all, the following tasks:

- The acquisition and evaluation of the UAV aerodynamic characteristics.
- Obtaining the traction characteristic $P$ of the engine as a function of the velocity of the incoming air stream $V$, the shaft speed $n$, and the geometry of the propeller.

Measurement of these characteristics is possible in wind tunnels, and also can be obtained by:

- The virtual purging method for determining the aerodynamic characteristics of the UAV using the Floworks module in the Solid-Works package (Telukhin and Raspopov, 2010).
- Using the PropCalc 3.0 program for determining the engine traction characteristics.

As an example of results of using the virtual purging method, Figure 10.5 shows the dependence of the drag coefficient $c_{xa}$ and the lift coefficient $c_{ya}$ on the angle-of-attack. Here,

$$X_a = c_{xa} S \frac{\rho V^2}{2} \quad \text{and} \quad Y_a = c_{ya} S \frac{\rho V^2}{2}$$

where $\rho$ is the air density, $V$ is the flight speed, and $S$ is the wing area for a UAV with a wingspan of 1.56 m and a NACA-0016 wing profile.

**Table 10.1** Technical characteristics of some SUAVs

| Name of UAV | Technical characteristics | | | | | | | | |
|---|---|---|---|---|---|---|---|---|---|
| | Take-off mass (kg) | Payload mass (kg) | Wing span (m) | Length (m) | Work range (km) | Flight height (m) | Velocity (km/h) | Flight duration (h) | Country |
| WASP III | 0.454 | — | 0.735 | — | 5.0 | — | — | 0.7–0.8 | USA |
| RQ—14 | 2.7 | — | 1.1 | 0.9 | <5.0 | 90–150 | 65 | 1.0 | USA |
| Aladin | 3.2 | — | 1.5 | — | 15.0 | 30–150 | 37–76 | — | Germany |
| RQ—11 | 17 | — | 1.5 | 0.96 | 10.0 | 5000 | 95 | 1.0–1.2 | USA |
| Orbiter | 6.5 | — | 2.2 | 1.0 | 15–50 | >5000 | — | 2.0–3.0 | USA |
| Boomerang | 7.0 | 1.2 | 2.75 | — | 15.0 | 500 | 55–110 | 2.5 | Israel |
| "Inspector 101" | 0.25 | 0.05 | 0.3 | — | 1.5 | 25–500 | 30–45 | 0.5–1.0 | Russia |
| "ZALA 421-11" | 0.79 | 0.1 | 0.4 | 0.4 | 5.0 | <2500 | 60–130 | 0.5 | Russia |
| "Inspector 201" | 1.3 | 0.15 | 0.8 | — | 5.0 | 50–1000 | 35–90 | 0.5–0.6 | Russia |
| "ZALA 421-08" | 1.7–2.1 | 0.2 | 0.81 | 0.425 | <15 | 50–3600 | 65–130 | 15 | Russia |
| T23 "Aileron" | 2.8 | — | 1.47 | 0.45 | 10–30 | <3000 | 65–105 | 1.25 | Russia |
| "BROTHER" | 3.0 | 0.3 | 2.0 | 1.0 | 10 | <5000 | 90 | 1.0 | Russia |
| "Irkut—2M" | 3.0 | 0.3 | 1.5 | 0.5 | 20 | 300–3000 | 65–105 | <1.5 | Russia |
| "Curl" | 3.5 | 0.6 | 2.0 | 0.95 | <25 | 50–3000 | 60–120 | 1.0 | Russia |
| "ZALA 421-12" | 3.9 | <1.0 | 1.6 | 0.62 | <40 | <3600 | 65–120 | 2.0 | Russia |
| T-3 | 5.0 | — | 1.8 | 0.7 | <25 | — | 60–120 | 1.0–1.5 | Russia |
| "Inspector 301" | 5.5 | 0.4 | 1.5 | — | 15 | <1000 | 55–150 | 0.75–1.5 | Russia |

(a)

(b)

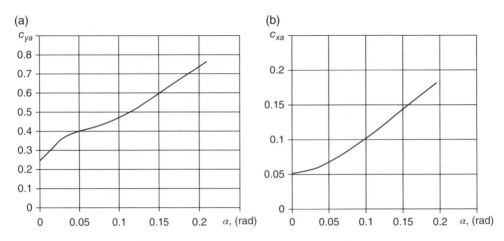

**Figure 10.5**   The dependence of coefficients $C_{ya}$ (a) and $C_{xa}$ (b) on the angle-of-attack

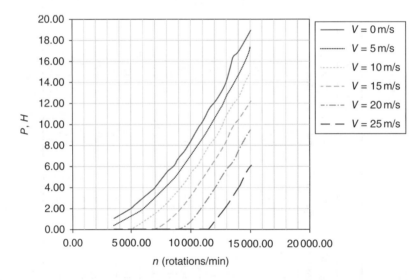

**Figure 10.6**   Traction characteristics of a three-bladed propeller type APC 7×5 for different flow velocities

The thrust performance of a fixed pitch propeller type APC 7×5 for various flight speeds obtained using PropCalc 3.0 is given in Figure 10.6.

The adequacy of the UAV movement as predicted by the Mathematical Model (MM) is checked by flight testing, which is performed as a result of:

- Comparison of the UAV free flight (i.e., without an autopilot) with the coordinates $K_p$ ($\psi$, $\vartheta$, $\gamma$, $H$, $V$) as registered in flight, and the UAVs MM reaction—$K_m$ is compared for the same coordinates in a rudder surface deviation $\delta_\kappa$.

- Comparison of the reaction of the UAVs autopilot with the coordinates $K_p$ as registered in flight, as well as those predicated by the MM of the UAV/autopilot for the given values $K_r$ ($\psi_r$, $H_r$, $V_r$, etc.).

A conformity assessment of the MM parameters with those of the actual UAV can be performed by using the relative errors of the $\varepsilon_c$ measurement and the averaged errors $\bar{\varepsilon}_c$ calculated for $n_c$ control time reports in the steady state.

$$\varepsilon_k = \frac{K_p - K_M}{K_p} \cdot 100\%, \quad \bar{\varepsilon}_c = \frac{1}{n_c} \cdot \sum_{i=1}^{n_k} \varepsilon_c.$$

Figure 10.7 shows the transient processes for the model aircraft TwinStar II via the angular coordinates, and in Table 10.2 the errors in their determination.

$K_m$ is the transient process of the MM.
$K_p$ is the transient process of the real model.
1 is the input action in the form of deviation of the respective rudders.

Table 10.2 shows that the average relative error in the motion parameters predicted by mathematical modeling does not exceed 10%, which testifies to the adequacy of MM in representing the real movement of the UAV, and confirms the efficacy of methods developed by applied research.

Taking the UAV as a complete control plant, the control actuators along with the onboard and ground-based equipment form the Automatic Control System (ACS) for that UAV.

Most modern ACSs use a three-channel scheme. Usually, the respective channels are called steering bodies: the elevator (stabilizer) channel, the rudder channel, and the aileron channel. There is a common functional separation channel management system as follows: the elevator (stabilizer) channel is used to control the pitch angle and altitude, the aileron channel is for roll direction and rate, and the rudder channel is to eliminate slip. Finally, there is a very important autothrottle system for controlling the speed.

The ACS performs the following tasks:

- Ensuring that the required handling characteristics are aimed at optimizing the dynamic properties of the aircraft under manual control.
- Stabilizing the angular position of the aircraft using information from the orientation system.
- Automating the trajectory control that can show the vehicle at a certain point of the route, or stabilizing the motion of the vehicle on a particular path at a predetermined speed. Note that the trajectory control requires the use of information from the navigation system.

## 10.4   UAV Navigation

For many tasks executed by modern UAVs it is necessary to ensure performances adequate for maintaining flight modes and flight paths that result in the vehicles reaching the required points on their routes and completing their missions. Current methods for achieving these requirements now follow.

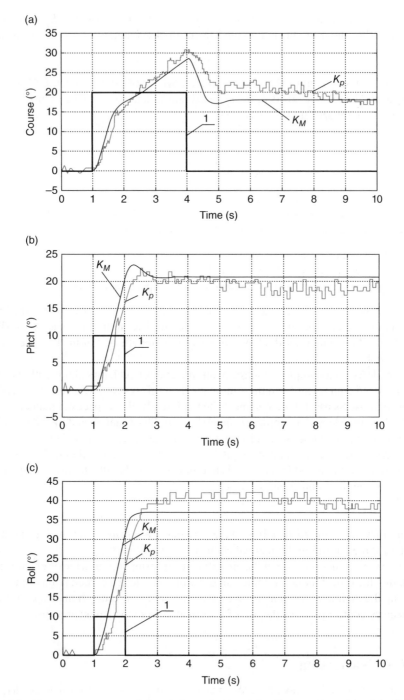

**Figure 10.7** Transient processes of the mathematical model and the real model of the TwinStar II airplane by course angle (a), pitch (b), and roll angle (c) at a speed of 18 m/s

**Table 10.2**   Errors in determining the flight parameters of mathematical models ($n_\kappa = 9$)

| Flight parameters (coordinate $K$) | Errors | |
|---|---|---|
| | max $\varepsilon_\kappa$ (%) | $\bar{\varepsilon}_k$ (%) |
| Course angle, $\psi$ | 22 | 9.5 |
| Pitch, $\vartheta$ | 19 | 10 |
| Roll angle, $\gamma$ | 14 | 10 |

## 10.4.1   Methods of Controlling Flight Along Intended Tracks

The flight path of a UAV is actually the curve in space taken by its center of mass. This space-time reference trajectory is represented by a flight navigation program that depends on the functional purpose of the UAV. Typically, a flight navigation program should include a checkpoint flight timetable, which is a sequence of individual elements of the flight that include the entry conditions at the points in space at which the main tasks will be carried out. Examples include the photographing of an area designated by the beginning and ending of photography along a flight path, or the dropping of a payload at a specific point.

The speed of the vehicle relative to the Earth is called the *Earth speed*, which is directed tangentially to the flight path, the horizontal projection of which is the *ground speed*.

Usually the trajectory of a route can be represented by the track and the flight profile. The route is the projection of the UAV flight on the Earth's surface, and its flight profile indicates its changes in altitude, which can be represented by a trajectory, or flight pattern, in the vertical plane. Reference points are usually allocated for the flight route, for example, the takeoff and landing points and also the points where routing changes such as turns occur. Points where flight mode changes occur are called *milestones*, and these include the beginning and end of objective function implementation for a given flight, the beginning and the end of the initial climb, etc.

UAV automatic control systems should ensure the precise execution of the route at the set altitude (flight level), the determination of navigational flight parameters, the enforcement of flight objectives, and the execution of approach procedures to the landing site at the correct time.

It is usual to distinguish between two-, three-, and four-dimensional navigation. Two-dimensional navigation implies the maintaining of a given route (track), three-dimensional (3-D) navigation adds the tasks of assignment and control of the flight profile, and four-dimensional navigation establishes that the flight and task performances are carried out according to a predetermined time schedule.

Motion control of aircraft, including UAVs, along a predetermined path (the *intended track*, or IT) is carried out by establishing sequential Turning Points Within The Route (TPR) in three ways: ground, course, or route (Figure 10.8). The LPP is the Line of Predetermined Path.

Motion control in the lateral direction is performed by means of establishing a Bearing Track For The Route (BTR) $\psi_w$. For flight along the Intended Track (IT) and the subsequent

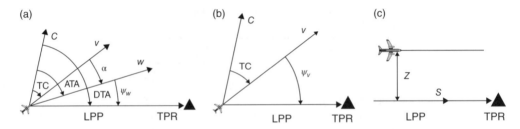

**Figure 10.8**  Methods of the flight control along an intended route: (a) path, (b) course, and (c) trip

motion of the vehicle along the BTR, the ground speed vector should be directed toward a specified point. To do this, the angle of the track bearing must be kept equal to zero:

$$\psi_w = DTA - ATA = DTA - (TC + \alpha) = 0, \qquad (10.2)$$

where DTA is the Defined Track Angle, ATA is the Actual Track Angle, TC is the True Course and $\alpha$ is the drift angle.

Condition (10.2) will provide flight to a given point on the shortest distance along the great circle passing through that given point and the TPR. This is an advantage of the track method, but when the vehicle deviates from the desired course, it does not provide a way to correct it. If the wind parameters are not known, then the UAVs avionics "believe" that the drift angle is zero, and the track method is converted into the course method.

During the course procedure, motion control in the lateral direction is executed by using the localizer bearing, which is kept equal to zero. In the absence of wind the vehicle will approach the BTR in the shortest distance, but under wind conditions it will follow a difficult path that does not coincide with the desired course. Hence, in some cases there can be significant deviations of the Actual Path of The Line (APL) from the desired course and significant deviations of the APL from the IT.

The trip flight mode along the desired course and the vehicle flight along the BTR are realized via the acquisition and continuous indication of the coordinates Z and S. The relevant problems are solved in a terrestrial coordinate system, one axis of which is the desired course and the second axis is perpendicular to it. The control parameter in the trip method is the linear lateral deviation Z from the desired course.

When Z = 0, the vehicle follows the desired course and ensures its access to the BTR. In the trip control method, the APL form is determined by the shape of the IT, and if the break points along the trip segments are related to a great circle, this method provides trip motion along that great circle. However, when deviating from the set trip the vehicle is flown along an appropriate restorative course, which is an advantage of the trip method.

The problem of optimizing the trajectory control using the trip navigation method is usually decided via methods of analytical construction of regulators. The primary regulator in UAV motion control system is autopilot, and the most rigorous solution is obtained by using the full nonlinear equations describing the spatial motion of the vehicle and the control system. However, in this approach it is difficult to take into account the hierarchy of governance, in particular the relationship of the navigation system (the "navigator") to the flight director (the "pilot") complexes, each of which has its own control loop. The navigation system is at a senior level to the piloting system that performs the actual flight execution, and its control part dictates action for that piloting system in the form of angles of roll, pitch and heading, and also speed and altitude.

It is precisely these effects that must be considered when solving the optimization problem, which is essentially the determination of changes in angles, altitudes, and speeds (all involving control laws) under which the system would be converted from a perturbed state into the correct one.

Under any method of flight control it is necessary to determine the coordinates of the points in space through which the UAV is intended to fly. This is the primary navigation task and currently it is solved mainly by positional and dead reckoning methods.

The Position Method consists of determining the coordinates of the vehicle location from the geometric relationships of the measured distances and angles of relative location of aircraft and known points such as landmarks, beacons, and lights. This method is based on astronomical techniques and both radio navigation and visual orientation. Radio navigation uses navigation satellites—this is Satellite Navigation System (SNS), which is widely used in UAVs. Electronic methods using beacons and visual methods with the help of video systems can be used for the automatic landing of UAVs.

Dead reckoning consists of calculating the trajectory of the vehicle by measuring the value and direction of its velocity and the coordinates of its starting point. The measurement of airspeed using absolute pressure sensors, along with Inertial Navigation System (INS) and SNS can be used. To measure the UAV course, magnetometer and an INS and SNS can be used.

## 10.4.2   Basic Equations for UAV Inertial Navigation

The following coordinates shown in Figure 10.9 are as follows.

The Inertial $OX_u Y_u Z_u$ coordinates start at the center of the Earth; the $OZ_u$ axis is directed along the axis of rotation of the Earth to the North Pole; and the $OX_u$ axis is directed along the line of intersection of the equatorial plane and the Greenwich meridian.

The Earth (equatorial) $O_0 X_0 Y_0 Z_0$ axes are fixed relative to the Earth, and the initial time coincides with the inertial coordinate system.

The Normal coordinate system is $OX_g Y_g Z_g$.

If a UAV is moving relative to the Earth at a linear speed $\bar{V}$ and at an altitude $H$, the changes of geodetic coordinates are determined by the following dependencies:

$$\dot{B} = \frac{V_{Xg}}{R_1}; \quad \dot{L} = \frac{V_{Zg}}{R_2 \cos \phi}; \quad \dot{H} = V_{Yg} \qquad (10.3)$$

where $V_{Xg}$, $V_{Yg}$, and $V_{Zg}$ are the projections of the velocity vector onto the axes $OX_g$, $OY_g$, and $OZ_g$ respectively; and

$$R_1 = \frac{a_{\mathrm{Kr}}\left(1-e^2\right)}{\left(1-e^2 \sin^2 B\right)^{3/2}} + H; \quad R_2 = \frac{a_{\mathrm{Kr}}}{\left(1-e^2 \sin^2 B\right)^{1/2}} + H$$

Here, $a_{\mathrm{Kr}} = 637\,8245\,\mathrm{m}$ and $b_{\mathrm{Kr}} = 635\,6863\,\mathrm{m}$ are the major and minor axes of the Krasovsky ellipsoid, respectively, and $e = \sqrt{a_{\mathrm{Kr}}^2 - b_{\mathrm{Kr}}^2} / a_{\mathrm{Kr}}$ is the eccentricity of the ellipsoid.

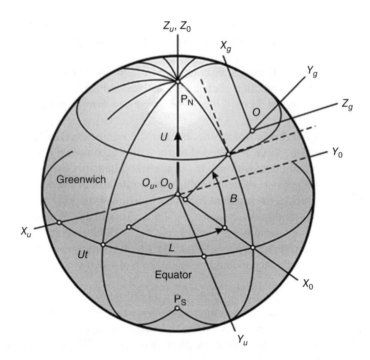

**Figure 10.9**   Relative positions of the inertial, equatorial, and normal coordinate systems

To implement the algorithm for a Strapdown Inertial Navigation System (SINS), the kinematic elements of the UAV motion are needed, that is, the projection of the absolute angular velocity of rotation of the trihedron $OX_gY_gZ_g$ on its axis and the projection of the absolute linear accelerations of its vertices. The projections of the absolute angular velocity of the coordinate system normal to the axis are defined as follows:

$$\begin{cases} \omega_{xg} = \left(U + \dot{L}\right)\cos B = U\cos B + \dfrac{V_{Zg}}{R_2}; \\[2mm] \omega_{yg} = \left(U + \dot{L}\right)\sin B = U\sin B + \dfrac{V_{Zg}}{R_2}\,tgB; \\[2mm] \omega_{zg} = -\dot{B} = -\dfrac{V_{Xg}}{R_1}, \end{cases} \tag{10.4}$$

where $U = 7.292116 \cdot 10^{-5}$ rad/s is the angular velocity of the daily rotation of the Earth.

The projections of the absolute linear acceleration are:

$$a_{xg} = \dot{V}_{Xg} + a_{xg}^a; \quad a_{yg} = \dot{V}_{Yg} + a_{yg}^a; \quad a_{zg} = \dot{V}_{Zg} + a_{zg}^a \tag{10.5}$$

where $a_{xg}^a$, $a_{yg}^a$, and $a_{zg}^a$ are the apparent acceleration components to be compensated.

Projections of apparent acceleration can be measured by fixed accelerometers mounted on board the UAV, that is, in the associated coordinate system. It is then necessary to

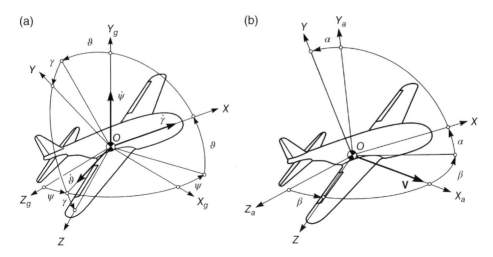

**Figure 10.10**  Coordinate systems: (a) normal and related coordinate system and (b) related and speed coordinate system ($\psi$, $\vartheta$, $\gamma$—heading, pitch, roll; $\alpha$, $\beta$—angle-of-attack and sideslip angle)

convert these accelerometer indications from the associated to the normal coordinate system. This is possible if the orientation of the associated coordinate system relative to the normal is known. To describe the relative positions of the associated and normal coordinate systems, the different kinematic parameters can be used: the Euler–Krylov angles, the direction cosines, the Rodrigues-Hamilton parameters, and the parameters of Kaylee Klein *et al.* (Bortz, 1971; Branets and Shmigleevskiy, 1973; Gundlach, 2011; Kuzovkov and Salichev, 1977; Matveev and Raspopov, 2009; Savage, 2010; Veremeenko *et al.*, 2009a).

Equations 10.4 and 10.5 are the key inertial navigation equations whose solution provides the location of the UAV coordinates.

Conversion from the $OX_gY_gZ_g$ coordinate system to the $OXYZ$ system can be made by successive turns at yaw angle $\psi$, pitch $\vartheta$, and roll $\gamma$ (Figure 10.10a).

The relative position of the normal and related coordinate systems can be described by the direction cosines and their own quaternions. A quaternion is a hypercomplex number:

$$\Lambda = \lambda_0 + \lambda_1 \mathbf{i} + \lambda_2 \mathbf{j} + \lambda_3 \mathbf{k}$$

whose elements are Rodrigues-Hamilton parameters $\lambda_n$ ($n = 1,2,3$), where $\mathbf{i}$, $\mathbf{j}$, and $\mathbf{k}$ are trihedron imaginary axes. Contact numbers of the final turn and the corresponding direction cosine matrix and quaternion are given in Table 10.3.

The resulting shift from the coordinate system by using a direction cosine matrix and quaternion is defined as follows:

$$A = A_\gamma A_\vartheta A_\psi = \begin{pmatrix} \cos\psi\cos\vartheta & \sin\vartheta & -\sin\psi\cos\vartheta \\ -\cos\psi\sin\vartheta\cos\gamma + \sin\psi\sin\gamma & \cos\vartheta\cos\gamma & \sin\psi\sin\vartheta\cos\gamma + \cos\psi\sin\gamma \\ \cos\psi\sin\vartheta\sin\gamma + \sin\psi\cos\gamma & -\cos\vartheta\sin\gamma & -\sin\psi\sin\vartheta\sin\gamma + \cos\psi\cos\gamma \end{pmatrix}$$

$$(10.6)$$

**Table 10.3**  The kinematic parameters characterizing the relative positions of the coordinate systems $OX_gY_gZ_g$ and $OXYZ$

| No. of final rotation | Angles Euler–Krylov | Directional cosines | Own quaternions |
|---|---|---|---|
| I | $\psi$ | $A_\psi = \begin{Vmatrix} \cos\psi & 0 & -\sin\psi \\ 0 & 1 & 0 \\ \sin\psi & 0 & \cos\psi \end{Vmatrix}$ | $\mathbf{P} = \cos\dfrac{\psi}{2} + \mathbf{j}\sin\dfrac{\psi}{2}$ |
| II | $\vartheta$ | $A_\vartheta = \begin{Vmatrix} \cos\vartheta & \sin\vartheta & 0 \\ -\sin\vartheta & \cos\vartheta & 0 \\ 0 & 0 & 1 \end{Vmatrix}$ | $\mathbf{Q} = \cos\dfrac{\vartheta}{2} + \mathbf{k}\sin\dfrac{\vartheta}{2}$ |
| III | $\gamma$ | $A_\gamma = \begin{Vmatrix} 1 & 0 & 0 \\ 0 & \cos\gamma & \sin\gamma \\ 0 & -\sin\gamma & \cos\gamma \end{Vmatrix}$ | $\mathbf{R} = \cos\dfrac{\gamma}{2} + \mathbf{i}\sin\dfrac{\gamma}{2}$ |

$$\Lambda = \mathbf{P} \circ \mathbf{Q} \circ \mathbf{R} = \lambda_0 + \lambda_1 \mathbf{i} + \lambda_2 \mathbf{j} + \lambda_3 \mathbf{k} \tag{10.7}$$

where the Rodrigues-Hamilton parameters are defined as follows:

$$\begin{aligned}
\lambda_0 &= \cos\frac{\psi}{2}\cos\frac{\vartheta}{2}\cos\frac{\gamma}{2} - \sin\frac{\psi}{2}\sin\frac{\vartheta}{2}\sin\frac{\gamma}{2} \\
\lambda_1 &= \cos\frac{\psi}{2}\cos\frac{\vartheta}{2}\sin\frac{\gamma}{2} + \sin\frac{\psi}{2}\sin\frac{\vartheta}{2}\cos\frac{\gamma}{2} \\
\lambda_2 &= \sin\frac{\psi}{2}\cos\frac{\vartheta}{2}\cos\frac{\gamma}{2} + \cos\frac{\psi}{2}\sin\frac{\vartheta}{2}\sin\frac{\gamma}{2} \\
\lambda_3 &= \cos\frac{\psi}{2}\sin\frac{\vartheta}{2}\cos\frac{\gamma}{2} - \sin\frac{\psi}{2}\cos\frac{\vartheta}{2}\sin\frac{\gamma}{2}
\end{aligned} \tag{10.8}$$

The matrix of the direction cosines (10.6) recorded in Rodrigues-Hamilton parameters is as follows:

$$A = \begin{pmatrix} 2\lambda_0^2 + 2\lambda_1^2 - 1 & 2\lambda_1\lambda_2 + 2\lambda_0\lambda_3 & 2\lambda_1\lambda_3 - 2\lambda_0\lambda_2 \\ 2\lambda_1\lambda_2 - 2\lambda_0\lambda_3 & 2\lambda_0^2 + 2\lambda_2^2 - 1 & 2\lambda_2\lambda_3 + 2\lambda_0\lambda_1 \\ 2\lambda_1\lambda_3 + 2\lambda_0\lambda_2 & 2\lambda_2\lambda_3 - 2\lambda_0\lambda_1 & 2\lambda_0^2 + 2\lambda_3^2 - 1 \end{pmatrix} \tag{10.9}$$

The yaw, pitch, and roll angles are determined by the Rodrigues-Hamilton parameters as follows:

$$\begin{aligned}
\psi &= \operatorname{arctg}\left(-\frac{A_{13}}{A_{11}}\right) = \operatorname{arctg}\left(-\frac{2\lambda_1\lambda_3 - 2\lambda_0\lambda_2}{2\lambda_0^2 + 2\lambda_1^2 - 1}\right) \\
\vartheta &= \arcsin\left(A_{12}\right) = \arcsin\left(2\lambda_1\lambda_2 + 2\lambda_0\lambda_3\right) \\
\gamma &= \operatorname{arctg}\left(-\frac{A_{32}}{A_{22}}\right) = \operatorname{arctg}\left(-\frac{2\lambda_2\lambda_3 - 2\lambda_0\lambda_1}{2\lambda_0^2 + 2\lambda_1^2 - 1}\right)
\end{aligned} \tag{10.10}$$

where $A_{mn}$ are the elements of the matrix (10.9).

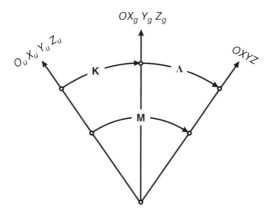

**Figure 10.11**  System transition scheme

To implement the algorithms of SINS, it is necessary to have the positional information (related to the coordinate system) relative to the normal at any given moment of time. This information can be obtained from the quaternion $\Lambda$ (10.7), which is an analog of the matrix of the direction cosines $A$ (10.6). While knowing $\Lambda$, the apparent acceleration of the UAV can be calculated, having been measured in the UAV-connected coordinate system, as well as in the normal coordinate system and being defined by the orientation parameters $\psi$, $\vartheta$, and $\gamma$ (10.10).

Figure 10.11 shows the transitions from the inertial coordinate system $OX_uY_uZ_u$ to the related $OXYZ$ and normal $OX_gY_gZ_g$ systems by means of quaternions $\mathbf{K}$, $\Lambda$, and the total transition through quaternion $\mathbf{M}$.

In accordance with Figure 10.11

$$\mathbf{M} = \mathbf{K} \circ \Lambda,$$

hence

$$\Lambda = \bar{\mathbf{K}} \circ \mathbf{M}, \tag{10.11}$$

where $\bar{\mathbf{K}}$-conjugate quaternion.

Differentiating the right and left side of (10.11) with respect to time:

$$\dot{\Lambda} = \dot{\bar{\mathbf{K}}} \circ \mathbf{M} + \mathbf{K} \circ \dot{\mathbf{M}}$$

which is converted to the form:

$$2\dot{\Lambda} = \Lambda \circ \Omega - \Omega_g \circ \Lambda, \tag{10.12}$$

where $\Omega, \Omega_g$ are hypercomplex displaying vectors of the absolute angular velocity of the normal as well as the related coordinate systems.

Equation 10.12 makes it possible to find quaternion $\Lambda$ and hence solve the problem of the orientation of the UAV relative to the normal coordinate system.

In the numerical solution of the Equation 10.12 there may be computational errors due to nonfulfillment of equality (of the so called standard quaternion):

$$\Lambda \circ \bar{\Lambda} = 1$$

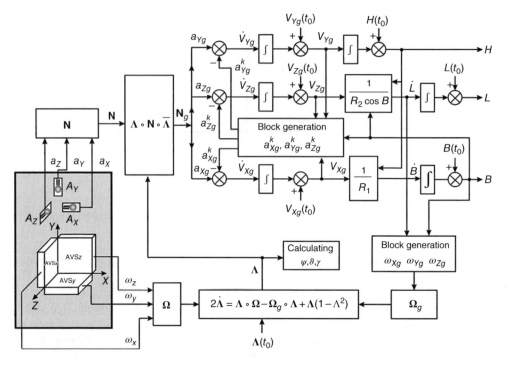

**Figure 10.12** Strapdown Inertial Navigation System (SINS) flowchart

For automatic quaternion normalization in place of Equation 10.12, it is necessary to solve that equation with a quaternion correction norm:

$$2\dot{\Lambda} = \Lambda \circ \Omega - \Omega_g \circ \Lambda + \Lambda\left(1 - \Lambda^2\right),$$

where $\Lambda^2 = \lambda_0^2 + \lambda_1^2 + \lambda_2^2 + \lambda_3^2$.

To convert the apparent acceleration from the coordinate system related to the normal use of equality redesign:

$$\mathbf{N}_g = \Lambda \circ \mathbf{N} \circ \overline{\Lambda}$$

where $\mathbf{N}$ represents the apparent acceleration vector, and $\mathbf{N}_g$ represents the vector in the normal coordinate system.

After the conversion of the orientation and navigation parameters into the normal coordinate system of the UAV, a navigation algorithm is implemented on the basis of (10.3) and (10.5). A flowchart for a continuous time SINS is shown in Figure 10.12.

A SINS flowchart represents the transformation of primary information about the apparent acceleration and angular velocity of the UAV into the coordinates of the center of mass and its angular position. Continuous SINS algorithms are closed; that is, to calculate the corresponding integrals and the transformation of variables from one coordinate system to another requires continuous time values for both the variables and their integrals and elements of the quaternions.

Since data processing via an onboard computer (microcontroller) is conducted in a discrete form, it requires approximations of continuous discrete signals, that is, the quantization of continuous signals in both time and level. To increase the accuracy of autonomous work, a strapdown orientation and navigation system is necessary to calibrate inertial sensors, which entails verifying various types of systematic error components.

## 10.4.3   Algorithms for Four-Dimensional (Terminal) Navigation

A terminal navigation system provides for a UAV flight from a point with coordinates $(x_0\ y_0)$ to a target (T) with coordinates $(x_T\ y_T)$ at a predetermined ground speed $Vgr$ in a definite time $t_f=t_T-t_0$. Frequently, flights take place within the action of large-scale wind flows such as the global jet streams, the average speeds of which are constant over considerable distances (units and tens of kilometers).

   Hence, it is possible to represent an SUAV flight trajectory in the form of a piecewise linear approximation, or Line Of Predetermined Path (LPP), in a terrestrial system of coordinates between the TPR, while taking into account a vector of air-disturbance speed (Figure 10.13).

   Navigational algorithms include an algorithm for an astatic Autothrottle (AT) to stabilize a given flight speed, an identification algorithm for large-scale wind disturbances, an algorithm of outputs to guide the UAV flight to the target in the set time, and an emergency mode "return" algorithm. These are as follows.

1. The Autothrottle Algorithm

Changes in a UAV ground speed $V_g$ with an autothrottle are described by a system of nonlinear differential equations:

$$\begin{cases} \dot{V}_g\left(t\right)=\dfrac{1}{m}\left(P-c_xS\dfrac{\rho V^2}{2}-G\sin\vartheta\right); \\[2mm] \dot{P}\left(t\right)=\dfrac{1}{T}\left[K\left(V_r,n_r\right)n_r-P\right], \end{cases} \tag{10.13}$$

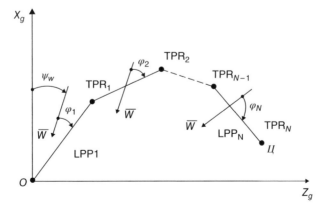

**Figure 10.13**   Piecewise linear approximation of a UAV flight trajectory

where $T$—engine time constant; $n$ and $n_r$—actual and required frequencies of rotation of the engine shaft, respectively; and $K$—functional communication parameters $V_r$, $n_r$ with traction $P$.

It is required to find equation for $n_r(t)$ that will describe the flight of a UAV with a given speed $Vgs(t)$ and with the limitation $V_g(t) - Vgr(t) = 0$, for minimizing the function:

$$I = \int_0^{t_K} \left[ V_g(t) - V_{gr}(t) \right]^2 dt. \tag{10.14}$$

In accordance with (10.13) the set points of traction $P_r$ and the rotational frequencies of the engine shaft $n_r(t)$ are defined by the following expressions:

$$\begin{cases} P_r(t) = \dfrac{m}{T_V} \Delta V + c_x Sq + G \sin \vartheta; \\[2mm] n_r(t) = \dfrac{1}{T_V T_n} \dfrac{T}{K(V_r, n_r)} m\Delta V + \dfrac{T}{T_n K(V_r, n_r)} c_x qS + \left(1 - \dfrac{T}{T_n}\right) n(t), \end{cases} \tag{10.15}$$

where $\Delta V = V_{gr} - V_g$; $T_V$, $T_n$—the time constants of the transition process on the UAV speed and frequencies of rotation of the engine shaft; $q = \rho V^2 / 2$.

For a UAV, including small-size $T_n \ll T_V$, maximum overshoots of speed and engine shaft rotation frequency should not exceed 10%.


## 2. The Wind Disturbance Identification Algorithm

A MM of the trajectory of a UAV center of mass during horizontal flight ($\omega_x = \omega_y = \omega_z = 0$) in the axis-related coordinate projection system, and under large-scale wind disturbances, is represented by the following system of equations, assuming that $\bar{W}_{yg} = 0$:

$$\dot{V}_{gx} = g(n_x - \sin \vartheta); \quad \dot{V}_{gy} = g(n_y - \cos \vartheta); \quad \dot{V}_{gz} = gn_z; \quad \dot{W}_{xg} = 0, \ \dot{W}_{zg} = 0, \tag{10.16}$$

where $n_x n_y n_z$—overload along respective axes (acceleration divided by the acceleration due to gravity).

The MM of the measuring system can be represented by the following system of equations for the instrument estimates of the velocity $V^{ins}$ and its components $V_{gx}^{ins}$ and $V_{gz}^{ins}$:

$$V_{gx}^n = V_{gx} + \Delta V_{gx}; \quad V_{gz}^n = V_{gz} + \Delta V_{gz};$$

$$V^n = \sqrt{\left(V_{gx} - W_{xg}\right)^2 + \left(V_{gz} - W_{xz}\right)^2} + \Delta V_V \tag{10.17}$$

where $\Delta V_i$—error in determining the respective speeds.

For a UAV, the identification of large-scale wind disturbances is generated as follows: its motion is described by (10.16) using measurements from (10.17), while considering the model of angular sensors $\alpha$ and $\beta$ (Figure 10.10b) and taking into account the estimation of the state vector:

$$X = \left( V_{gx}, V_{gy}, V_{gz}, W_{xg}, W_{zg} \right)^T.$$

On the basis of the Kalman filter, the equation for estimation appears in continuous form and has the following form (Stepanov, 2009; Valavanis and Vachtsevanos, 2015):

$$\hat{X}(t) = K(t) \left[ z(t) - h(\hat{x}) \right],$$

where $K(t)$ is the mathematical gain; $z(t) = \left[ V_{gx}^{ins}, V_{gy}^{ins}, V_{gz}^{ins}, V^{ins}, \alpha^{ins}, \beta^{ins} \right]^T$ is the measurement vector; and $h(\hat{x})$ is a prediction vector measurement obtained by known vector estimates.

In the presence of some number $(i)$ in the normal coordinate system, $(OXgYgZg)$, the expression for the sensor readings becomes:

$$V_{gxi}^{ins} = V_{gxi} + \Delta V_{gxi}; \quad V_{gyi}^{ins} = V_{gyi} + \Delta V_{gyi}; \quad V_{gzi}^{ins} = V_{gzi} + \Delta V_{gzi};$$

$$V_i^{ins} = \sqrt{\left( V_{gxi} - W_{xgi} \right)^2 + \left( V_{gyi} - W_{xyi} \right)^2 + \left( V_{gzi} - W_{xzi} \right)^2} + \Delta V_i;$$

$$\alpha_i^{ins} = \arcsin \left( -\frac{V_{gyi} - W_{yi}}{\sqrt{\left( V_{gxi} - W_{xi} \right)^2 + \left( V_{gyi} - W_{yi} \right)^2}} \right) + \Delta \alpha_i;$$

$$\beta_i^{ins} = \arcsin \left( \frac{V_{nzi} - W_{zi}}{\sqrt{\left( V_{gxi} - W_{xi} \right)^2 + \left( V_{gyi} - W_{yi} \right)^2 + \left( V_{gzi} - W_{zi} \right)^2}} \right) + \Delta \beta_i,$$

where $\Delta V_{gxi}, \Delta V_{gyi}, \Delta V_{gzi}, \Delta V_i, \Delta \alpha_i, \Delta \beta_i$ are in general the functional errors of measurement of the vector component representing discrete random processes.

In the simplest (or elementary) case, while assuming that the LPP is maintained—and using angles of roll and pitch close to zero—the model vector of speed and direction is defined by the following equation:

$$W = \sqrt{W_{xh}^2 + W_{zg}^2}, \quad \psi_W = \arctan \left( \frac{W_{zg}}{W_{xg}} \right).$$

3. The Algorithm Output Over the Route

Over a predetermined distance $L_p$ from a starting point to an end point and turning at $N$ points on the route (TPR$_i$, $i = 1, 2, \ldots, N$), a UAV will take a flight time $t_f$, whence the required average velocity of flight without taking into account the time of turns is equal to:

$$V_r = \frac{L_p}{t_f}. \tag{10.18}$$

The possible minimum ($t_{min}$) and maximum ($t_{max}$) time to the end point, assuming that the UAV is capable of accelerating and decelerating within the limits of $a_{max}^a, a_{min}^d$, are defined by following equations:

$$t_{min} = \frac{L_p}{V_{g\,max}} + \frac{V_{g\,max} - V_{gi}}{a_{max}^a};$$

$$t_{max} = \frac{L_p}{V_{l\,min}} + \frac{V_{g\,min} - V_{gi}}{a_{min}^d},$$

where $V_{gi}$, $V_{g\,max}$, $V_{g\,min}$ are the initial, maximum, and minimum ground velocities, respectively.

After identification of wind parameters $W_i$, $\psi_{Wi}$ for each TPR$_i$, the wind direction $\varphi_i$ (Figure 10.13) on each $i$-th flight section is known. The air speed on each section of the flight is $\bar{V} = \bar{W} + \bar{V}_g$; therefore, the demanded velocity for each section will be:

$$\bar{V}_d = \bar{W}_i + \bar{V}_r \tag{10.19}$$

where $W_i = |\bar{W}|\cos\phi_i$ is the projection vector of the wind speed at the LPP$_i$, the $i$-th section of the flight.

A function representing the current difference between the demanded route $L_p$ and the route actually taken by the UAV moving at a velocity $V_r$ over the remaining time to the end point appears as follows:

$$I = L_p - L - V_r\left(t_f - t\right), \tag{10.20}$$

where $L$ and $t$ are the current trip distance and time, respectively.

Hence, the algorithm output over the flight comprises:

- identification of wind parameters $W_i$, $\psi_{Wi}$, and $\varphi_i$ in real time for each $i$-th section of the flight;
- determination of the demanded average ground velocity $V_r$ using formula (10.18);
- calculation of the required air velocity for each section of the flight using formula (10.19);
- a feasibility check of the required air speeds:

$$V_{min} \leq V_d\left(t\right) \leq V_{max}$$

and in the case of unfeasibility, a correction in time $t_f$ must be made and followed by a return to step 2.

Determination of the set rotation frequency of the engine shaft via formulas (10.15) and (10.20) is accomplished by:

$$n_r^*\left(t\right) = \beta_0\beta_1\frac{T}{K\left(V_r,n_r\right)}m\left(V_{gr} - V_g\right) + \frac{\beta_0 T}{2K\left(V,n\right)}c_x S\rho V^2 + \left(1 - \beta_0 T\right)n\left(t\right) + \Delta n_T, \tag{10.21}$$

where $\beta_0 = 1/T_n$, $\beta_1 = 1/T_V$, $\Delta n_T = K(V,n)I$; and $K(V,n)$ is the coefficient of recalculation for the functionality of $I$ (10.20) in an increment of shaft rotational frequency.

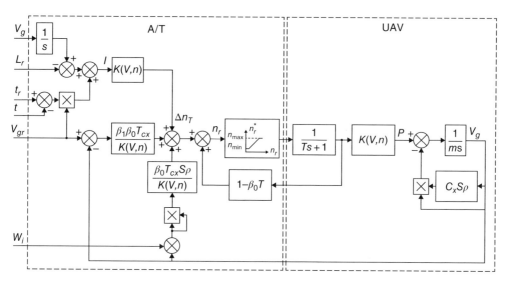

**Figure 10.14**   Block diagram for a UAV-AT (UAV Autothrottle) system for the terminal navigation mode

The set rotational frequency of the power unit is restricted as follows:

$$n_{\min} \le n_r^* \le n_{\max}$$

Given the assumption that $\vartheta = 0$, it is possible to write the differential Equation (10.13) in an operator form to give expressions for the UAV traction and ground speed:

$$P = \frac{K(V,n)n_r}{Tp+1}; \quad V_g = \frac{1}{mp}\left(P - \frac{CxS\rho V^2}{2}\right) \tag{10.22}$$

where $p$ is a differentiation operator.

Using (10.21) and (10.22), a block diagram of a UAV autothrottle system can be drawn as in Figure 10.14 for mode terminal navigation (the UAV performance at a given point at a set time and at a set speed).

Figures 10.15 and 10.16 show the results of the mathematical modeling of a UAV-A/T system for a flight through four control points ($TPR_1$ ... $TPR_4$) and with a return to the start point ($TPR_1$). The total length of a route $L_p = 3200\,m$, the set time-of-flight (ToF) $t_f = 140\,s$, and the ground speed of arrival $Vg_T \approx 23\,m/s$ ($82\,km/h$). The wind speed in the western direction $W_{west} = 2\,m/s$, and the wind speed in the northern direction $W_{north} = 2\,m/s$.

## 10.5   Examples of Construction and Technical Characteristics of the Onboard Avionic Control Equipment

At the heart of any ACS lies a sequence of actions: measuring the state of the system, comparing the current state with that desired, and the subsequent generation of control actuation for compensating any deviation of the current state from the desired one. The determining factor in that sequence is the measurement of the current state, and this function is implemented by means of an Integrated Navigation Complex that consists essentially of several navigation

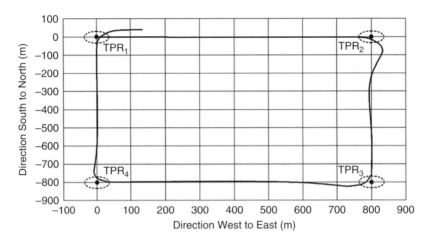

**Figure 10.15** Flight trajectory of a UAV in terminal navigation mode taking into account the influence of wind conditions

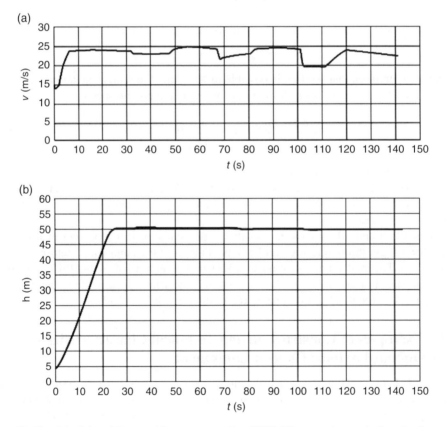

**Figure 10.16** Schedules of the transition processes for a UAV-AT system in terminal navigation mode: (a) ground speed, (b) height, (c), (d), and (e) angles for course, pitch, and roll, respectively

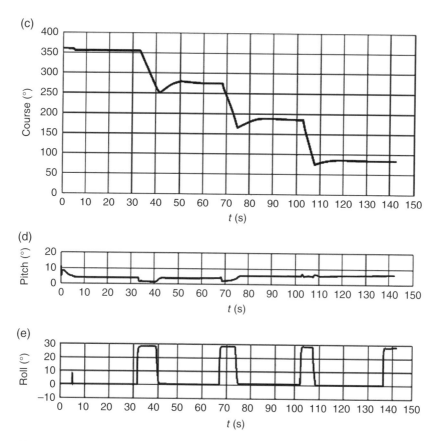

**Figure 10.16** (Continued)

systems (inertial, magnetometric, satellite, etc.), each of which has its own principle of operation and characteristic information renewal rate. A "master filter" chooses the best condition received from any of the systems, depending on the current state of motion (Alaluev, Ladonkin, Malyutin *et al.*, 2011; Veremeenko *et al.*, 2009b).

The principle used in the LLC "TekNol®" product for assessing the state of the SUAV INS and for setting the autopilot coefficients for specific navigation and control flight modes is illustrated by the functional diagram of Figure 10.17 (Lozano, 2010; Stepanov, 2009; Voronov, 2006).

The onboard complex is a full-function navigation system for a UAV configuration controller that performs the following tasks:

1. Determination of navigational parameters, orientation angles, and UAV motion parameters (angular rates and velocities, and accelerations)
2. Navigation and management of the UAV in flight along a predetermined path
3. Stabilization of the orientation angles of the UAV in flight
4. The generation of readouts for the telemetry transmission channel carrying data on navigation parameters and UAV orientation angles and
5. Programmable control taking into account the payload.

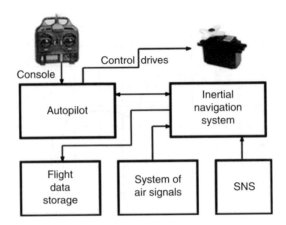

**Figure 10.17**   A functional diagram of the LLC TekNol avionics complex for SUAV navigation and control

The complex consists of three major modules: a receiver for a SNS/GPS, an INS, and an Autopilot (AP).

The chosen GPS module for the SNS is the Trimble® Lassen® iQ, which determines the coordinates and transmits the data to the INS. This INS contains sensors (accelerometers and gyroscopic sensors for angle rates and angular velocities) and is adjusted according to the SNS data and built-in barometric altimeter. It then measures the navigation and pilotage parameters of the UAV motion and transfers these to the autopilot module.

When the AP receives data from the INS, it produces control commands on the basis of predetermined control laws and sends them in the form of PWM signals to the UAV control-surface actuators/servo units. The AP module is also compatible with a manual control system that can be switched on or off by command. For this purpose, the radio channel uses Pulse-Code Modulation (PCM) via a standard remote control center.

In its basic configuration, the flight management is carried out through channels relevant to ailerons for bank, elevators for pitch, and hence altitude control; a rudder for directional control; and an engine controller.

In one embodiment, the UAV "Dozor" developed by the Design Bureau (DB) "Transas" JSC of St. Petersburg was installed with the "TekNol"® system, though later replaced with an ACS of its own development. The accuracy of the UAV flight parameters provided by "TekNol" LLC avionics is shown in Table 10.4. Here, the parameters are specified as values of the mean-square error/integrated square error ($1\sigma$) determined in comparison with data from standard instruments based on the results of flight tests.

In addition to "TekNol" LLC, JSC company RDC Rissa of Moscow has positioned itself as a developer of avionics and a mass-producer of the STA-32® ACS for UAV applications. It is capable of adaptation by the user for particular types of UAV (Topekhin, 2006).

The second and third Moscow International Forums and Exhibitions entitled "Unmanned Multipurpose Complexes/Vehicle systems" (2008/2009) included presentations on the "Ptero"® LLC systems for monitoring and diagnostics, which are capable of a wide range of applications (Valiev, 2007). Other UAV facilities include an airborne avionics complex for the Canadian Micropilot autopilot, which comprises GPS-receivers, radio modems,

**Table 10.4**   Accuracy of the UAV flight parameters provided by the Integrated Navigation Complex

|  | Integrated navigation complex mode | Autonomous inertial mode |
| --- | --- | --- |
| Coordinates (integrated solution) | 6 m | 500 m (5 min after loss of GPS) |
| Altitude | 2 m | 6 m |
| Ground speed | 0.2 m/s | 5 m/s (5 min after loss of GPS) |
| Vertical speed | 0.25 m/s | 0.3 m/s |
| Orientation angles (roll, pitch) |  |  |
| Straight flight[a] | 0.2°...0.3° | 0.3°...0.4° (indefinite time) |
| Maneuvering[b] | 0.3°...0.5° | 0.5°...0.7° (indefinite time) |
| Highly maneuverable flight[c] | 1.3° | 1.5° (indefinite time) |
| Course/heading (course angle)[d] | 0.4° | 3.0° (5 min after loss of GPS) |

[a] Straight flight is determined as the absence of intentional maneuvers for heading/course, roll, and pitch.
[b] Roll and pitch angles do not exceed 45°.
[c] Roll and pitch angles are in the range from 45 to 75°.
[d] The autonomous/standalone (without SNS) definition of heading is possible when conducting compensation for magnetic deviation.

microcontrollers for servo control, parachute release, payload and power equipment that control the electric propulsion system.

At the heart of the Ptero-E® (an upgraded version of the Ptero®) UAV control system lies the principle of measuring relative altitude by Topcon® precision GPS receivers. The UAV has four antennas: one at the center of mass, one in the nose, and two on the wingtips (Figure 10.18). By measuring the height differences between the four antennas it is possible to obtain the roll and pitch angles at any UAV attitude. Moreover, GPS makes it possible to determine the coordinates of the UAV irrespective of attitude, altitude above sea level, ground velocity, and heading angle.

The error in determining the heading angle does not accumulate over time and does not restrict the flight time of the UAV. Because the measurements of the angles of roll and pitch are relative to signals from the same types of antennas and the same types of receivers, the accuracy of angle determination does not depend on the GPS signal over a given time or its reception level. Thus, the determination error of the coordinates common to all GPS receivers has no effect on the determination accuracy of the roll or pitch angles. However, not all the receivers are suitable for use in the airborne avionics of the Ptero-E® model.

A disadvantage of this type of avionics is its dependence on channel licenses granted by foreign states. Obtaining licenses for precision channels is controlled by the US DoD, in addition to which, the cost of four precision GPS receivers exceeds the cost of even classical avionics working on the basis of inertial elements.

Foreign developers offer quite a wide range of autopilots. One of the most popular is the Kestrel V2.22® autopilot as installed on the Kestrel "Rabbit" 3000 Series® microcontroller; and the Micropilot MP2128®, both shown in Figure 10.19.

The Kestrel V2.22® autopilot contains three accelerometers, three gyroscopic angular velocity sensors, a three-axis magnetometer, and a GPS decoder. Stable operation of the

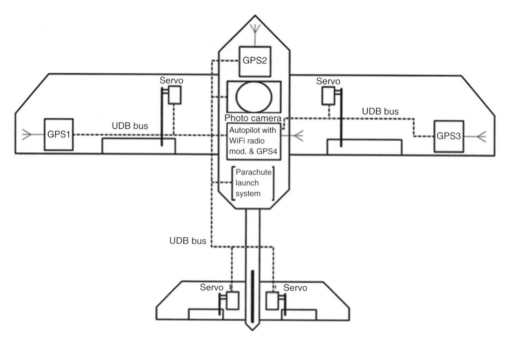

**Figure 10.18**   Layout of airborne avionics using a "Ptero"® UAV control system

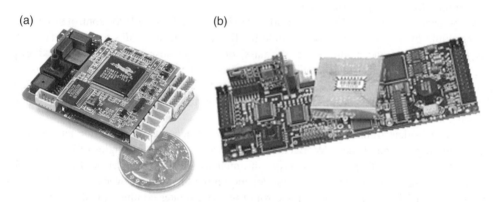

**Figure 10.19**   Autopilots: (a) Kestrel V2.22® and (b) MP2128®

module is ensured by a built-in 20-level temperature compensator and a switching voltage regulator (3.3 and 5V; 500 mA). The processor operates at a frequency of 29 MHz and has 512 kB of RAM. There are four serial ports, four built-in and eight external servo controllers. There are three analog 12-bit inputs and 12 digital channels for information exchange (six bidirectional, three inputs, and three outputs). All this provides for the control of voltage and current, self-tuning, duplication of the data transmission system with error correction, stabilization, and video camera control.

The MP2128® autopilot contains three-axis MEMS gyroscopes and accelerometers, a barometric altitude sensor, an air speed sensor, a module for GPS satellite navigation, and a

**Figure 10.20**   General view of the UAV Navigation autopilot AP04®

system for collecting and storing data. The computing core is a RISC-microprocessor capable of performing 150 Million Operations Per Second (MOPS). Whereas the physical size of the module is only $10 \times 4$ cm, its mass is only 28 g. The MP2128® provides full control of the UAV from the moment of takeoff to landing.

The UAV Navigation AP04 autopilot can be used on a SUAV with a wingspan less than 1 m and on a medium size UAV with a wingspan up to 4 m, as well as on rotary UAVs. The AP contains a Strapdown Attitude Reference System (SARS) (Figure 10.20).

The AP04® is a fully integrated autopilot offering optional manual UAV control via a built-in radio channel and payload control. It provides automatic takeoff, flight over a specified route and automatic landing. It can also be configured during the flight—the built-in radio channel allows data transfer up to a distance of about 100 km.

Automatic flight over a route is provided in the form of control points having three dimensions (latitude, longitude, and altitude), and it can be used for UAVs having the following ranges of air speed (km/h): lower: 25–150, normal: 35–250, and upper: 45–450.

The manual operation system allows a ground-based operator to take full control of the UAV using a standard joystick or equipment intended for model airplane control. The AP04 also contains a redundant microprocessor that ensures a high level of safety and resiliency.

Further features of the AP04® include payload control of a parachute and a gyro-stabilized video camera. An RS232 port can be used to communicate with a ground station, and an interface allows engine control and for tracking the parameters of its operation, which can also be transmitted to that ground station. In all, the autopilot allows control of up to 16 servo units or other peripherals.

## 10.6   Small-Sized Unmanned WIG and Amphibious UAVs

Wing-in-ground-effect, or WIG, vehicles include manned Ekranoplanes for application over water surfaces, but here, only unmanned aerial vehicles—WIG UAVs—are considered, along with unmanned surface vehicles (WIG USVs).

## 10.6.1  Emerging Trends in the Development of Unmanned WIG UAVs and USVs, and Amphibious UAVs

Significant developments have already taken place on UAVs and USVs, and these have paved the way for developing unmanned WIG versions that integrate many properties, characteristics, and functions of both normal UAVs and USVs. The relevant tasks are in many cases simpler and more advantageous than those for standard UAVs (Nebylov, 2013a, 2013b; Nebylov and Nebylov, 2014, 2015; Nebylov and Wilson, 2002).

A few types of conventional UAVs are able to perform very low-altitude flight close to water surfaces, but stabilization and control at extremely low altitudes are extremely complex tasks. However, WIG vehicles are specially intended for performing such low-altitude motion and if appropriately designed, stability is considerably enhanced by the dynamic air cushion near the surface. There are also many other important advantages, the most significant being the ability to integrate many properties, characteristics and functions of both marine craft and aircraft in a single transport vehicle.

The WIG is an interesting and peculiar physical phenomenon with multilateral characteristics, having both positive and negative influences on flight in "WIG-mode." Small-sized unmanned WIG vehicles and unmanned amphibious vehicles face peculiar challenges in their control systems and navigation complexes compared to standard small-sized UAVs.

In order to make full use of the WIG-effect and to provide good functional characteristics for amphibious platforms as transport vehicles, the following features usually distinguish them from conventional UAVs:

- They use a small aspect ratio wing that is attached relatively low on the body or the "flying-wing" configuration
- Boundary plates on wing tips are used for enhancing the wing aerodynamics when moving close to the supporting surface, often doubling as float plates
- The tail assembly consists of a high fin (or fins) with the rudder, the horizontal stabilizer, and the elevator attached as high as possible and
- They incorporate special equipment for expediting takeoff and landing from water.

Trouble-free motion close to an underlying surface can be ensured only by:

- Providing precise control of the altitude above a surface with errors of not more than 3–10 cm
- Ensuring the vehicle's stability in spite of transient nonlinear aerodynamic effects due to its close proximity to the surface and
- Providing noncontact measurement, tracking, and estimation of coordinates and biases of sea waves for improving motion control efficiency (Figures 10.21 and 10.22).

Autonomy technology that will become important in the development of Unmanned Small-Sized WIG vehicles will include (Nebylov et al., 2014) the following:

- Developments in sensor systems for combining information from different onboard sensors.
- The handling of communication and coordination between multiple agents in the presence of incomplete and imperfect information.

**Figure 10.21**  Small-sized unmanned amphibious vehicle prototype (under test at the St. Petersburg State University of Aerospace Instrumentation)

- Motion planning (also called "path planning") for determining optimal flight paths while meeting certain objectives and constraints such as obstacles.
- Trajectory generation for determining the optimal control maneuvers for following given paths between locations.
- Task allocation and scheduling for determining the optimal distribution of tasks among a group of agents under time and equipment constraints.
- Cooperative tactics formulation for the optimal sequencing and spatial distribution of activities between agents in order to maximize the chance of success for any given mission scenario.

The first requirement of physical autonomy is the ability of an unmanned vehicle to take care of itself without causing a threat to its environment and while still trying to fulfill its mission success rates. According to a study carried out in International Institute for Advanced Aerospace Technologies, St. Petersburg (IIAAT), the following tasks are important for implementation by an embedded software for unmanned Ekranoplanes.

- Embedded electronics real-time solutions
- Development of an advanced system configurable for both the WIG-mode and the free flight mode
- Provision of waypoint navigation via GLONASS, GPS, D-GPS, and INS
- Development of a fail-safe navigation mode
- Considerations relevant to dynamic real-time gains, limits, and adjustments
- Provision of dual extended Kalman filtering for precision navigation
- Capability for compressed digital image transfer

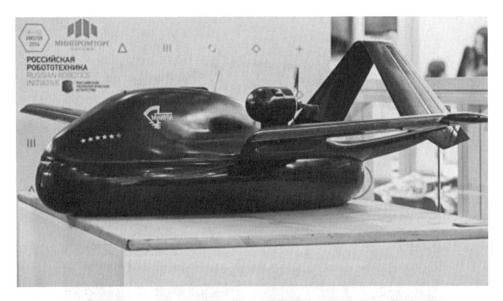

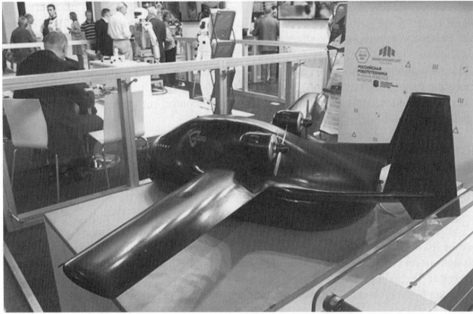

**Figure 10.22**   The Russian "Chirok" amphibious UAV model on public display at the Innoprom International Technology Exhibition, 2014

- Capability for autonomous takeoff and landing from land, water, or ice surfaces and autonomous flight in different modes including floating on water surfaces
- A return-home mode in the event of communication loss, GLONASS/GPS signal loss, etc.
- A communication relay mode
- Customizable software filters for sensor noise reduction

- Automatic target coordinate detection via INS
- Fault-tolerant embedded software and
- Special programs for supporting cooperative operations by many unmanned Ekranoplanes.

Computational Intelligence required the following:

- Neural networks
- Fuzzy systems
- Evolutionary computation and
- Hybrid intelligent systems.

Special requirements:

In order to fulfill operations and missions successfully, the following points must be given proper attention, and the allocation of systems such as algorithms and programs must be carefully considered by taking into account the various modes of motion. These considerations include the following:

- Robust and optimal control methods and algorithms
- Special automatic control methods and flight parameter monitoring systems
- Artificial intelligence
- Preprogrammed flight control
- The use of GLONASS and GPS SNS for navigation
- Limited ground station control for special modes and missions
- Asynchronous learning environments
- Network-centric operations
- Robust acquisition of relocatable targets
- Smart antennas
- Antenna arrays
- Beam forming
- Integrated systems and
- Space-time adaptive processing.

Control systems for unmanned WIG vehicles, including unmanned amphibious seaplanes flying in ground effect for long periods, have inherent peculiarities compared with those for normal UAVs. These include a marked nonlinear dependence on all aerodynamic coefficients, the character of their correlation with a relatively low flight altitude $h_r$, and the effect of wave disturbances (Nebylov et al., 2015).

When flying relatively high above the supporting surface, a WIG vehicle or Ekranoplane can, like an airplane, has longitudinal stability only if its center of gravity (CG) is ahead of its aerodynamic center. Given correct CG positioning, the aerodynamic center in airplane flight (which depends slightly on the angle-of-attack) fulfills this requirement within certain margins.

Within the supporting surface action zone, the longitudinal stability can be disturbed because the aerodynamic force depends not only on the angle-of-attack but also on the altitude. Furthermore, the aerodynamic center position may vary depending upon several factors under the influence of the supporting surface. When the altitude decreases, the aerodynamic center moves backward due to pressure increment at the back edge area of the wing under positive angles of attack, and moves forward under zero and negative angles of attack.

During pitch angle variation, the drag and hence the flight speed changes, so the presence of a velocity stabilization system is essential. Therefore, all the channels in the control complex must participate substantially in the maintenance of the demanded motion of the vehicle in the longitudinal plane. The synthesis of control laws can be carried out using several criteria, but the general structure of these laws turns out to be quite similar in the majority of cases. Estimations of the vehicle stabilization errors, linear as well as angular rates and the effects of wave disturbances, must be accurately filtered.

Effective attempts for solving the problem of accurately measuring low altitudes have been made at IIAAT SUAI, St. Petersburg by developing phase Radio Altimeters (RA) specially designed for WIG vehicles (Nebylov, 2013a, 2013b). A complete flight parameter measuring system has been designed for controlling an experimental WIG vehicle and recording its flight parameters under the following conditions:

- Flight altitudes up to 5 m with 5 cm accuracy
- Speeds up to 180 m/s with an accuracy of 0.1–0.2 m/s
- Roll and pitch angle measurement with 0.1–0.2° accuracy and
- Vertical overloads up to 3 g with 0.06 g accuracy.

The primary sensors included three special radio altimeters, a vertical reference system, and a multiantenna DGPS receiver. These were integrated with the aim of improving accuracy and providing fault-tolerance properties.

### 10.6.2   Radio Altimeter and Inertial Sensor Integration

The dependence of functional properties on the altitude in the range equal to units of meters is practically absent. So, the most effective method for low-altitude flight parameter measurement (for small-sized unmanned WIG vehicles or unmanned amphibious craft) is the use of an active radar, which is integrated along with inertial sensors for geometrical altitude and vertical acceleration. The most important advantages conferred by this method are its comparative simplicity and the availability of appropriate instruments. The output signal of a radio altimeter installed on a vehicle flying at low altitude above a strongly disturbed sea surface is:

$$x_{LV}(t) = h(t) + \sigma_{LV}(t) = h(t) + y(t) + \Delta h(t), \tag{10.23}$$

where $h(t)$ is the true flight altitude relative to the average level of the disturbed sea surface, $\sigma_{LV}(t)$ is the resulting error $h(t)$, $y(t)$ is the wave profile ordinate for the point situated in the center of the irradiated area, and $\Delta h(t)$ is the location error $h(t) + y(t)$ of the altimeter.

When the vehicle itself undergoes vertical motion, this can be measured by an accelerometer with a vertical sensitivity axis, which is:

$$\delta(t) = \delta_1(t) + \delta_2(t) + \delta_3(t), \tag{10.24}$$

where $\delta_1(t)$ is an error caused by sensor gain factor while deviating from its nominal value, $\delta_2(t)$ is a slow shift in the sensor scale zero, and $\delta_3(t)$ is an error caused by inaccurate stabilization of the sensor's vertical axis sensitivity.

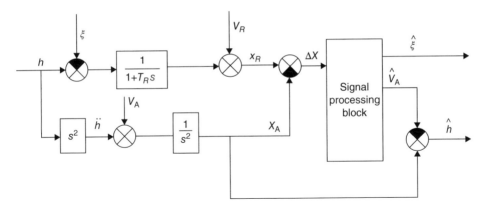

**Figure 10.23**   Block diagram for an integrated radio-inertial system

An increase in the effectiveness of the integrated measuring system may be achieved by the technique of separating one sensor error from the background error of another while considering their frequency differences. The separated error can then be subtracted from the overall output signal to compensate this for an error.

A generalized block diagram for a radio-inertial measuring system is shown in Figure 10.23.

Improvements in such a measurement scheme can be made by using current information about sea wave movement at several points under the vehicle. To do this, the measurement system requires a multichannel design involving angular position estimation on the basis of additional information processing using sensors situated at various points on the vehicle.

An accelerometer with vertical sensitivity provides the basic information for low-altitude flight control, and three radio-altimeters and accelerometers, spatially distributed along the vehicle body, make possible good estimations of altitude and angular acceleration (Obviously, the minimum number of radio-altimeter/accelerometer sensor pairs necessary for roll and pitch angle measurement is three).

An experimental installation was developed at IIAAT SUAI for assisting in the design of a dedicated 3-D ToF technique using PMD® imaging sensors for applications in motion control systems. This 3-D ToF technique is particularly appropriate for seaplane landing assistance under harsh conditions and has exhibited very promising results for installation of WIG vehicles. Methods for determining spatial position using vision-based methods for vehicles above water surfaces have also been developed.

A generated 3-D image is a matrix whose size is dependent on the containment chamber dimensions. The output level of each element in the matrix is dependent on its distance from the corresponding site on the surface. Usually, ranges are measured in radial coordinates from the center of the camera but can be converted to the Cartesian form if so required.

The described features of the 3-D ToF camera for the measurement of distance above a water surface are advantageous because the camera automatically measures distance close to a perpendicular, that is, the shortest distance. However, because the water surface is usually disturbed, the reflected signal comes back from a comparatively wide area. Each pixel in the resulting area at the camera has an individuality that can provide an estimate of the distance above the corresponding site on the water surface. Accordingly, it is necessary to process the 3-D images with the purpose of providing an estimate of the shortest perpendicular distance.

The results obtained using 3-D ToF PMD cameras for height measurement above a water surface confirm this as an alternative and unconventional method, which is especially effective for low—and extremely low—altitude motion above a sea surface. The reflected signal has a nontrivial form but does offer a method for providing stable height estimation. In particular, experiments have shown that at heights of about 4.5 m above a water surface, the r.m.s. measurement error was around 0.05 m. Further experimentation and research have indicated that the technology has the potential for measuring not only true height but also the angular orientation of the vehicle and the characteristics of the underlying sea surface disturbances during close motion close to it.

The use of specially designed radio-altimeters in addition to the 3-D ToF technique provides greater reliability and efficiency for automatic control systems along with the capability of supporting low-altitude flight for WIG vehicles. It is also a valuable tool for assisting the landing of amphibious UAVs during wavy surface conditions.

## 10.6.3   Development of Control Systems for Unmanned WIG Aircraft and Amphibious UAVs

Algorithm structures for the complex filtration of radio-altimeter measurements and vertical accelerometer outputs are installed for the measurement of height and also for pitch and bank angles.

An algorithm block diagram for the filtration of radio altimeter and vertical accelerometer measurements (Figure 10.24) includes a block for the recalculation of radio-altimeter measurements from the point of inertial unit (IU) installation, and also the recalculation of the altitude estimate from the point of installation of an IU at the CG. In the diagram, Filter 1 is for altitude and Filter 2 is for vertical accelerations.

An integrated system for the measurement of motion parameters close to a sea surface can be built using three radio-altimeters and a compact INS. This involves three angle-rate sensors and three linear accelerometers along with a calculator and temperature transmitter for the compensation of temperature drift in them all.

In the diagram, Unit 1 is for the recalculation of altimeter outputs relevant to the point of IU installation. Unit 2 is for the recalculation of altitude estimates at the CG and also at the points of altimeter installation.

This measurement system allows for the tracking of sea wave profiles $\xi_n$, $\xi_l$, $\xi_r$ at three points corresponding to the points of radio-altimeter installation at the nose and both sides of the vehicle and with an accuracy of 10 cm at Seaway #4. This is important for optimizing a mode of approach for landing and splashdown. In addition, the problem of automatically estimating the general direction of sea waves has been solved, this being important for the optimization of landing on water. The problem of automatically estimating the general direction of sea wave propagation with the use of three radio-altimeter outputs must also be highlighted (Figure 10.25).

Recalculation of the altimeter outputs relevant to the CG may be executed according to:

$$h_{GC\_K} = h_k - x_k \psi + z_k \theta - y_k,$$

where the index $k=n, l, r$ ($n$—nose altimeter, $l$—left-side altimeter, and $r$—right-side altimeter).

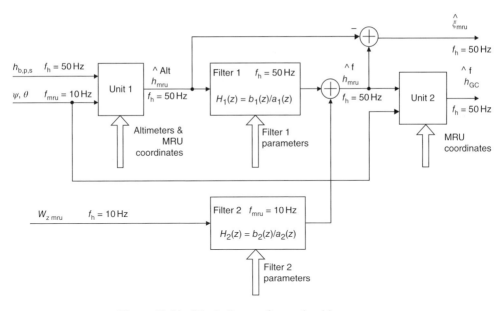

**Figure 10.24**　Block diagram for an algorithm structure

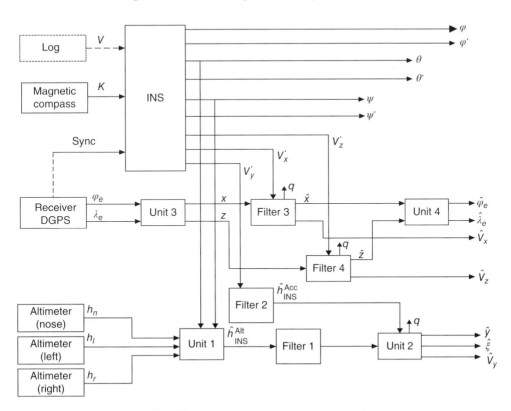

**Figure 10.25**　Block diagram of the integrated measuring system

For recalculation of altitudes from the CG to the point of INS installation, the relationship:

$$\hat{h}_{INS}^{Ail} = med\left(h_{GC\_k} + x_{INS}\psi - z_{INS}\theta + y_{INS}\right)$$

is used, where med (.) is the median definition operation.

The formula for recalculation of the filtered value of an altitude from the point of IU installation to the CG (Unit 2) is:

$$\hat{h}_{GC}^{f} = \hat{h}_{INS}^{f} - x_{INS}\psi + z_{INS}\theta - y_{INS}.$$

The altitude and vertical acceleration filters have the following transfer functions:

$$H_1(s) = \frac{\tau^2 s^2 + 2\tau k_3 s + k_3}{s^3 + \tau^2 k_3 s^2 + 2\tau k_3 s + k_3}$$

$$H_2(s) = \frac{s}{s^3 + \tau^2 k_3 s^2 + 2\tau k_3 s + k_3}$$

where $k_3 = 0.035s^{-3}$ and $\tau = 1.32 / \sqrt[3]{k_3} = 4.035s$.

In discrete time, the structure of the filters is described by the formulas:

$$H_1(z) = \frac{b_2^1 z^2 + b_1^1 z + b_0^1}{z^3 + a_2^1 z^2 + a_1^1 z + a_0^1}$$

$$H_2(z) = \frac{b_2^2 z^2 + b_1^2 z + b_0^2}{z^3 + a_2^1 z^2 + a_1^1 z + a_0^1}$$

## 10.6.4 The Design of High-precision Instruments and Sensor Integration for the Measurement of Low Altitudes

Specially designed and constructed for low-altitude measurement, isotope altimeters use the principle of measuring $\gamma$-beam radiation reflected from an underlying surface (Nebylov, 2013a). They have the great advantage of constructional simplicity but the intensity of the measured $\gamma$-radiation depends not only on a vehicle's altitude but also on the slope steepness of the underlying surface, which leads to errors. For the same reason there will be further errors when used above any rough surface. Isotope altimeters exhibit many other disadvantages, too, whereas for radio altimeters the information parameter of a radio signal is not energy, so there is a tremendous opportunity for increasing measurement accuracy via stabilized directional antennas.

In principle, high-altitude measurement accuracy can also be achieved by the laser altimeter via its capability for producing a very narrow beam. However, its main disadvantage is unreliability when working in difficult meteorological conditions, such as fog. Furthermore, modern laser altimeters usually involve pulse operation modes in order to realize powerful beams with high-quality modulation, and the effective division of direct and return beams presents technical difficulties that limit the minimal altitudes, which can be measured.

Methods for algorithm synthesis for the processing of signals from several radio altimeters, accelerometers, vertical gyros, and GPS receivers have been developed at IIAAT SUAI (St. Petersburg). For the estimation of critical parameters for low-altitude flight above the sea and for the characteristics of wave disturbance, synthesis using the Kalman and robust filtration has been developed. These ensure good estimation quality even in the presence of incomplete *a priori* information about errors in the primary sensors, along with allowances for diversity in the vehicle motion modes.

Such measuring systems allow the tracking of sea wave profiles $\xi_n$, $\xi_l$, $\xi_r$ at three points corresponding to radio altimeter installation at the nose and at the left and right sides of the vehicle, respectively, with accuracies of 10 cm at sea state number 4.

The problem of automatic estimation of the general direction of sea wave distribution with the use of three radio altimeter outputs is still moot at the time of writing, and this is important for optimization of the landing approach mode and splashdown for amphibious seaplanes and WIG aircraft.

# References

Alaluev P.V., Ladonkin A.V., Malyutin D.M. *et al. Microsystems for Orientation of UAVs*. Edited by: Raspopov V.Y.: Mashinostroenie, Moscow, 2011, 180 p. (in Russian).

Austin Reg. *Unmanned Air Systems: UAV Design, Development and Deployment*. Wiley, Hoboken, NJ, 2010, 372 p.

Beard R.W., McLain T.W. *Small Unmanned Aircraft: Theory and Practice*. Princeton University Press, Princeton, NJ, 2012, 300 p.

Bortz J.E. A new mathematical formulation for strapdown inertial navigation, *IEEE Aerospace and Electronic Systems Magazine*, 1971, AES-7(1), 61–66.

Branets V.N., Shmigleevskiy I.P. *Application of Quaternion in Problems of Solid Body Orientation*. Nauka, New York, 1973, 320 pp. (in Russian).

Gundlach J. *Designing Unmanned Aircraft Systems: A Comprehensive Approach (AIAA Education Series)*, American Institute of Aeronautics & Astronautics, 2011, 805 p.

Kuzovkov N.T., Salichev O.S. *Inertial Navigation and Optimal Filtrations*. Mashinostroenie, Moscow, 1977, 144 p. (in Russian).

Lozano R. (Ed.) *Unmanned Aerial Vehicles: Embedded Control*. Wiley-ISTE, 2010, 352 p.

Matveev V.V., Raspopov V.Y. *Fundamental Principles for Design of Non-Platform Inertial Navigation Systems*. Edited by: Raspopov V.Y.: CSRI Elektropribor, St. Petersburg, 2009, 300 p. (in Russian).

Nebylov A.V. (Ed.) *Aerospace Sensors*. Momentum Press, New York, 2013a, 350 p.

Nebylov A.V. WIG-Craft Flight Control Systems Development. Fifth European Conference for Aeronautics and Space Sciences (EUCASS), Munich, Germany, July 1–5, 2013b, pp. 1–6.

Nebylov A., Nebylov V., Fabre P. WIG-Craft Flight Control above the Waved Sea. ACNAAV 2015: Workshop on Advanced Control and Navigation for Autonomous Aerospace Vehicles, Seville, Spain, June 10–12, 2015, pp. 102–107.

Nebylov A.V., Nebylov V.A. Wing-in-Ground Effect Vehicles Flight Automatic Control Systems Development Problems. AEROTECH V: Progressive Aerospace Research, Applied Mechanics and Materials, vol. 629. Trans Tech Publications Inc., Pfaffikon, October 2014, pp. 370–375. Available at: http://www.ttp.net/978-3-03835-232-7/6.html (accessed on January 22, 2016).

Nebylov A.V., Nebylov V.A. WIG–Craft Motion Control Concept. Proceedings of the Third CEAS Euro GNC 2015 Conference, Tu/PMB/Tr.1. Toulouse, France, April 13–15, 2015.

Nebylov A., Nebylov V., Sharan S. Development of New-Generation Automatic Control Systems for Wing-in-Ground Effect Crafts & Amphibious Seaplanes. Advances in Control and Optimization of Dynamical Systems, Indian Institute of Technology, Kanpur, India, March 13–16, 2014, pp. 1–8.

Nebylov A., Sukrit S. Perspectives for Development of Autonomous & Intelligent WIG Crafts and its Peculiar Control Problems. IFAC "AGNFC" Workshop Proceedings, Samara, Russia, June 30–July 2, 2009, pp. 1–6.

Nebylov A., Sharan S., Rumyantseva E. Comparative Analysis of Design Variants for Low Altitude Parameters Measuring System. Automatic Control in Aerospace-IFAC Symposium Proceedings, Toulouse, France, June 25–29, 2007, p. 46.

Nebylov A., Wilson P. *Controlled Flight Close to Surface*. WIT Press, Southampton, 2002, 312 p.

Raspopov V.Y. *Micro-System Avionics*. Tula: Grif K, 2010. 247 pp. (in Russian).

Savage P. G. Coning algorithm design by explicit frequency shaping. *Journal of Guidance Control and Dynamics*, 2010, 33(4), 1123–1132.

Stepanov O.A. *Basic Theory of Estimation for Application to Problems Concerning Navigation Information Processing*. St. Petersburg: CSRI Elektropribor, 2009, 496 pp. (in Russian).

Topekhin A. UAV systems "NTC Rissa". *Aerospace Kurier*, 2006, 6(48), 67. (in Russian).

Telukhin S.V., Raspopov V.Y. Determination of aerodynamic coefficients of UAV, *"Vestnik" Journal of Computer and IT*, 2010, 2, 17–22 (in Russian).

Valavanis K.P., Vachtsevanos G.J. *Handbook of Unmanned Aerial Vehicles*. Springer Publication, Dordrecht, 2015, 3022 p.

Valiev A. "Ptero" system distance control and diagnostics, *Aerospace Kurier*, 2007, 5(53), 42–43 (in Russian).

Veremeenko K.K. *et al. Guidance and Control of Unmanned Aerial Vehicles on the Basis of Contemporary Information Technologies*. Edited by: Krasilschikov M.N., Sebryakov G.G.: Fizmatlit, Moscow, 2009a, 280 pp. (in Russian).

Veremeenko K.K., Zheltov S.Y., Kim N.V. *et al. Modern Information Technologies in Navigation and Guidance of UAVs*. Edited By: Krasilschikov M.N., Sebryakov G.G.: Fizmatlit, Moscow, 2009b, 556 pp. (in Russian).

Voronov V. Onboard systems for UAV control, *Aerospace Kurier*, 2006, 6(48), 62–63 (in Russian).

Unmanned Aircraft Systems. The Global Perspective. 2008/2009. Available at: www.wikipedia.org (accessed on November 19, 2015).

# Index

*Aerospace Navigation Systems*, First Edition. Edited by Alexander V. Nebylov and Joseph Watson.
© 2016 John Wiley & Sons, Ltd. Published 2016 by John Wiley & Sons, Ltd.